History of Hydrogeology

INTERNATIONAL CONTRIBUTIONS TO HYDROGEOLOGY

28

Series Editor: Dr. Nick S. Robins
Editor-in-Chief IAH Book Series, British Geological Survey, Wallingford, UK

INTERNATIONAL ASSOCIATION OF HYDROGEOLOGISTS

History of Hydrogeology

Editors

Nicholas Howden
Department of Civil Engineering, University of Bristol, Bristol, UK

John Mather
Emeritus Professor, University of London, UK

CRC Press
Taylor & Francis Group
Boca Raton London New York

CRC Press is an imprint of the
Taylor & Francis Group, an **informa** business

A BALKEMA BOOK

Published by:
CRC Press/Balkema
P.O. Box 447, 2300 AK Leiden, The Netherlands
e-mail: Pub.NL@taylorandfrancis.com
www.crcpress.com – www.taylorandfrancis.com

First issued in paperback 2020

ISBN 13: 978-0-367-57662-2 (pbk)
ISBN 13: 978-0-415-63062-7 (hbk)

This book contains information obtained from authentic and highly regarded sources. Reasonable efforts have been made to publish reliable data and information, but the author and publisher cannot assume responsibility for the validity of all materials or the consequences of their use. The authors and publishers have attempted to trace the copyright holders of all material reproduced in this publication and apologize to copyright holders if permission to publish in this form has not been obtained. If any copyright material has not been acknowledged please write and let us know so we may rectify in any future reprint.

Visit the Taylor & Francis Web site at
http://www.taylorandfrancis.com

and the CRC Press Web site at
http://www.crcpress.com

Typeset by MPS Limited, Chennai, India

Library of Congress Cataloging-in-Publication Data

History of hydrogeology / editors, Nicholas Howden & John Mather. — 1st ed.
      p. cm. — (International contributions to hydrogeology ; 28)
   Includes bibliographical references and index.
   ISBN 978-0-415-63062-7 (hardback : alk. paper)
   1. Hydrogeology — History.   I. Howden, Nicholas.   II. Mather, John Russell, 1923-
III. Series: International contributions to hydrogeology ; v. 28.
   GB1003.H57   2012
   551.4909—dc23

                                                    2012031741

Table of contents

Preface

I was first introduced to the history of hydrogeology project at the 2007 IAH congress in Lisbon. I had arrived a day early and was eating breakfast when Nick Robins introduced me to John Mather. Before I finished my coffee, John persuaded me to attend the special session on the History of Hydrogeology later that morning and, by lunchtime, had me signed up to help with chapter reviews and editing. At the time I don't think I really appreciated the magnitude of the task before us but, looking back, can see that it would never have been achieved without the help and support of numerous people. First we would like to thank the contributing authors for their work and their patience. Many have waited several years since draft chapters were first produced, and all have been supportive of our work to bring the book to press. Second, we must thank Ian Holman, Harriet Nash and John Chilton for their help in editing various chapters – sometimes through several iterations. Their support has been invaluable in bringing the book together, and has often involved a large amount of work at relatively short notice. We would also like to add thanks to the series editor Nick Robins who has been a constant support during organised sessions at the IAH meetings in Lisbon and Toyama, the editing process over the past few years, and in recent months during the final stages of preparing the book for publication. In addition we are grateful for the support from Taylor Francis in bringing the book to press.

Nicholas Howden
June 2012, Bristol

Biographies

Nicholas Howden is Senior Lecturer in Water in Queen's School of Engineering at the University of Bristol. He studied Engineering at Durham University and hydrology, hydrogeology and geochemistry at Imperial College London, and then became a consultant in the mining, oil and gas industries. In 2007 he joined the UK's National Soil Resources Institute (NSRI) at Cranfield University as Lecturer in Soil and Water Engineering and moved to Bristol in 2010. At Bristol he is Director of the MSc in Water and Environmental Management and is also Head of the Water and Environment Research Group in the Faculty of Engineering. Nicholas was awarded the Young Hydrogeologists' Prize by the Geological Society in 2003, and has been a member of IAH since 2004. He is the author of 60 articles in the scientific and technical literature.

John Mather retired from the position of Lyell Professor of Geology in the Department of Earth Sciences at Royal Holloway, University of London in 2001. He currently lives near Exeter in south-west England where he makes functional stoneware pottery and writes on the history of hydrogeology. Educated at Liverpool University he joined the British Geological Survey in 1966, rising to the position of Assistant Director before moving to academia in 1990. He has been a member of IAH since 1973 and was chairman of the British Committee from 1991 to 1995 and co-chair of the 27th Congress held in Nottingham in 1997. He has undertaken research projects in the UK, Romania, China, West Indies, Central America and the Middle East and advised the European Community, OECD, IAEA and UK and US agencies on various aspects of groundwater development and waste management. During a career spanning nearly 50 years he has specialised in the geochemical aspects of hydrogeology and is the author of some 150 articles in the scientific and technical press.

The history of hydrogeology project

Nick S. Robins

Editor-in-Chief, IAH Book Series

INTRODUCTION

From time to time it is useful for practitioners to look back over the historical developments of their science. Lessons can be learnt from the past, otherwise 'wheels' can so easily be reinvented. Indeed, much of the pre-digital technical literature is at risk of being dismissed by today's workers if it cannot be accessed at a desktop computer. With this in mind, John Mather, Emeritus Professor at the University of London, initiated, in 2001, the History of Hydrogeology Project with the objective of producing this volume. The project has taken much longer to fruit than was originally anticipated, but it has embraced two well attended international workshops, one held at the IAH Congress in Lisbon in 2007 and one at the IAH Congress held at Toyama the following year. The workshops helped to keep people informed how the Project was progressing and were also a means of attracting additional material. Two of the presentations from the workshops, neither of which fitted the geographical framework adopted in this volume, have already been published elsewhere.

The science of hydrogeology has developed from humble beginnings into the broad church of modern day applications we know today. Early understanding of some groundwater processes was demonstrated by the Arab peoples and by the Romans, working respectively in the arid zones of the Middle East and the more humid climates of Europe; by the peoples of South Asia; by the American Indians, and by others. Since the early days of numerical hydrogeology, which started with the realisations of Henry D'Arcy (1803–1858) and others, the science has developed into a complex series of investigatory procedures which collectively form the modern-day hydrogeologist's tool box, from which he or she can pick and choose tools that best address the complex issues we face today. Hydrogeology remains, as always, a branch of the over-arching science of geology; hydrogeology is, after all, the study and investigation of all aspects of the mineral that is water, or as Joseph Lucas originally defined it (Lucas, 1877):

> Hydrogeology ... takes up the history of rainwater from the time that it leaves the domain of the meteorologist, and investigates the conditions under which it exists in passing through the various rocks which it percolates after leaving the surface.

One significant change in our approach to hydrogeological investigation has occurred in the recent past. Whereas hydrogeology was initially a science that could provide solutions to groundwater problems without the need to talk to other disciplines, hydrogeology has now become, and quite rightly, just one component of study in the bigger picture of the water budget at sub-catchment, catchment and regional scale. Site specific issues, such as point source and diffuse pollution problems need to be viewed in the context of catchment scale understanding. Contemporary hydrogeology, therefore, provides analysis of just one component of the water cycle but it is now pursued within a holistic application that provides solutions that may relate equally to water supply, ecological services and the overall management, not least of change, of available surface and groundwater resources. Indeed, we have now got to the stage where we can conjoin a groundwater flow model running at one discrete time step with a surface water model running at another time step; computer programmers it seems, know no bounds. Indeed, the future demands of hydrological analysis to make sensible forecasts, within the constraints of current climate change and demand predictions, are becoming the next challenges of the hydrogeologist. And it will be a task for a future generation to review the historical worth of these efforts.

A number of articles have been written in the past on the history of hydrogeology, but none were intended to be comprehensive. Fetter (2004a; 2004b) wrote a useful two part summary on the evolution of hydrogeology as a science and others have written histories on specific aspects of hydrogeology, for example, Worthington & Gunn (2009) provided a valuable historical overview of the development of understanding in carbonate aquifers, while Vegter (1987) reviewed progress in groundwater development in South Africa (on the science of geohydrology as it is locally known), Mather (2004) offered a selection of papers on the history of hydrogeology from a British standpoint and several other authors have presented a range of historical summaries for a variety of specific geographical regions or specific disciplines within the science of hydrogeology.

This volume, *History of Hydrogeology*, is a first attempt at bringing the story of the evolution of the science of hydrogeology together from a country specific viewpoint. It does not cover history to the present day but rather takes the story, based on a country or region, up and until about the period 1975 to 1980, i.e. when hydrogeology was still an independent science and was evolving and developing, and in some areas doing so quite rapidly at a time when the personal computer was still a figment of the imagination of science fiction writers. This cut-off in time also avoids the contentious issue of setting the work of people who are still alive within an appropriate historical context and of trying to identify the modern framework that is clearly still evolving.

THE HISTORY OF HYDROGEOLOGY PROJECT

The context of the History of Hydrogeology Project is itself worth recording. The initiative started at the IAH Cape Town Congress in 2000 where discussions were held between various members. The timing was considered appropriate for two reasons. The first was the perception that the science of hydrogeology was becoming a mature science in that its development had slowed down while the practitioners were increasingly providing routine applications. That being said, important advances in understanding were, and are still, being made, not least using the computer power now available to every hydrogeologist in the modern era. Hydrogeological methodologies based on a thorough understanding of the key processes were, nevertheless, advanced and few, if any, radical new discoveries are likely in the future. The second reason was less philosophical and was based simply on the likelihood of losing insight into the historical development of hydrogeology as each of its elder statesmen cease communicating with fellow scientists.

A discussion paper was prepared by John Mather (UK) with input from Philip Commander (Australia), Co de Vries (Netherlands) and John Moore (USA) and presented at the Munich Congress in 2001. The outline for the project comprised three parts:

1   Pre-1800: looking at early development in hydrogeology on a unified historical basis.
2   1800–1980: looking at concepts and ideas and how they have developed in different countries in response to local pressures – on a regional or country basis.
3   Highlights: teasing out the crucial developments in various sub-disciplines within hydrogeology.

In the event, the project focussed on Part 1, Prehistory of hydrogeology, a project given to Michael Knight (Australia) which has since been taken out of IAH auspices, and Part 2. Potential authors for Part 2 were identified and requested to submit chapters by September 2003. However, authors seemingly had little appetite for the project and by 2005 only 13 of the invited 25 papers had been submitted. Significantly, many of those countries where important hydrogeological advances had been made over the last 200 years were among the absentees.

In 2004 the book *200 Years of British Hydrogeology* (Mather, 2004) was published by the Geological Society of London. This book describes various facets of the evolution of hydrogeology in Britain and Ireland. It was hoped that the publication of this volume would provide a fillip to the global project, but the submission rate still remained tardy.

Unlike *200 Years of British Hydrogeology*, for which many of the key authors were known to the project co-ordinator and could be targeted directly, likely and willing authors for the History of Hydrogeology had to be internationally sourced and could not always be identified so easily. Messages were sent to senior members of a variety of government institutions (many a persuasive beverage was bought at bars at various IAH congresses) and, little by little, promises that more country papers would be produced slowly emerged.

One of the early decisions taken within the project was whether to provide potential authors with a template. This could have taken the form of a prescription with headings such as national hydrogeological setting, key groundwater investigations, groundwater governance history, key individuals, key methodologies, etc. This approach was not adopted, both because it would inhibit and constrain authors and because it would produce a book, albeit one which might have greater value as a reference source than this volume, which would inevitably be a tedious read. Nor were authors told that there had to be a diagram on every other page to make the book look pretty or that the length had to be in the range of a set number of words, although they were told that the target was 10 000 words; the form of the paper and its focus on hydrogeological history was left to the authors.

Respective country authors have taken three distinctive routes: one is to focus on the people themselves, another on the concepts that have been developed in their country and the third is to consider the country's achievements. Some, of course, combine all three, but most tend to focus on just one theme making for a richly diverse and varied set of chapters. Countries faced with specific issues focus on those issues, for example, Serbia concentrates on karst and caves, the Netherlands on groundwater control and Japan on subterranean dams. Others have the potential to confuse the reader through past changes in territorial ownership and location of political borders, for example, throughout much of Eastern Europe. Some have even dwelt on political pressures, although this is one aspect that the editors have tended to inhibit.

Although the History of Hydrogeology Project was started in 2001 it has been unable to produce until 2012. Why then did it take so long? There are several answers to this question and many lessons have been learnt along the way:

1   Coverage – there are a number of countries that have made significant contributions to the science of hydrogeology that are (still) not represented in this book. Time was needed to allow contributions to formulate in order to attain best representation as possible, at least in the present volume.
2   Editorial – contributions have needed both light touch editorial and a more hands on approach, in some cases with multiple iterations and a lengthy gestation period. Editorial services have been required to an extent that a team of people, whose mother tongue is English, were co-opted – these include John Chilton, Ian Holman and Harriett Nash.
3   Editorial management – in the event, has been a much larger task than originally anticipated and it required Nicholas Howden to be co-opted onto the project to assist John Mather to ensure that the project deliverable became reality.

The first lesson, the issue of coverage, was always going to be a difficult one. France might say Henry D'Arcy was the founder of hydrogeology, Russia might say it was first at map making, the Balkan states will tell you that they defined karst hydrogeology, while America and Canada could tell of their important role in developing numerate analysis of groundwater issues, not least with regard to pollution. But what of all the other countries, Australia with its vast aquifers and problems of salinisation, Britain and its focus on Chalk hydrogeology, small island problems and the salt water interface and many others. The aim of the History of Hydrogeology Project was to include as many of the key countries in the volume as was feasible. The corollary to the issue of incompleteness is that this volume is now seen as a first volume in a series of many books that will in the future take up the thread and present the history of hydrogeology in Australia, Canada, much of the Middle East, Africa and the whole of South America, to name just a few.

The second lesson, the editorial problems that have been encountered, was addressed by a long, hard and protracted team effort. The problem, of course, stemmed from the absence of any initial steer being given to potential authors and resulted in paper lengths varying from just four pages of script to fifty pages, some of which were amply illustrated while others had no graphics, and there were those boldly written in English by an author who rarely needed to practice his knowledge of this language and which required extensive editorial support.

Figure 1 Coverage of History of Hydrogeology by country or region.

The third lesson learnt was that the magnitude of the editorial management of the project was far greater than anyone had anticipated. Although John Mather had single-handedly carried the project to a stage where about eighteen chapters had been received, he was aware that he alone could not hope to deal with bringing them to publication without additional help. In 2007, Nicholas Howden, University of Bristol, who had taken an interest in the project through the Lisbon Workshop, offered to help and became co-project manager and principal editor.

The result is a book which, although entitled *History of Hydrogeology*, is in fact the history of hydrogeology viewed from the perspective of twenty one country chapters (see geographical coverage, Figure 1).

We owe a debt to both John Mather and Nicholas Howden for having produced this book for future generations of hydrogeologists to enjoy and to learn from. We also owe a major debt to the many contributing authors, several of which have patiently waited nearly ten years to see their work in print. The final product has been worth the wait, a product which we should consider as the first in an occasional series of volumes recording the history of hydrogeology.

THE NEXT PHASE OF THE HISTORY OF HYDROGEOLOGY PROJECT

Some countries that have contributed to the development of the science of hydrogeology are missing from this book, while other countries with specific tales to tell are also absent. This should not detract from this first ambitious attempt at beginning to pull the overall story together. There will be another volume on the history of hydrogeology, a volume that could follow the national divisions adopted here or could perhaps look specifically at themes, for example, the evolution of test pumping analysis, the conceptual flow model, and other specific areas of work. There is also the volume on prehistory that is yet to be published (the original Part 1 proposal of this project) which will describe how the arid lands of the Middle East were developed with innovative use of self-draining qanats, how the Romans developed their recreational hot groundwater baths and even the role of groundwater in worship and cleansing adopted by the Celtic tribes and the American Indians; there is a story here with lessons for all practising hydrogeologists. Whatever the form of the next 'History of Hydrogeology' volume, it too will be a welcome addition to the background literature on the science of hydrogeology.

REFERENCES

Fetter, C.W. (2004a) Hydrogeology: a short history, Part 1. *Ground Water*, 42 (5), 790–792.
Fetter, C.W. (2004b) Hydrogeology: a short history, Part 2. *Ground Water*, 42 (6), 949–953.
Lucas, J. (1877) Hydrogeology: one of the developments of modern practical geology. *Transactions of the Institution of Surveyors*, 9, 153–184.
Mather, J.D. (ed.) (2004) *200 Years of British Hydrogeology*. Geological Society, London, Special Publications, 225.
Vegter, J.R. (1987) The history and status of groundwater development in South Africa. *Environmental Geology*, 10 (1), 1–5.
Worthington, S.R.H. & Gunn, J. (2009) Hydrogeology of carbonate aquifers: a short history. *Ground Water*, 47 (3), 462–467.

The history of hydrogeology in Australia

W.H. Williamson
Ryde, New South Wales, Australia

ABSTRACT

Australia is a land of extremes subject to an adverse and unreliable climate. The original inhabitants lived mainly in coastal zones without permanent settlements. European settlers, who arrived from 1788, founded towns from which they penetrated into the drier interior following rivers or sinking wells. There was pressure for Government to provide guidance in obtaining water supplies. The various States began to establish Geological Surveys in the 1850s and, although their prime objective was to advise on minerals they also advised on groundwater. Artesian water was discovered in 1878 and the boundaries of the Great Artesian Basin were established by around 1900. Declining yields and the long-term reliability of supply resulted in numerous conferences and committees. Following World War II geologists were appointed by organizations outside the Geological Survey. From the mid-1960s to 1980 intensive groundwater exploration took place and hydrogeology became accepted as a discipline in its own right.

INTRODUCTION

With the exception of Antarctica, Australia is by far the world's driest continent. It is sometimes referred to as the world's oldest continent – not because of its geological formations but rather because of its predominantly ancient landscape. This enormous continental island of some 7.7 million km² has very low relief. About half has an altitude of less than 300 m above sea level; only about 5% exceeds 600 m, practically all of which is concentrated in a zone along the eastern margin. Its highest point, Mt Kosciusko, in the far south-east, is only 2230 m in height.

Australia is a land of extremes. Climatic zones range from tropical in the north, arid in the interior, to temperate in the south. Its general aridity is evident in that 30% of it has a median annual rainfall of less than 200 mm, and half less than 300 mm. Less than 12% receives more than 800 mm, and these areas are restricted to the extreme north, east, south-east, and far south-west (Figure 1). The climatic picture is compounded by the great majority of the continent being subjected to rainfall variability ranging from moderate to extreme. High temperatures and heat waves can also cause difficulties. Evaporation too can be extreme, with average annual Class A pan rates ranging up to over 4000 mm in central Western Australia. The net result of these factors is a land subject to adverse and generally unreliable climatic conditions – a land particularly susceptible to prolonged and devastating droughts, as well as to bushfires and, almost incongruously, to floods. The tropical north is also subject to destructive cyclones.

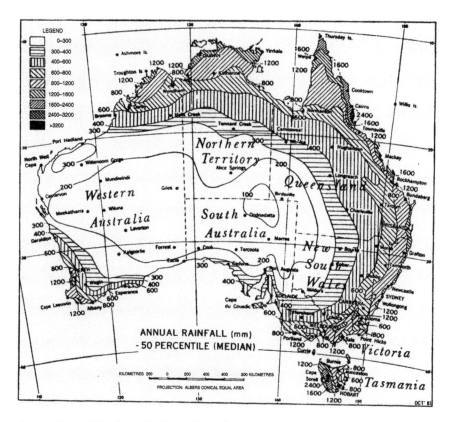

Figure I Annual rainfall in Australia. Taken from: *The Climate of Australia,* Commonwealth Bureau of Meteorology, AGPS, 1989. Commonwealth of Australia copyright reproduced by permission.

In such an environment, where the importance of groundwater resources is obviously enhanced, one might expect a rapid development in the application of hydrogeology. While this indeed happened, it took place predominantly from about 1960.

The development of hydrogeology in any country requires, normally, that there is first substantial use of groundwater. In this latter regard, a number of factors put Australia in a very retarded position compared with most major countries in the early 1800s. At that time, not only had the settlement of Europeans in Australia just started, its purpose was to serve as a penal colony for the British! European settlement commenced at Sydney with the arrival of the First Fleet, and its cargo of convicts, in 1788.

There was, of course, an indigenous population, the Australian aborigines, who were later shamefully treated. They lived mostly in the relatively well-watered coastal zones, but they also sparsely populated arid areas. Being astute observers, they were able to survive in such areas, where Europeans would have perished. They are known to have used wells, for example between some sand ridges and in dry creek beds, as well as springs, soaks and rock holes. But they were essentially food gatherers and hunters, moving in small groups. They did not have permanent settlements, nor did they cultivate crops or engage in animal husbandry.

The new colony of New South Wales (NSW) experienced considerable privation over the early years, including coping with the harshness of the new environment and limited availability of good quality land and water. Population gradually increased, mostly from continuing transportation of convicts, but settlement also expanded as land grants were made to military personnel and time-expired convicts. However, expansion was constrained by the Blue Mountains, which run parallel to, and within about 70 km of, the coast. It was not until 1813 that explorers found a way across them, and the young colony was able to burst its bonds and gain access to the hinterland. In the meantime, the then Van Diemen's Land, now Tasmania, had been claimed by the British in 1803 and a penal colony established there. Today, Australia comprises seven States (Figure 1). Of these, South Australia is the only one to have had free settlers from the outset, commencing in 1837.

By the 1850s, European settlement of Australia was still essentially in its infancy. The main centres of population were the few towns which are now the capitals of their respective States, and then, to a lesser extent, along the coastal zones which were relatively well-watered. As is normally the case, the initial spread of settlement inland mainly followed reliable surface water supplies, but as settlers penetrated into the drier inland grazing country, or if they did not have riparian access, recourse was soon made to tapping groundwater with wells.

With a rapidly expanding wool industry, and the need for more pasture land, settlers eagerly followed up the discoveries of early explorers and quickly spread inland. Further impetus was given by the announcement of payable gold in NSW in 1851, and by 1856 the population had more than doubled to 929 000. Enterprise quickened in every direction, but concern was also developing about water resources and the need for conservation measures. It was groundwater that attracted the first Government expenditure on rural water-conservation works. In 1866, £2900 was allocated to the sinking of wells to render practical a stock route from the Darling to the Lachlan and Warrego Rivers in NSW.

BUILDING THE FOUNDATIONS

In NSW, although there had been some earlier activities, including visits of various scientific expeditions, the Rev. W.B. Clarke, known as "the father of Australian geology", was prominent in geological activities prior to meaningful Government appointments. Soon after his arrival in 1839, he combined geologising with his duties as a cleric and mapped relatively large tracts of country (Vallance & Branagan, 1968). His geological expertise was also recognised from a groundwater viewpoint. In 1850, the Governor directed that an Artesian Well Board be formed "for the purpose of considering and reporting as to the best means of conducting an undertaking for endeavouring to obtain an abundant supply of pure water for the City of Sydney, by boring on the Artesian principle within the walls of Darlinghurst Gaol," and Clarke was appointed Chairman of the Board (Clarke, 1850). The attractions of Darlinghurst Gaol were that it was in a reasonably elevated position from which water could be distributed by gravity, and that the hard-labour convicts could be used on a tread-wheel as the motive power for the boring plant. Clarke was not optimistic of success, and estimated it would take three years to bore to 150 m. The bore was commenced in November 1851, but had an ignominious ending, being abandoned at 23 m as a result of sabotage (perhaps not surprisingly as it must have been purgatory on a tread-wheel in Sydney's summer temperatures!). However, Clarke's initial report (Clarke, 1850) would appear to be Australia's first groundwater report.

With population expanding, and settlers venturing further inland, inevitably there was pressure for Government to provide guidance and assistance in obtaining water supplies, including groundwater. Of major importance was the establishment of Geological Surveys in the various States, for they gradually provided geological maps, a basic requirement for hydrogeologists. The first such Survey was formed in Victoria in 1856, though it had a Geological Surveyor from 1852. However, in all cases they were part of Departments of Mines, and the prime object in setting them up was in relation to minerals such as gold and coal, rather than water. Nevertheless, most of the early Government Geologists had a special interest in groundwater, notably A. R. Selwyn in Victoria, C. S. Wilkinson in NSW, R. Logan Jack in Queensland, H. Y. L. Brown in South Australia, and A. Gibb Maitland in Western Australia. With time, as Surveys were able to carry out more intensive mapping, the locations of bores and wells were included on maps, and available details incorporated in reports. Eventually, prior to World War II, some reports specific to the groundwater resources of various mapped areas were being published. Johns (1976) gives a valuable outline of the history and role of the various Surveys, as well as descriptions of the activities and personal traits of their early leaders.

The apparent plethora of Geological Surveys in Australia arose from the Australian Constitution, adopted in 1901, under which the States are responsible for their own resources, such as forests, soils and minerals, including water. With the exception of Northern Territory (NT) and the Commonwealth, all the Surveys were in operation before the Constitution, so it was only formalising what was already in practice. In any event, considering the enormous area of most of the States, the remoteness of the main settlements from each other, and that in the early 1800s the geology of Australia was virtually unknown, there was justification for the individual State Surveys.

The early activities of the various Surveys were oriented mainly towards mineral prospects and mining, but in some cases, particularly in relation to alluvial gold in shallow or deep leads, they had significant relevance to hydrogeology. Examples in Victoria are given by Smythe (1869) and in NSW by Andrews (1910), Jones (1940), and Rayner (1940).

DISCOVERY OF THE GREAT ARTESIAN BASIN

A tremendous boost was given to the search for adequate water supplies by the discovery of artesian water in 1878 (in Australia, the term "artesian" implies that a bore will flow naturally). It was probably this discovery, more than anything else, which triggered the movement towards hydrogeology becoming a discipline in its own right in Australia, even though this happened over many decades.

Scientists of the day had already been pondering on the fate of the water disappearing from the Murray-Darling Rivers system. Rawlinson (1878) had appealed for an enquiry "into the cause of the disappearance of the vast bodies of river water which collect on the inner water-shed of the coast ranges of Australia" and considered that the interior of Australia would "ultimately be proved to be the storage reservoir where are conserved the rain and river waters which other theories fail to account for". The Government Astronomer claimed that less than 1.5% of the rain falling in the upper Darling River catchment flowed past Bourke, whereas for a corresponding point on the Murray, the discharge appeared to be of the order of 25% of the catchment rainfall Russell (1879). In the same year, Professor Tate examined mound springs near Lake Eyre in South Australia, pronouncing them to be natural artesian wells and predicting that boring for artesian water would be successful.

In fact, Professor Tate's prediction had already been proven, for in 1878 a flowing supply had been obtained on the remote Kallara Station in New South Wales. Finding the supply obtained by reconditioning a 33 m deep well to be inadequate, the manager had an auger hole drilled on from the bottom. At 43 m a flowing supply was obtained – the first flowing bore in what was to prove to be one of the world's largest artesian basins. Its site was about 170 km WSW of Bourke, and 30 km from the Darling River. Also it was about 120 m W of Wee Wattah mud springs, which were doubtless the result of shallow artesian conditions in the area.

The recognition of the bore and mud springs indicating artesian conditions led to further boring and within a few years small flows had been obtained from shallow bores near such springs in Queensland, NSW, and South Australia. These results encouraged deeper exploration in areas remote from any springs and, within a decade of the Kallara bore, large artesian flows were being obtained from depths of over 300 m. The first of these in NSW was in 1887 on Kerribree Station, about 80 km WNW of Bourke.

In Queensland, following one of the then-familiar droughts in the west, in 1881 the Government created a Water Supply Department, under J. B. Henderson as Hydraulic Engineer, with the aim of providing more adequate water supplies for inland towns and on main coach roads and stock routes. However, the drilling plant then available was suitable only for shallow depths, and the groundwater encountered was mostly too saline for human consumption. Also in 1881, the then Government Geologist, R. Logan Jack, had speculated on the possibility of encountering artesian water in inland Queensland. So in 1885, following yet another drought, when he and Henderson were asked to advise on the prospects of obtaining artesian water in the western interior, he reported favourably, and Henderson recommended the importation of plant and operators for deep drilling.

The first of the deep bores to be completed was on Thurralgoonia Holding, near Cunnamulla, a flow of 363 m³/d being obtained at 393 m. This discovery had great impact in Queensland and there was rapid expansion of drilling. By 1889, 34 artesian bores had been completed, with 524 by 1899. In 1888, the drilling of 13 more bores for town water supply was approved, and their waters markedly improved the living conditions and health of the western communities. By 1900, artesian bores for 24 inland towns had been completed or commenced (Dept. Coordinator-General of Public Works, 1954).

It is significant that, in a collaborative report to the Government in 1893, Henderson and Jack expressed considerable concern about the number of bores, in which the flows were inadequately or not controlled, with consequent wastage of water, and the difficulty in obtaining adequate information on the bores. Their recommendations included that (a) the Crown assume control of artesian supplies in Queensland; (b) control of drilling activities be exercised; (c) the waste of water be combated; and (d) the State asset be conserved. The commendable foresight of this report resulted in legislation being passed in the Assembly to control artesian water, but frustratingly this was nullified by the Legislative Council, said at the time to be because of "vested interests". Clearly, the pastoralists did not want the constraints of controls, and they had the political power to offset such attempts. The matter was not rectified until the passing of the Water Act of 1910, but this was after the artesian water situation had deteriorated much further.

In the meantime, in NSW, with the introduction of the Public Watering-places Act in 1884, the Government invited tenders for the construction of artesian bores on some of the far-western stock routes. The successful completion of many of these provided safe and permanent water supplies on stock routes which in some cases had previously been little used. Development was further promoted by the Artesian Wells Act in 1897, which enabled groups of settlers to obtain Government assistance to construct an artesian bore to serve their collective properties, the water being distributed to them by open drains.

During these early years of development in the Great Artesian Basin, geologists of the various State Geological Surveys were mapping large areas, assessing the new information coming to light from bores, and advising on the prospects of boring in various areas. It is a tribute to the geologists of the day that, in spite of the enormous size of the Basin (some 1.7 million km^2, or 22% of the Australian mainland) its limits were broadly established by about 1900 (Figure 2). It is indeed fortunate that geological conditions have provided this water-source in regions in which rainfall is often so low and unreliable that otherwise much of the pastoral industry would be impracticable.

In NSW, a Royal Commission was appointed in 1884 "to make a diligent and full enquiry into the best method of conserving the rainfall, and of searching for and developing the underground reservoirs supposed to exist in the interior of the Colony, and also into the practicability, by a general system of water conservation and distribution, of averting the disastrous consequences of the periodical droughts to which the Colony is from time to time subject." The Commission was extant for four years and its lengthy hearings were published in three reports, in 1885, 1886 and 1887, respectively, but it failed to achieve much direct action.

The most tangible outcome of the Royal Commission was the establishment of a Water Conservation and Irrigation Branch, attached to the Department of Mines and Agriculture, but later, in 1896, transferred to the Public Works Department. Initially, this Branch was concerned with stream gauging and related activities, and assessing the potential for irrigation developments. However, in 1913 it became a Commission in its own right. In relation to artesian bores, it continued with the previous practice of having Government bores constructed under private contract, but in its first year it purchased two "Combination Cable Rig and Hydraulic Rotary Boring Plants" for drilling bores to a maximum depth of 900 m. The intention was to operate these plants with "day labour" under the supervision of officers of the Commission. For its shallow boring (maximum 150 m) activities, it then had only one plant, and insufficient funds to acquire others that year (Water Conservation and Irrigation Commission Annual Report, 1913). This was the humble beginning to activities of that Commission which were later to play a significant role in the development of hydrogeology in Australia.

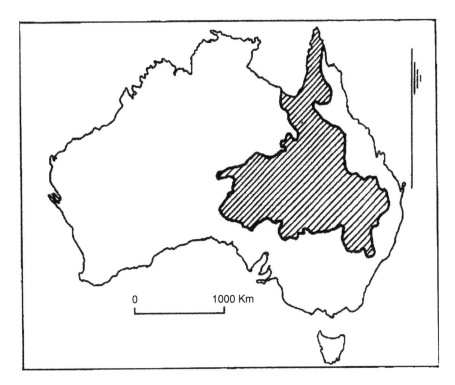

Figure 2 The Great Artesian Basin. Adapted from: *Review of Australia's Water Resources,* Australian Water Resources Council, AGPS, 1975. Commonwealth of Australia copyright reproduced by permission.

PROBLEMS DEVELOP IN THE GREAT ARTESIAN BASIN

With an apparently inexhaustible supply issuing from artesian bores, attention was also given to utilising the water for small scale irrigation projects, starting in 1891, near Bourke, NSW. It was not long before difficulties were experienced. An irrigation expert reported "as the soil cakes on the surface after irrigation, and the water from this and all other bores contains alkali, the necessity for keeping the surface broken up and for using the water sparingly is obvious" (Home, 1897). So here, another aspect of water chemistry was coming into the picture, the apparently unexpected residual alkali problem, which proved to be the undoing of this irrigation scheme. Over most of the Great Artesian Basin, the waters of the main artesian aquifers are essentially sodium bicarbonate waters and for this reason, although not unduly high in total salts, the waters are generally unsuitable for irrigation. The unfavourably high ratio of sodium to calcium plus magnesium ions has an adverse effect on soil structure, ultimately rendering the soil impervious. Even so, Cox (1906) considered that it was possible to cope with the alkali problem, and advocated extensive use of the artesian waters for irrigation, and of the artesian head for the production of mechanical and electrical power. Symmonds (1912) details considerable experimentation with materials such as gypsum and nitric acid to try to offset the ill effects of the alkali, but an economic and practicable solution was not forthcoming. Surprisingly, it was fourteen years before the irrigation scheme was finally abandoned.

In spite of this setback, here was a major water resource underlying an enormous area of semi-arid and arid lands. It was potable, being successfully used for town water supply, and served the pastoral industry. Apart from the irrigation aspect, it appeared to be the panacea of water supply problems for some 22% of Australia. But was it?

Soon other problems started to develop. With bores tapping aquifers at depths of 300 to 1200 m, construction and completion techniques were generally inadequate for proper control of the high-pressure water. It was found that the flows, although often initially large (flows in excess of 9000 m^3/d were not uncommon), soon showed a marked reduction, some bores even ceasing to flow. Various theories were advocated to explain this, e.g. leakage into upper formations, caving of the aquifer, choking of the bore, loss of gas pressure, and regional loss of head because of droughts; but none was sufficient to be of general application. In some areas, too, difficulties were being experienced because of severe corrosion, steel casing in the bores sometimes being eaten through in a matter of months. Furthermore, with regard to the potential of the Basin's water resources and their appropriate development, many conflicting views were raised. Even opinions as to the source of the water were not unanimous, a matter which raised an interesting controversy.

Well before 1900, the origin of water in artesian basins, or, for that matter, virtually all groundwater, was recognised as being of meteoric origin, i.e. from rainfall. However, J. W. Gregory, then Professor of Geology at Melbourne University (and later Glasgow University) caused a scientific stir by advocating a plutonic origin, from deep-seated igneous rocks. His views were first aired in a lecture in 1901, and were refuted by Knibbs (1903). Gregory developed his theories further in his book "The Dead Heart of Australia" (1906), basing his case mainly on the anomalies in temperature and pressure and the chemical composition of the water. He claimed that the rise of the artesian water was due to the combined effects of the temperature of the igneous rocks from whence the waters originated, and the pressure of the overlying rocks. Pittman (1907), then NSW Government Geologist and active in the study of the Basin for many years, attacked Gregory's theories. However, not to be outdone, Gregory criticised Pittman's arguments and maintained his views. The final broadside by Pittman (1914) replied in considerable detail, but Gregory remained unconvinced. At that time, no one fully understood the hydraulics of such a large artesian system. Even Pittman (1901), in trying to explain the diminishing flow of one of the bores, stated "... seeing that the drought has lasted between 3 and 4 years when the diminishing flow was observed, there cannot be much doubt as to the cause of the deficiency".

This controversy had ramifications beyond simply being a scientific argument. If Gregory was right, there were serious implications as to the continuity of the water resource. But even apart from this controversy, the marked reductions and cessations of flow caused considerable alarm, particularly among pastoralists, as it was feared that this invaluable resource might have a relatively short life. It was evident that the whole matter required careful investigation and review, and to this end, in 1908 the NSW Government invited other States to form a consultative Board. Unfortunately, only South Australia was willing, and the matter lapsed. It was revived again in 1911, this time successfully, and the First Interstate Conference on Artesian Water was held in Sydney in 1912. There were five such conferences, the last in Sydney in 1928. At the latter, the need to extend the activities of the conference to cover all groundwater in Australia was included, and it was resolved that "the whole body of underground water, whether under pressure or not" should come within the scope of the conference. Unfortunately, although it was proposed that the next meeting be in 1930, this proved to be the last of this series of Conferences on Artesian Water. Even so, they had been of enormous benefit. They clarified many problems, allayed many fears, instigated the systematic collection and interpretation of data on Australian artesian basins, and stressed the need for controlled development of artesian water resources.

Although there were no further Artesian Water Conferences, the movement towards having a national forum for the consideration of all groundwater was not forgotten. At a Brisbane meeting of the Australian Agricultural Council in 1937, the national aspects of water conservation and irrigation schemes were stressed and a joint Commonwealth and State enquiry was recommended. An Interstate Conference on Water Conservation and Irrigation was subsequently convened in Sydney in 1939, but the Commonwealth was not represented. The matters considered by this conference were of very wide scope but it is significant that the first resolution in its report recommended that the States and the Commonwealth participate in "a national investigation into the question of underground water supplies and the matters relating thereto".

The conference also resolved that "a permanent advisory committee or council, consisting of representatives from each State and from the Commonwealth, should be formed to consider and advise on those problems relating to the conservation and utilisation of water, which are of national or interstate interest and importance". It was recommended that this matter be referred to a proposed further conference "for recommendations as to the constitution, powers and duties of such committee or council". Unfortunately, World War II burst onto the scene, and it was to be many years before these proposals were implemented.

However, there were still concerns, particularly by pastoralists, about declining yields of artesian bores, cessations of flows, and the long-term reliability of supply. In response the Queensland Government set up a Committee for its own official investigation in 1939. An enormous amount of data had accrued over the years, especially since the start of the Interstate Artesian Conferences. Furthermore, Theis's non-equilibrium equation and the theory of elasticity of aquifers had been developed since the last of those conferences in 1928. Unfortunately this Committee's work was interrupted by World War II, and not resumed until 1946. Its final report was not presented to Parliament until 1954, but it provided an important collation of the state of knowledge of the Queensland portion of the Great Artesian Basin at that time (Dept. Coordinator-General of Public Works, 1954). In relation to uncontrolled and leaking bores, it concluded (p. 49) that stringent conservation measures were not warranted, on the grounds that "the resultant benefits were insufficient to outweigh the cost and the many difficulties involved in the implementation of a programme which could be expected to meet with strenuous opposition from the majority of the owners who would be required to carry it out". This controversial conclusion prompted a scathing minority report from committee member C. E. Parkinson, which included the disturbing statistics (p. 77): "Of the flowing bores in the State, 26.4% have no control valves; 16.9% have casing that has chemically deteriorated; 10.2% have leaking casing; 4.2% run to swamps; and 28.2% run to watercourses". Perhaps the only bright note was that, in

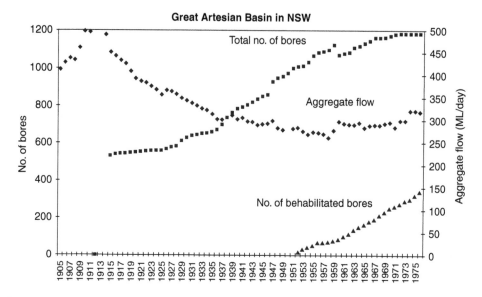

Figure 3   Effects of bore rehabilitation on the aggregate flow of artesian bores in NSW (Department of Land and Water Conservation, NSW).

1954, Queensland introduced legislation requiring that new artesian bores be properly controlled by valve and the water from them distributed only by pipeline.

In contrast, in NSW, a programme of rehabilitating uncontrolled bores had been commenced in 1952 and continued until 1976. It was not long before beneficial effects became apparent. By 1959 the aggregate available flow of artesian bores in NSW showed an upward trend for the first time since 1911 (Figure 3). In South Australia too, where there were about 300 artesian bores, rehabilitation of uncontrolled bores began in 1977.

POST WORLD WAR II TO THE 1960's

With the end of WWII there began a period of increased emphasis on water developments, including groundwater. In NSW, this was not only due to the resurgence to be expected when a nation can return to normal pursuits, but also because it had been experiencing a succession of droughts from the mid-1930s until 1947. At that stage, hydrogeological matters were dealt with by the Geological Survey and the then Water Conservation & Irrigation Commission, among other things, constructed artesian and sub-artesian bores. The Survey serviced the Commission in matters such as determining from drilling samples whether bedrock had been reached in the Great Artesian Basin, and to some extent in the selection of bore sites outside the Basin. However, only one or two geologists could be made available for this work, and then only on a part-time basis.

A significant post-war development in NSW, serving to help re-establishment, and as an anti-drought measure, was the Farm Water Supplies Act, in 1946. Under this Act, primary producers could obtain technical advice and financial assistance on matters relating to water supply, including groundwater. This scheme proved very popular, particularly in drought periods, when there was great emphasis on groundwater.

In 1950, the Commission, until then essentially an engineering organisation, decided it should have its own geological staff to specialise on groundwater. At that stage, apart from the pressures arising from the Farm Water Supplies Act, the Commission was operating twenty boring plants throughout the State, and it needed geological expertise on hand. The author was appointed to the Commission in May 1950, but immediately a problem arose. The then Government Geologist advised that if the author were classified as a geologist he would lodge an official objection on the grounds that the Geological Survey should service other government departments in geological matters. To overcome this impasse, and since he was to specialise in groundwater, the author suggested he be classified as a hydrogeologist. This was accepted, and appears to be the first usage of the designation of Hydrogeologist in Australia.

At that time there were no courses in Australia to provide training in hydrogeology, so it became a matter of gaining in-service experience and much self-education from overseas literature. Learning the capabilities of various types of boring plant, and the design of various types of bores, was gained mainly through considerable discussion and field work with the Engineer for Boring, and Boring Superintendents, as well as from literature. Later, finding that the results of test-pumping of completed bores were being treated in a rather rudimentary and pragmatic way, the author introduced time-drawdown and time-recovery measurements to allow mathematical analysis of the data. For high-yielding bores, such as for irrigation or town water supply, step-drawdown tests were also applied with the object of determining bore efficiency and to allow extrapolation for the effects of long term pumping at rates higher than those of the test-pumping.

At that time too, belief in water divining was rife in rural areas. Sometimes a land-holder would seek technical advice to check the prospects of a divined site, or conversely get a diviner to check the hydrogeologist's site. Even today, there are still some adherents to divining, especially in areas where there is no difficulty in obtaining groundwater. Perhaps the most telling evidence against divining is the statistics accrued by the then NSW Water Conservation & Irrigation Commission over the period 1918 to 1945 from its drilling of 3638 bores, shown in Table 1. About half were on divined sites, and in spite of the "advantage" of having been divined, there were twice as many failure bores on such sites than on sites not divined. Prior to the introduction of new regulations in 1947, the Commission was committed to boring on whatever site the landholder wanted, and many of the undivined sites were not favourably located. However, with the advent of hydrogeological advice, failure bores became rare.

During the 1950s, NSW, Queensland and South Australia started to undertake regional exploratory drilling programs for groundwater supplies. In NSW, the first of these programmes commenced in 1953 in the Hunter Valley (Williamson 1958). In 1957, attention turned to the much more extensive inland drainage systems, initially the Lachlan (Williamson 1964), and later, in 1963, the Namoi (Williamson 1970). The investigations in these inland systems revealed a far higher groundwater potential than was formerly known, and led to a major increase in the use of groundwater for

Table 1   Comparative results of water divining. Water Conservation and Irrigation Commission NSW. Annual Report for 1945, p. 21.

	Divined		Not divined	
	Number sunk	Per cent	Number sunk	Per cent
Bores in which supplies of serviceable water estimated at 100 gallons per hour or over were obtained	1291	70.4	1516	83.9
Bores in which supplies of serviceable water estimated at less than 100 gallons per hour were obtained	185	10.1	96	5.3
Bores in which supplies of unserviceable water were obtained	87	4.8	61	3.4
Bores – absolute failures, no water of any kind obtained	269	14.7	133	7.4
Total	1832	100	1806	100

irrigation and town water supply. Queensland started an exploratory programme in the Dumaresq Valley in 1958 (Queensland Irrigation and Water Supply Commission 1965), while South Australia carried out such drilling in an area south of Adelaide in 1950 and 1955–56 (O'Driscoll & Shepherd 1960), and in the Willochra Basin in the mid-1950s (O'Driscoll 1956). The Victorian Department of Mines had a long history of stratigraphic drilling, particularly in the Otway Basin. This was continued in that Basin in the 1950s (Wopfner & Douglas 1971) and also in the Murray Basin (Johns & Lawrence 1964). Western Australia, Northern Territory, and Tasmania were not able to undertake regional groundwater exploration programmes in this post-World War II period, and had to await developments in the 1960s in order to have the necessary funds and resources. Indeed, these factors were severely handicapping all of the States in their groundwater investigations.

POST – 1960 DEVELOPMENTS

Events moved swiftly in the 1960's. An Underground Water Conference of Australia was formed, and had its first meeting in Canberra in 1961. More importantly, an Australian Water Resources Council (AWRC) was formed soon afterwards, and held its first meeting in March 1963. The Council was a Ministerial body, serviced by a Standing Committee comprising heads of relevant authorities, and this in turn was serviced by Technical Committees. In 1964, the Underground Water Conference was reconstituted as the Technical Committee on Underground Water (TCUW). An initial task was to have each State prepare a review of its surface water and groundwater resources as in 1963, and these were collated and published as "*Review of Australia's Water Resources, 1963*" (AWRC, 1965). A major contribution in the Review was the first meaningful map of Australia's groundwater resources.

The Review showed there were glaring deficiencies in information on both surface water and groundwater resources. To assist in achieving more complete knowledge in these regards, the Commonwealth passed the State Grants (Water Resources) Act, 1964. This had the object of encouraging the States to implement accelerated programmes of research and investigation into surface water and groundwater resources, and finance was made available by subsidy, initially for a three-year period. This was a major step in the development of hydrogeology in Australia. At last it had become politically acceptable for finance to be made available to increase hydrogeological staff numbers in the under-resourced relevant State authorities, and to embark on systematic exploration and investigation programmes. There was, of course, an immediate problem – the recruitment and training of requisite staff. There were still no appropriate training courses in tertiary educational institutions in Australia. Hence, to augment in-service training, TCUW organised a series of Groundwater Schools; intensive courses of two weeks duration. The first was held in Adelaide in 1965. Apart from the lectures and workshop content, the Schools provided an important forum for discussion and exchange of experience, adding markedly to their value. The Schools also brought out the multi-disciplinary nature of hydrogeology, attracting both geologists and engineers. However, there was a severe shortage of geologists, and this difficulty was compounded by a synchronous boom in mining, starting in the mid-1960s. Not only were additional geologists needed in relation to mining, but also hydrogeologists in relation to associated groundwater matters.

Probably the major transitional factor to elevate hydrogeology to a discipline in its own right at this time was that it had become quantitative. Prior to this, groundwater reports were almost entirely qualitative insofar as the resource was concerned. Usually there were neither sufficient data nor staff to do otherwise. This change came at an opportune time, for in every State there were already areas in which there was concern either about over-development of groundwater resources or urgent need for additional resources.

Although there were initial problems in obtaining requisite staff and equipment, the mid-1960s heralded the start of greatly accelerated progress in the investigation and assessment of the nation's groundwater resources. An indication of the progress is given by two landmark publications: *Groundwater Resources of Australia* (AWRC, 1975); and *Review of Australia's Water Resources* (AWRC, 1976). The latter provides the first quantitative assessment of Australia's water resources, both surface water and groundwater. Even so, much remained to be done, and the State Grants Act to provide Commonwealth subsidies to the States was extended for a number of three-year periods.

The mid-1960s to about 1980 was the period of the most intensive groundwater exploration. As groundwater resources were proven and developed, there was a gradual transition in emphasis from exploration to management. Thus, as investigations proceeded, the need for quantitative assessments became paramount, particularly with regard to how the systems would respond to development. It was clear that the closed mathematical or analytical solutions used in relation to the behaviour of an individual bore or well would not lend themselves to considering the response of a whole aquifer system to multiple extractions, so recourse must be made to other approaches. Fortunately, there had been important advances overseas in these regards during the 1960s. Various analogue models had been developed, initially physical and then electrical. From the outset, it was evident that physical models would have very limited application, but electrical analogue models appeared to offer more scope. Lawson and Turner (1970) built a resistance network and McPharlin (1970) reported on the use of the more sophisticated resistance-capacitance network models on two projects. With the introduction of computers Pilgrim (1970) reviewed the use of digital models, based on both finite difference and finite element methods, and correctly foresaw their advantages would increase as the technology and capacity of computers improved. At last, here was a management tool that lent itself to assessing the long-term effects of various development strategies and which was also more amenable both to setting up the model and adjusting hydrologic parameters during the process of calibration. The first reported Australian study of a computer-based simulation of a substantial aquifer system appears to be that by Kalf and Woolley (1977) in relation to an investigation started in 1971 in the Murrumbidgee Valley, NSW.

A particularly ambitious project was carried out by the then Bureau of Mineral Resources (BMR) over 1971 to 1979 to model the Great Artesian Basin. This was at the instigation of the then TCUW, and involved not only an overview of the geology and hydrogeology of the Basin, but also modelling to simulate its hydrodynamics. The comprehensive overview was prepared by Habermehl (1980), and the model GABHYD developed by Seidel (1980) over 1975 to 1978. To deal with 3-D transient groundwater flow, a finite difference model was chosen. The geometry and hydrogeology of the Basin was drastically simplified, both horizontally and vertically, in order not to exceed the capacity of the computer used. The multiple aquifer system was reduced to only two groups, with each group represented by one layer of grid nodes in the model. Horizontally, a square grid of 25 km spacing was used. Naturally, the model lends itself only to regional predictions. To complement the project, in 1974, a major study of the isotopic hydrology of the Great Artesian Basin was jointly commenced by six relevant organizations (including three from the USA), in conjunction with the then BMR.

Green (1965) had carried out earlier isotopic studies, commencing in 1962, using Carbon-14 dating in the NSW part of the GAB, but the age limit of about 35 000 years before present restricted its application to marginal zones. However, the ages determined were in accord with those derived by Hind and Helby (1969) from hydraulic data – a rate of water movement of about 1.6 m/year. Hind and Helby (1969) also presented basement contours in the NSW part of the Basin, as well as thermal gradients and their relationship to basement lithology. Later, Polak and Horsfall (1979) dealt with geothermal gradients over the whole Basin.

Other substantial developments occurred during the 1960–1980 period. One of these was the formation of the first Australian consultant group specifically for hydrogeological activities. Although there were a few consultant engineering groups carrying out some groundwater projects, such projects were generally subordinate to their main interests. Hence the formation of Australian Groundwater Consultants in 1966 by two hydrogeologists from the author's staff was a significant move. The contemporaneous rapid expansion of groundwater investigations by State departments and the onset of a mining boom after the mid-1960s provided an ideal opportunity for consultancies. Both required hydrogeological expertise, but the economic bias in favour of mining developments led to the movement of some experienced departmental staff into the consulting field.

A second specialist consultancy, Stephen Hancock and Associates, formed in Victoria in 1967, and both of these were drawn into mining developments in Western

Australia, working as partners on some projects. They amalgamated as Australian Groundwater Consultants Pty. Ltd. in 1969. In the early 1970s, several other consulting or contracting companies formed and/or entered the field of specialised hydrogeological services, and others came from overseas.

Further scope for hydrogeological consultants came in the 1970s because of the emphasis that had finally developed on environmental issues. Most States had legislation in this regard in place by the early 1970s, e.g., Clean Waters Acts in NSW in 1970, Western Australia in 1971, and in Tasmania in 1973; and the Underground Water Preservation Act in South Australia in 1969. Contamination of groundwater systems poses a particular problem because of its long-term effects, and although State departments became heavily involved, the demand for hydrogeological input became such that it also increased the demand for consultants. The spectre of radioactive contamination around uranium mines and their associated plant and tailings ponds roused much early concern. Soon, the principles of contaminant containment, fixation, management and monitoring were being applied with greater intensity to all coal and metalliferous mining and mine waste management, as well as potentially contaminating industries, including power stations, petroleum refineries, and metallurgical and chemical works.

In mining projects, conflicts of interest are not uncommon. In an important example in NSW, for many years the Sydney Water Board had resisted proposals by four coal-mining companies to allow mining to extend under five major water storage reservoirs owned and operated by the Board. Matters were finally brought to a head when a Public Inquiry was instituted in 1974, and in the course of this the author was made available to the Department of Mines in relation to hydrogeological aspects. The Inquiry sittings were held in 1976, and it was concluded that, subject to certain conditions, mining should be permitted under the reservoirs but not under the dam structures (Williamson, 1978).

Another prominent example is the open-cut mining of major brown coal deposits of the Latrobe Valley in the Gippsland Basin, Victoria. Winning of the coal requires the lowering of artesian pressures in sand aquifers beneath the coal to prevent heaving of the open cut floor and to aid batter stability. Dewatering commenced at Morwell Open Cut in 1960 and has resulted in lowering of the potentiometric surface by up to 130 m, the effects being evident within a large area and at distances up to 50 km. The water is used in power generation and for industrial purposes, and the dewatering has resulted in surface subsidence of up to 2 m. The water being extracted is at about 50°C, and the thermal gradient is as high as 5 m per degree through the thick brown coal seams. Thompson (1978) showed that the high groundwater temperature is essentially due to the insulating effect of the coal beds on local areas of high geothermal flux, with the presence of artesian aquifer systems within the coal sequence modifying the simple thermal conductivity relationships.

Most major groundwater investigations in Australia have been directed towards unconsolidated sediments, or, to a lesser extent, sedimentary rocks. Although fractured rock systems have received relatively little attention, this is not to say that they are not important, for they provide the only groundwater source over about half the continent. Prior to the 1960s, bores were used to obtain water for stock and/or domestic purposes. Such supplies normally put limited demand on the system, so there was little need for quantitative investigations. Also, general principles and the suite of data from the many thousands of landholder's bores gave a broad picture of groundwater availability in regions of fractured rocks.

In many areas, particularly in those of low rainfall, the groundwater is brackish or salty, or the yields are low. An example of the outcome of an exploratory drilling programme in such an area is given by Lord (1971). In a period of severe drought in SW Western Australia in 1969, the Government instituted a drought relief programme, with bore sites selected by geologists of the Geological Survey. The area is underlain by Precambrian granite and gneiss, and was known to have poor groundwater potential. In a crash programme, of 2639 bores drilled, only 10% were classed as successful. However, the criteria of "success" were a yield of only 4.5 Kl/day, with salinity less than 11 000 mg/l!

Although the yield of most bores in fractured rock tended to be less than 2.5 l/s, in some areas much higher yields were available. This is expected in calcareous rocks, where solution channelling is common, but rocks normally less favourable were also involved. Hillwood (1967), in reviewing the problems and results of bores in hard rock areas in South Australia, gives many examples of yields of the order of 12.5 l/s. Down-hole hammer drilling was then coming into vogue in Australia, and it was significant that the yields being obtained from hard rock bores drilled by this method were consistently higher than those drilled by cable-tool. This is believed to be due to the clean hole being maintained by the hammer method.

Considerable interest was taken in groundwater in fractured rocks after a mining boom commenced in the mid-1960s, particularly in relation to mines in arid and semi-arid areas. Of concern were the dewatering of mine workings, the supply of water to the mines for needs such as ore processing, dust suppression, or general water supply, and commonly to meet water supply requirements of small townships set up to service mines. These townships often had to be located at a substantial distance from their respective mines because the salinity criteria for their water supplies were more stringent than for the mine. O'Driscoll (1979) reviewed the significance of groundwater to mining developments in Western Australia from the late 1800s to the 1970s, and the majority of mines were in fractured rock environments.

At the other end of the spectrum, a particular problem was where groundwater yields from fractured rocks were so low that bores did not warrant developing even for stock and domestic supplies. This had been of concern to the author for many years in NSW, and in 1978–79 he led an AWRC-funded research project involving the use of hydraulic techniques to stimulate the yields of such bores. At that time, the NSW Water Resources Commission was using combination rotary and down-hole hammer plants for these bores, and the project approach was to use the mud-pump of the plant, in conjunction with a booster auxiliary pump, to pressurise the aquifer zone isolated by packers. Up to six-fold increases in yield were obtained, thus converting what had formerly been taken as failure bores to viable stock and domestic water supply bores. An important finding was that sufficient inter-block propping was achieved by injecting only water, and no advantage was gained with viscosity-inducing additives and propping agents as used in oilfield technology (Williamson & Woolley 1980).

In the 1960s to 1980s, there were also some important Australian developments in the application of geophysical techniques to hydrogeological work. Earlier applications had been almost entirely derived from overseas methods. Some of the earliest geophysical reports relevant to groundwater were by staff of the then Bureau of Geology, Geophysics and Mineral resources (BMR), established in 1946. Examples are resistivity investigations in five areas in WA (with special attention to the granite country) by Wiebenga (1955), and Dyson and Wiebenga (1957) in relation to an investigation in 1956 regarding the Alice Springs water supply. Seismic refraction, resistivity traverses and depth soundings were used in this latter investigation, but the results proved unsatisfactory. An important advance in using seismic refraction techniques was developed by Hawkins, with Wiebenga and Dyson. It was used as a standard procedure in the Canberra area in 1956, and termed the Reciprocal Method (Hawkins, 1961). It was this method that the then NSW Water Conservation & Irrigation Commission adopted when, in 1965, it incorporated a program of seismic refraction surveys to track the buried "valley-in-valley" in its investigations of the extensive Lachlan Valley, and subsequently in other major inland valleys in that State. Additional developments of the method were made by the Commission's geophysicists in the course of that work, including the incorporation of blind zones (Merrick *et al.*, 1978).

A further significant enhancement to seismic refraction surveying and interpretation was the Generalised Reciprocal Method developed by Palmer (1980) during the 1970s. This combined many of the better features of previous methods, including the standard reciprocal method. It added the concept of an "optimal offset" to accommodate significant irregularity in the subsurface refractor being mapped. The method has gained wide acceptance internationally as the most effective available for seismic refraction interpretation.

Following the establishment in 1964 of the Technical Committee on Underground Water, in the Australian Water Resources Council, an early concern was the problem of obtaining reliable quantitative interpretations of the various geophysical methods. Later, when funds became available for approved research projects, one such project, carried out during 1969 to 1971, was on geophysical well logs in water bores in unconsolidated sediments. The report highlighted the difficulties involved in gaining quantitative interpretations, and made recommendations for further research (Emerson & Haines, 1974).

In the Great Artesian Basin, a difficulty in effecting stratigraphic correlations was that only driller's logs were available for the great majority of bores. To obtain more specific data, the BMR conducted a geophysical well logging programme in the Basin from 1960 to 1975. In all, wire-line logs were run in 1250 wells. Many could not be accessed because of complicated headworks or internal obstructions. Other operational difficulties included high water temperatures, commonly in the range of 30°C to 50°C, but in some areas in Queensland up to 100°C at the well head. Since the wells were cased, the logging was mainly restricted to nuclear logs. Natural gamma logs were run in all wells and in some cases also neutron-gamma logs. From many wells, temperature, differential temperature, and casing collar locator logs were acquired, and from some flowing wells, flow-meter logs were also obtained. Where more than 100 m were uncased, spontaneous potential, resistivity, and calliper logs were run in those sections. This programme provided a valuable suite of complementary data for subsequent studies of the Basin.

Of surface methods, electrical resistivity, either by traversing or depth sounding, had long had problems. Unfortunately, reports were commonly being made without benefit of ground-truth data, and subsequent drilling had often found them to be misleading. In NSW, geophysicists N. Merrick and D. O'Neill, in the author's Hydrogeological Section, contributed important developments in this field. One of the difficulties in Schlumberger depth sounding was the data shifts that often occurred when potential electrode spacing had to be expanded. An advance was made in this regard by Merrick (1974) with his Pole-Multidipole Method. This acquired simultaneous sounding data for a multi-electrode array, and also offset the ambiguities created by data shifts. It was a forerunner to the multi-electrode soundings and imaging arrays now in popular use.

In spite of the number of electrical resistivity surveys carried out, it would be fair to say that until 1971 the interpretation of the data was a tedious process and of dubious reliability. A major constraint was the difficulty of modelling the theoretical response of the fundamental model consisting of known layer thicknesses and resistivities. Overseas, Ghosh (1971) produced a major breakthrough by applying linear filter theory to the problem, providing a basis for simple, rapid and inexpensive modelling. However, Ghosh's "digital filter" was rudimentary and limited. In Australia, O'Neill (1975) refined Ghosh's technique by developing a more generally applicable filter for the widely used Schlumberger electrode array. Subsequently, the filter became a widely adopted standard for resistivity modelling that largely eliminated the constraints of Ghosh's original filter. O'Neill's filter was a catalyst for the subsequent development of resistivity "inversion" (Merrick, 1977) that finally paved the way for more reliable resistivity interpretation. Later, additional filters were developed by O'Neill and these were incorporated by O'Neill and Merrick into a methodology for resistivity modelling for a generalised (four electrode) array, eliminating the need to adhere to conventional arrays and providing a mechanism for efficient and flexible resistivity surveying. In 1973–74, these staff also developed the first automatic inversion method of interpretation that ran on a micro-computer (Merrick, 1977).

CONCLUSION

From the foregoing review, it will be evident that the 1960–1980 period was particularly important in the development and application of hydrogeology in Australia. 1980 has been taken as the as the cut-off point, but it is clear that by then hydrogeology had been accepted in Australia as a discipline in its own right. In this dry continent, groundwater inevitably played an increasingly important role, so that the future augured well for the hydrogeologist. Hydrogeology had really arrived!

ACKNOWLEDGMENTS

The author acknowledges with thanks information from Philip Commander, Stephen Hancock, Peter Jolly, Peter Dillon, Noel Merrick, David O'Neill, and Michael Williams, and assistance from Sue Irvine, Librarian, Dept. of Land & Water Conservation, NSW. Don Woolley kindly reviewed the text, and made helpful suggestions.

REFERENCES

Andrews, E.C. (1910) The Forbes–Parkes Goldfield. *Mineral Resources Report Geological Survey, NSW*, 13.

Australian Water Resources Council. (1965) *Review of Australia's Water Resources, 1963*. Canberra, Department National Development.

Australian Water Resources Council. (1975) *Groundwater Resources of Australia*. Canberra, Department of Environment and Conservation, AGPS.

Australian Water Resources Council. (1976) *Review of Australia's Water Resources, 1975*. Canberra, Department of National Resources, AGPS.

Clarke, W.B. (1850) *Artesian Well Board. Report to Colonial Secretary*. Papers of NSW Legislative Council, 6, 595–596.

Cox, W.G. (1906) *Irrigation with Surface and Sub-Surface Waters, with Special Reference to Geological Development and Utilisation of Artesian and Sub-Artesian Supplies*. Sydney, Angus and Robertson.

Department of the Coordinator-General of Public Works. (1954) *Artesian water supplies in Queensland*. Report of the Artesian Water Investigation Committee. Brisbane, Government Printer.

Dyson, D.F. & Wiebenga, W.A. (1957) *Geophysical investigation of underground water. Alice Springs NT*. Final Report. BMR Australian Record 1957/89.

Emerson, D.W. & Haines, B.M. (1974) The interpretation of geophysical well logs in water bores in unconsolidated sediments. *Bulletin Australian Society of Exploration Geophysicists*, 5, 89–118.

Ghosh, D.P. (1971) The application of linear filter theory to the direct interpretation of geoelectrical resistivity sounding measurements. *Geophysical Prospecting*, 19, 192–217.

Green, J.H. *et al.* (1965) University of New South Wales radiocarbon dates I. *Radiocarbon*, 7, 162–165.

Gregory, J.W. (1906) *The Dead Heart of Australia*. London, John Murray.

Habermehl, M.A. (1980) The Great Artesian Basin, Australia. *BMR Journal of Australian Geology and Geophysics*, 5, 9–38.

Hawkins, L.V. (1961) The reciprocal method of routine shallow seismic refraction interpretation. *Geophysics*, 26, 806–819.

Hillwood, E.R. (1967) Problems of locating water supplies in hard rock areas. *Mining Review*, 126, 80–85.

Hind, M.C. & Helby, R.J. (1969) The Great Artesian Basin in New South Wales. *Journal Geological Society of Australia*, 16, 481–497.

Home, F.J. (1897) *Report on the Prospects of Irrigation and Water Conservation in New South Wales*. Sydney, Government Printer.

Johns, M.W. & Lawrence, C.R. (1964) Aspects of the geological structure of the Murray Basin in north-western Victoria. *Geological Survey of Victoria, Underground Water Investigation Reports*, 10.

Johns R.K. (ed.). (1976) *History and Role of the Government Geological Surveys in Australia*. South Australia, Government Printer.

Jones, L.J. (1940) The Gulgong goldfield. *Geological Survey of NSW, Mineral Resources Reports*, 38.

Kalf, F.R. & Woolley, D.R. (1977) Application of mathematical modelling techniques to the alluvial aquifer system near Wagga Wagga, New South Wales. *Journal Geological Society of Australia*, 24, 179–194.

Knibbs, G.H. (1903) The hydraulic aspects of the artesian problem. *Journal Royal Society of NSW*, 37, 24–44.

Lawson, J.D. & Turner, A.K. (1970) Groundwater analogues. *Proceedings of the Groundwater Symposium, 1969. Reports Water Research Laboratory, University of NSW*, 113, 105–124.

Lloyd, A.C. (1934) Geological survey of the Dubbo district with special reference to the occurrence of sub-surface water. *Annual Report Department of Mines, NSW.* 89.

Lord, J.H. (1971) Underground water investigation for drought relief in Western Australia 1969/70, Final Report. *Geological Survey of WA. Annual Report* for 1970. 11–14.

McPharlin, D. (1970) Hydrogeologic electric analogue models constructed by the South Australian Mines Department. *Proceedings of the Groundwater Symposium, 1969. Reports Water Research Laboratory, University of NSW,* 113, 125–132.

Merrick, N.P. (1974) The Pole–Multidipole method of geoelectrical sounding. *Bulletin Australian Society Exploration Geophysicists,* 5, 48–64.

Merrick, N.P. (1977) A computer program for the inversion of Schlumberger sounding curves in the apparent resistivity domain. *Water Resources Commission, NSW. Hydrogeological Report* 1977/5.

Merrick, N.P. *et al.* (1978) A blind zone solution to the problem of hidden layers within a sequence of horizontal or dipping refractors. *Geophysical Prospecting,* 26, 703–721.

O'Driscoll, E.P. (1956) The hydrology of the Willochra Basin. *Report of Investigations, Geological Survey of South Australia,* 7.

O'Driscoll, E.P. (1979) Groundwater and its importance to the mineral industry. In: Prider, R.T. (ed) *Mining in Western Australia.* Perth, University of WA Press, pp. 167–177.

O'Driscoll, E.P. & Shepherd, R.G. (1960) The hydrology of part of County Cardwell in the upper south-east of South Australia. *Report of Investigations, Geological Survey of South Australia,* 15.

O'Neill, D.J. (1975) Improved linear filter coefficients for application in apparent resistivity computations. *Bulletin Australian Society Exploration Geophysicists,* 6, 104–109 [*Errata,* 7, 48].

Palmer, D. (1980) *The generalised reciprocal method of seismic refraction interpretation.* Tulsa, Oklahoma, Society Exploration Geophysicists.

Pilgrim, D.H. (1970) Digital models in regional groundwater studies. *Proceedings Groundwater Symposium, 1969. Reports Water Research Laboratory, University of NSW,* 113, 133–151.

Pittman, E.R. (1901) *Mineral Resources of New South Wales.* Sydney, Government Printer.

Pittman, E.R. (1907) *Problems of the Artesian Water Supply of Australia with Special Reference to Professor Gregory's Theory.* Sydney, Government Printer.

Pittman, E.R. (1914) *The Great Artesian Basin and the Source of its Water.* Sydney, Government Printer.

Polak, E.J. & Horsfall, C.L. (1979) Geothermal gradients in the Great Artesian Basin, Australia. *Bulletin Society Exploration Geophysicists,* 10, 144–148.

Queensland Irrigation and Water Supply Committee. (1965) *Progress report on groundwater investigations of Dumaresq River alluvium, AMTM25 to AMTM110.* Groundwater Group Report, 409 [unpublished].

Rawlinson, T.E. (1878) Subterranean water supply in the interior. *Transactions Philosophical Society of Adelaide South Australia,* 124–126.

Rayner, J.M. (1940) Magnetic prospecting of the Gulgong deep lead. *Geological Survey of NSW. Mineral Resources Reports,* 38, Part 2.

Russell, H.C. (1879) The River Darling – The water which should pass through it. *Journal Royal Society,* 13, 169–170.

Seidel, G. (1980) Application of the GABHYD groundwater model of the Great Artesian Basin, Australia. *BMR Journal Australian Geology and Geophysics,* 5, 38–45.

Smythe, B.R. (1869) *The goldfields and the mineral districts of Victoria.* Melbourne, Government Printer.

Symmonds, R.S. (1912) *Our Artesian Waters: Observations in the Laboratory and in the Field.* Sydney, NSW Government Printer.

Thompson, B.R. (1978) On the thermal waters of the Gippsland Basin and the problems associated with the study of high temperature aquifers. *Hydrogeology of Great Sedimentary Basins,*

Memoirs of the International Association of Hydrogeologists Conference, Budapest, 1976. IAH Memoirs, 11, 492–509.

Vallance, T.G. & Branagan, D.F. (1968) NSW geology — Its origins and growth. In: *A Century of Scientific Progress.* Centenary Volume, Royal Society of NSW, 9.

Wiebenga, W.A. (1955) Geophysical investigation of water deposits of Western Australia. *BMR Australian Bulletin,* 30.

Williamson, W.H. (1958) *Groundwater resources of the Upper Hunter Valley, New South Wales.* Sydney, Government Printer.

Williamson, W.H. (1964) The development of groundwater resources of alluvial formations. *Water Resources Use and Management. Proceedings of Australian Academy of Science Symposium, Canberra, 1963.* Victoria, Melbourne University Press, Victoria, pp. 195–211.

Williamson, W.H. (1970) Groundwater in unconsolidated sediments. Recent developments in NSW. *Proceedings of the Groundwater Symposium, University of NSW, August 1969. University of NSW Water Research Laboratory Reports,* 113, 1–12.

Williamson, W.H. (1978) Hydrogeological aspects of coal mining under stored waters near Sydney, Australia. In: *Water in Mining and Underground Works.* Granada, Spain, SIAMOS, 1, pp. 309–328.

Williamson, W.H. & Woolley, D.R. (1980) Hydraulic fracturing to improve the yield of bores in fractured rock. *AWRC Technical Paper,* 55, Canberra, AGPS.

Wopfner, H. & Douglas, J.G. (eds.). (1971) *The Otway Basin of south–eastern Australia.* Special Bulletin Geological Surveys of South Australia and Victoria.

Island hydrogeology: Highlights from early experience in the West Indies and Bermuda

N.S. Robins

British Geological Survey, Maclean Building, Wallingford, Oxfordshire, UK

ABSTRACT

Until the 1950s the development of groundwater resources in the Caribbean was limited to occasional drilling on volcanic islands and an understanding that use of shallow dolines and shallow wells in the low-lying limestone islands was preferable to deeper exploitation using boreholes. Although the theory behind fresh water lenses was developed in the late nineteenth century a practical understanding was only formalized in the 1970s. Some important island-wide studies were then carried out by engineering consultancies as part of national development plans. This early work in the West Indies and Bermuda has contributed significantly to the understanding of the hydraulics of the fresh water lens and of the water balances of small islands.

INTRODUCTION

The Caribbean archipelago mainly comprises low elevation limestone reef islands and high elevation volcanic islands. The diversity of relief and geology has enabled many groundwater resource development techniques to be applied in these islands, albeit with varying degrees of success. The present day understanding of the fresh water lens beneath low-elevation karst limestone islands stems from work carried out in Bermuda, Cayman Islands and Bahamas, whilst work on other islands in the region has contributed to understanding of groundwater flow in young volcanic terrain.

With hindsight, there are also some Caribbean examples of poor practice. For example, the detrimental effect of the Grand Lucayan Waterway upon the fresh water lens on Grand Bahama proved to have an irreversible impact upon the groundwater resource potential of the island. It begs the question whether the abandonment of the Trans-Florida barge canal project in the 1970s took heed of the lessons then being learnt in the Bahamas in the wake of the Grand Lucayan Waterway experience on Grand Bahama? There are several examples of abstraction regimes inducing seawater intrusion; that in Cuba during the introduction of rice cultivation in the late 1940s is probably the worst (United Nations, 1976).

Rainfall in the Caribbean is largely controlled by orographic and rain-shadow effects – hence the Windward and Leeward Islands. It is affected by prevailing winds and by land elevation; contrast the hot, dry, low-islands of the Turks and Caicos group with the higher elevation rainforests of the volcanic islands of Grenada and St. Vincent and on a smaller scale that of Nevis (Figure 1). Rainfall over the interior of these high

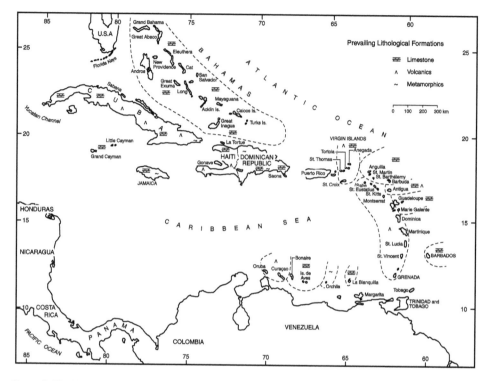

Figure 1   The larger Caribbean Islands showing the main lithological formations (after United Nations, 1976).

topography islands drains, in part underground, through pyroclastic and reworked volcanic deposits to the arid coastal lowlands where it augments otherwise sparse surface resources. Nevertheless, few surface streams actually reach the sea except at times of spate.

Carlozzi and Carlozzi (1968) used vegetation characteristics to emphasise the effects of orographic rainfall and rain shadow within a threefold classification of islands in the Lesser Antilles (Table 1).

It is notable that few of the 'Rain Forest Islands' contain any karst limestone. In contrast nearly all the more arid 'Seasonal Forest Islands' and all the dry 'Tropical Desert Islands' are sustained by karst aquifers. This simple classification belies the economic and social importance of karst aquifers in the Caribbean.

The maritime climate of the region, located between latitudes 12 and 27°, is dominated by the prevailing westerly Trade Winds that have traversed the tropical Atlantic Ocean. The moisture laden Trade Winds are strongest during the winter months; cumulus cloud development mainly occurs during the summer when the inversion layer is higher in elevation or absent altogether (Nieuwolt, 1977). Consequently, rainfall maxima tend to occur between July and October. Easterly depressions are most frequent between July and September and these disturbances bring further rainfall to much of

Table I   Island types (after Carlozzi and Carlozzi, 1968).

Rain Forest Islands:

Dominica	Grenada
Guadeloupe*	Martinique
Montserrat	Nevis
Saba	St. Kitts
St. Lucia	St. Vincent
Tobago	Trinidad*

Seasonal Forest Islands:

Anguilla*	Antigua*
Barbados*	Barbuda*
St. Barthelemy*	St. Eustatius
St. Marten*	

Tropical Desert Islands:

Aruba	Bonaire*
Curacao	

*Islands with significant karst aquifers

the region within these months. In addition the region may suffer the effects of up to eight hurricanes per year.

Given this environmental setting and the development of colonial interests in the islands during the last two centuries, access to groundwater has been pivotal to the economic development of the region. Initially, however, groundwater exploitation was primarily based on drawing water from natural 'slobs', dolines in the limestones, and from traditional hand dug wells excavated into volcanic detritus around the periphery of the volcanic islands (Earle, 1924). Sugar was the main cash crop until the 1960s, and this required fresh water for irrigation and milling. The water supply demands of the industry resulted in the drilling rig being introduced to some islands, although little was actually accomplished in the way of aquifer development by boreholes until the 1950s when the sugar industry was at its peak. Initially groundwater investigation and development was at best ad hoc, but during the ensuing 25 years detailed studies added greatly to the understanding of the fresh water lens beneath low-lying karst limestone islands.

The earliest surveys of the relationship between geology and groundwater occurrence were undertaken in Carriacou (Lehder, 1935), Barbados (Serin, 1946), in Puerto Rico (McGuinnes, 1948) and St Croix in the US Virgin Islands (Cederstrom, 1950). Little is reported from the 1950s other than a number of strategic groundwater surveys for military purposes (Klein et al., 1958). In the late 1960s and through the 1970s a number of detailed surveys were commissioned for some of the islands (e.g. Guyton, 1966; Halcrow, 1966; Tahal, 1971; Stanley Associates, 1978). The United Nations (1976) report on groundwater in the western hemisphere included 13 reports on the occurrence of groundwater in major Caribbean islands. Nevertheless, Robins et al. (1990) concluded that little of the hydrogeological investigation that had been carried out in the region had been reported in the open literature and what was present in public records was not easy to access. Notwithstanding, the reports indicate that a deeper understanding of island water balances was available for many Caribbean islands in the 1970s than was the case in comparable environments in much of the rest of the world at that time.

THE VOLCANIC ISLANDS

An early example of interest in groundwater is a map delineating areas of saline and fresh groundwater in Antigua (Tempany, 1914).

The London based drilling contractor Isler & Company deployed a percussion rig to the region to drill boreholes for public supply on some of the volcanic islands from the early 1930s onwards. Latterly, Drilling & Petroleum Operations of Trinidad undertook a number of water drilling contracts on a variety of islands in the 1940s.

The earliest documented survey of water supply on the volcanic island were carried out by Martin-Kaye (1954; 1956) in his role of Government Geologist to the Leeward Islands following his appointment in 1951. Martin-Kaye described the water supply in Antigua focussing on the more permeable alluvial deposits:

> Possessing no adequate permanent streams public supplies are derived from impounded rainy weather stream flow, a few small springs, and from wells. Drilling in the last few years (early 1950s) in the alluvium has added about 2300 m^3 per day to the island's water supply picture.

An early example of abstraction exceeding the renewable resource in a volcanic island was voiced by Martin-Kaye (1959) when reviewing the prospect of a new supply borehole to support the town of Basseterre in St. Kitts. He concluded that proposals to increase the supply of groundwater to 4500 m^3 per day would draw one third of the long term potential, but the basis of his calculations are not reported. Two wells then provided 1470 m^3 per day with additional requirements drawn from ephemeral streams. On the adjacent island of Nevis, a 150 mm diameter borehole was drilled to 55 m in 1945 to supplement a high elevation spring draining from a catchment in the wet and wooded central part of the island. The borehole was only commissioned in 1951 after an earthquake broke the delivery pipe from the spring.

THE LIMESTONE ISLANDS

The concept that fresh water is lighter than seawater and can float above it was recorded in the writings of Pliny the Elder (23–79) some 1900 years ago. Joseph Du Commun, teaching at the West Point Military Academy from 1818–1831, picked up the same theme (Carlston, 1963). The hydrostatic relationship between immiscible fresh water and salt water bodies was later investigated by Badon Ghyben (1888/89) and Herzberg (1901), and subsequently described as a dynamic equilibrium by Hubbert (1940). In the Caribbean, work by Mather and Buckley (1973), Mather (1975), LRD (1971/7) and UNDP (1977) consolidated this concept. However, it had long been the practice on New Providence Island in the Bahamas to develop fresh water sources from a network of shallow boreholes and wells to reduce the drawdown from any single well, thereby limiting upconing of the fresh water/saline water interface (Riddel, 1933; Stubbs and Langlois, 1954).

The landmark report by Vacher (1974) and subsequent paper (Vacher, 1978), working in Bermuda, independently championed the concept of developing fresh water lenses overlying salt water.

The high secondary hydraulic conductivity of karst island aquifers is balanced only by low hydraulic gradient in order to retain groundwater within the aquifer. The areal size of a karst aquifer is, therefore, critical to the volume of fresh water that it can retain. In the small island aquifer, recharge may effect a temporary rise in water level, but this quickly subsides as throughflow to creeks and the coast displaces much of the recharging volume of groundwater to the sea.

The importance of the dynamic mixing zone was recognised in the Caribbean at an early stage. Electrical resistivity soundings on Exuma and other islands were used by LRD (1971/7) to reveal the characteristic salinity curves of the lens (Figure 2). The thickness of the mixing zone in North Andros and Grand Bahama ranges from 12 to 20 m at the coast to just 1 m at the centre of the islands some 4 km inland.

For the most part in the Caribbean, the Ghyben-Herzberg principal has been found only to apply in part because the caveats of this principle include porous medium, infinite homogeneous aquifer, and many of the other standard Theisian controls. Given the widespread occurrence of karst conduit systems in the islands at the present sea level as well as at a lower fossil sea level, and its associate karst development, the application of the Ghyben-Herzberg principle is fraught in many island situations. Conduit flow destroys the pressure balance, and blue holes or cenotes allow sea water dymanic access to the interior of an island. In any case, the ratio of the elevation of

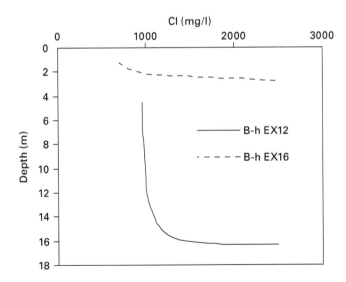

Figure 2   Resistivity soundings depicting the dynamic mixing zone beneath Exuma; EX12 shows a shallow near shore mixing zone whereas EX16 represents the sharper and deeper interface found further away from the shore (after LRD, 1971/7).

the water table to the depth of the saline interface beneath sea level is normally much less than 1:40, perhaps only 1:20 or less.

The difficulties inherent in determining the volume of a fresh water island lens are well illustrated by the varied estimates made over the years for long term yield of the Little Valley groundwater lens on Grand Cayman. The original estimate declined some fifteen-fold over a period of only 15 years (Table 2).

Predictions of lens volume with island size and shape have been made by Mather and Buckley (1973) and developed by Cant and Weech (1986) who demonstrated a clear relationship between the volume of freshwater stored in an island groundwater lens and the areal size of the island (Figure 3). The fresh water lens thickness varies approximately with the square root of the lens area and the cube root of the lens volume, but it is more particularly dependent on aquifer transmissivity and frequency of recharge events. Furthermore, the more extensive lenses only occur where annual rainfall exceeds 1150 mm. Rainfall less than this amount was demonstrated to be ineffective against the prevailing evapotranspiration, and brackish groundwater may

Table 2 Yield estimates (m³ per day) for the Little Valley groundwater lens on Grand Cayman (after Hukka and de Waal, 1983).

Consultant	Year	Yield
Black, Crown & Eidness	1966	4500
Reid, Crowther & Partners	1968	3800
Institute of Geological Sciences	1971	1300
Wallace Evans & Partners	1974	1100
Richards & Dumbleton	1975	270
Richards & Dumbleton	1980	310

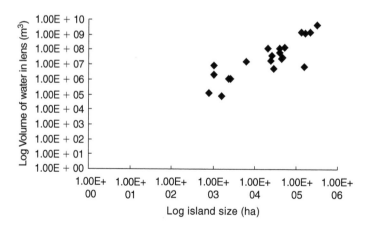

Figure 3 Log of island size versus volume of fresh water lens for 21 Bahamian islands (after Cant and Weech, 1986).

occur with hypersaline ponds at the surface. Shallow water table conditions in which the water table is within the root zone, for example on Great Inagua which is otherwise a relatively large island, also promote saline ground water.

Predictions of the variations of lens volume with island size and shape were made by Budd and Vacher (1990). Overestimates of the volume of fresh water are common particularly in low-lying islands where brackish and salt groundwaters are present at shallow depth and which reduce the effective 'hydraulic' area of the island. 'Fossil' karst levels occur at depth within the saline part of the aquifer and these tend to modify the influence of the hydraulic conductivity and hydraulic gradient in the fresh water zone itself.

Seasonal variation in water level within island freshwater lenses in karst limestone vary from a few centimetres in Anegada, in the British Virgin Islands, to 1 m on the smaller islands of the Bahamas and 3 m on some of the larger islands. These seasonal increases in storage do not reflect the total input from recharge minus abstraction, because the greater part of the balance is lost to throughflow to the sea during and immediately after the recharge event.

The delicate nature of the fresh water lens is illustrated by the Grand Lucayan Waterway Project. The Grand Lucayan Waterway on Grand Bahama was dug in 1977. A number of marina type developments have been cut into various islands, but few have had the dramatic effect on groundwater resource potential as this development. The main channel is up to 250 m wide and 2 m deep; it was cut in 1977 through the coastal barrier sands and into the karst limestone for a length of 4.5 km. Construction was undertaken in dewatered sections of canal with pumping from the then fresh water lens at rates up to 3 m^3 per second. This caused massive upconing of saline water. Opening of the ends of the canal to the coast allowed ingress of salt water; happily a further 5 km of waterway, now brackish due to open water evaporation, has never been connected to the sea. Nevertheless, the project destroyed the former 12 m thick freshwater lens replacing it with saline or at best brackish water over a 3 km wide swathe across the width of the island.

Considerable effort has been made in Jamaica in order to determine the regional storativity of the extensive White Limestone Group karst aquifer. White (1985) discusses various techniques, including laboratory measurements of porosity on core (0.006 to 0.03); aquifer tests (0.01 to 0.04); analogue determinations (in which storativity was not found to be a critical factor in simulating field observations); and by water balance determinations which relate the amount of water coming out of storage with the resultant reduction in saturated volume (0.03). Versey (1959) reported declines in water quality in the fresh water store as early as 1948, recovering with intensive rains in the early 1950s and declining again over the next eight years indicative of over pumping the lens.

HISTORY OF NASSAU WATER SUPPLY

The context of historical water supply development in the Caribbean region is illustrated by the water supply for Nassau. Over 60% of the population of the Bahamas live in Nassau, which is the capital city and is situated on New Providence. Of the remainder, 15% live in Grand Bahama, and the rest are distributed between the many

Table 3 Sources of supply to New Providence, Bahamas in 1980 (after Cant, 1980).

Source	Production ($10^3 m^3$ per day)	Percentage
Public supply wellfields	11	25
Private supply wellfields	6	15
Desalination	2	5
Barge imported water from Andros	4	10
Private, wells, roof catchments, and private desalination	18	45

other islands, only 14 of which have land areas greater than 50 km². New Providence was supplied solely by groundwater from the fresh water lens until 1961; not surprisingly supplementary supplies are now derived from other sources. By 1980 the supply to New Providence was still dominated by groundwater (Table 3) but anthropogenic pollution and increased demand has seen a reduction in the groundwater component.

Urban development of Nassau began in the early Nineteenth Century; St. Matthew's Parish Church was built by 1804, the Royal Victoria Hotel by 1861. Early water supply to the town derived from hand dug wells and roof top catchments. The first group of eight boreholes on New Providence was put into the Blue Hill Wellfield in 1928. This was replaced in 1937 by four trenches and the site was closed down altogether in 1943 having gone brackish (it was later reopened in 1967). Better locations were found with thicker fresh water lenses: ten trenches and 7 boreholes inaugurated the Prospect Wellfield in 1938, a further 100 boreholes, each equipped with windmills, followed in the 1940s and forty more in the 1950s. The Perpalls and Windsor wellfields were developed throughout the 1940s and 1950s with over 500 boreholes and numerous trenches. The biggest is the Cousins Wellfield which possesses over 1000 boreholes developed from 1963 onwards. The design criterion was that drawdown be limited to between 0.03 m and 0.30 m (the latter revised to 0.07 m in later years) depending on the available thickness of fresh water.

A major review of the government wellfields and their future potential was carried out by Peter Hadwin (UNDP, 1977). This study revealed a comprehensive understanding of the dynamics of the freshwater lens on New Providence. Hadwin reported:

> Government and commercial wellfield rose from 905 to 2025 ha between 1965 and 1968; total production (including desalination) doubled between 1966 and 1973. Density of well development is much too great in some wellfields and insufficient in others, but not all unpumped sources should be brought back into production. Some 30% of the wells are too deep and draw in saline water; many others are in poor condition, and about 10% are caved in, blocked or too shallow.

Growing concern in the 1980s about rising chloride levels on New Providence led to a realisation that the presumed recharge rate of about 0.3 m per year should, according to revised meteorological calculations, only be about 0.2 m per year (Watson, 1984). Given this new recharge estimate, abstraction figures for Government wellfields in New Providence showed that groundwater had been exploited between 1958 and

1983 at a rate some 50% greater than the theoretical renewable resource. The effect of this had been a slow increase in chloride concentrations which to date has not been rectified. Currently 20% of supply is derived from desalination. Desalination in the region is typically from saline water drawn from beneath the fresh water lens which provides constant salinity and temperature and a scavenging effect on the salt water interface.

SOME SPECIAL TECHNIQUES

During the late 1960s and early 1970s, experiments were conducted on the so called 'aquavoir' system of preventing the lateral movement of fresh water from a coastal aquifer at a site at Coral Harbour, New Providence. The system provided an impermeable vertical membrane which contains an area of aquifer into which infiltrating rainwater will gradually displace brackish groundwater, the latter escaping under the bottom of the membrane. Controlled abstraction may then take place from the contained fresh groundwater.

The small pilot installation proved encouraging, although this site was overrun by a hurricane before full use could be made of the facility. The Coral Harbour site was established on a small isthmus of oolitic limestone, in which groundwater with a chloride concentration of $7000 \, \text{mg} \, \text{l}^{-1}$ stood at a depth of 1 to 2 m below ground level. The porosity of the limestone was as high as 44%. A group of six circular membrane units were constructed, each with a ground surface areas of up to 0.5 ha; arrays of observation boreholes were also installed. The membranes penetrated up to 3 m below mean groundwater level (Howard Humphreys, 1971).

Operational records of the aquavoirs included rainfall, potential evapotranspiration estimates, tidal and groundwater level monitoring as well as monitoring the specific electrical conductance of groundwater in the observation boreholes. Observations showed a reduction in chloride concentration of between 55 and 88% over one year, and between 53 and 88% over two years; one cell actually deteriorated in year 2, and another remained steady. These inconsistent outcomes were likely the result of tidal mixing.

Although the technique was patented by a consulting engineer in Nassau with corresponding applications filed in forty two countries, the aquavoir did not survive to become a household name on tropical islands. The limitations of product quality and quantity along with high initial capital cost greatly reduced the attraction of the aquavoir and its development did not proceed.

Use of trenches or covered infiltration galleries has been made in certain areas in order to spread the stress on a shallow fresh water lens linearly. On North Andros and New Providence, shallow trenches about 1 m deep and 300 m long intersect the water table to facilitate abstraction from the fresh water lens. These trenchfields lose water to evaporation and increases in salinity may occur even if they are not pumped; pumping throughout the dry season may effect an increase in chloride concentration of about $70 \, \text{mg} \, \text{l}^{-1}$, but non-pumping may also see an increase of up to $40 \, \text{mg} \, \text{l}^{-1}$.

In Cuba, canals have been cut into some of the karst aquifers in order to promote artificial recharge by intercepting surface runoff. Artificial recharge wells are also used in some basins in Cuba to inject surplus surface water underground. This development resulted from the introduction of rice cultivation in 1948 (Gonzáles-Báez, 1987). Over a five year period the salt water interface was pulled inland by about 20 km with conse-quent salinisation of both soil and aquifer due to continued irrigation with increasingly brackish water. The situation was only rectified by complete cessation of pumping for a period of 12 years and the subsequent introduction of artificial recharge.

THE CARIBBEAN INFLUENCE

Hydrogeological investigation for the development and management of groundwater resources in the Caribbean has been at the forefront of much of the understanding of the hydraulics of small tropical islands over the last fifty years. The classic work on the fresh water lens carried out in a number of low-lying karst limestone islands is well known, and workers such as Richard Cant, Philip Weech and Leonard Vacher are synonymous with this area of study. The work reported from Bermuda by Vacher (1978) remains a classical hydrogeological paper. In addition the major national studies carried out by a number of consulting organisations, not least that of LRD in the Bahamas, have provided considerable insight into the management of fresh water lenses.

Work in the volcanic islands has also contributed to the present day understanding of small-island hydrogeology, particularly regarding the issue of the island wide water balance. However, much of this work has never been presented in the literature and is not easy to locate, e.g. work in St. Kitts by Halcrow (1966) and subsequent work by Christmas (1977), both these classic studies of recharge and water balance, remain largely unknown.

The key impact from the hydrogeological development in the Caribbean is under-standing of the fresh water lens. Perhaps also the vision of Martin-Kaye in the 1950s, ashore from his schooner for a day of fieldwork, is equally lasting in its own picturesque way. Many hydrogeologists from diverse backgrounds have enjoyed work-ing in the region, long may this region remain a seat of groundwater investigation and innovation.

ACKNOWLEDGEMENTS

The author is grateful to Joe Troester and the late Bill Back, formerly with the USGS, to regional water engineers such as Brian Kennedy and Gary Penn, and to Philippe Barragne-Bigot, formerly with the UN Caribbean Small Oceanic Islands Water Project, all of whom helped unravel the groundwater development history for the region. John Mather is also thanked for pointing towards various documents and papers which describe the development of groundwater understanding in the region.

REFERENCES

Badon Ghyben, W. (1888/89) Nota in Verband met de Voorgenomen Putboring nabij Amsterdam. Tijdschrift van het koninklyk Instituut van ingenieurs. The Hague.
Budd, D.A. & Vacher, H.L. (1990) Predicting freshwater lenses in carbonate palaeo-islands. *Journal of Sedimentary Petrology*, 61, 43–53.

Cant, R.V. (1980) The occurrence and development of groundwater resources in the Bahamas. UN/CSC Seminar, Barbados.

Cant, R.V. & Weech, P.S. (1986) A review of the factors affecting the development of Ghyben–Herzberg lenses in the Bahamas. *Journal of Hydrology*, 84, 333–343.

Carlozzi, C.A. & Carlozzi, A.A. (1968) Conservation and Caribbean regional progress. St Thomas, USVI, The Caribbean Research Institute.

Carlston, C.W. (1963) An early American statement of the Badon–Herzberg Principle of static fresh-water–salt-water balance. *American Journal of Science*, 261, 88–91.

Cederstrom, D.J. (1950) Geology and ground water resources of St Croix, Virgin Islands. *US Geological Survey Water Supply Paper 1067*.

Christmas, J. (1977) *Hydrogeology of an unconfined coastal aquifer system, Basseterre Valley, St Kitts*. PhD Thesis, University of West Indies, Jamaica.

Earle, K.W. (1924) *Reports on the Geology of St Kitts-Nevis, BWI and the geology of Anguilla, BWI*. The Crown Agents for the Colonies.

Gonzales-Baez, A. (1987) Case history No 10, exploitation of open coastal aquifers in Cuba. In: *Groundwater Problems in Coastal Areas*. Paris, UNESCO.

Guyton. (1966) *Report on groundwater investigation on part of Great Abaco Island, Bahama Islands*. William F. Guyton & Associates.

Halcrow. (1966) *Report on the water resources of St Kitts, Nevis and Anguilla and on their development*.London, Sir William Halcrow & Partners.

Herzberg, A. (1901) Die Wasserversorgung einiger Nordseebader. *Journal Gasbeleuchtung und Wasserversorgung*, 44, 815–819.

Humphreys, H. (1971) *Report on research into the conservation of rainwater by underground storage at sea level*. Reading, Howard Humphries & Sons.

Hubbert, M.K. (1940) The theory of groundwater motion. *The Journal of Geology*, 48 (8), 785–944.

Hukka & de Waal. (1983) Unpublished report, United Nations Department of Technical Co-operation and Development.

Klein, H., Roy, N.D. & Sherwood, C.D. (1958) *Geology and groundwater resources in the vicinity of the Auxiliary Air Force Bases, BWI Report*. US Geological Survey.

Lehder, E. (1935) *Report on the possibilities of establishing an artesian water supply for the island of Carriacou, with appended notes on the general geology of Carriacou*. St George, Grenada, Government Printing Office.

LRD. (1971/7) *Land Resources of the Commonwealth of the Bahamas*. Report in six volumes. UK, Land Resources Division.

Martin-Kaye, P.H.A. (1954) *Water supplies in the British Virgin Islands*. Report to Government, British Guiana.

Martin-Kaye, P.H.A. (1956) *The water resources of Antigua and Barbuda*. Report to Antigua Government.

Martin-Kaye, P.H.A. (1959) *Reports on the geology of the Leeward and British Virgin Islands*. Reports to the Governor of the Leeward Islands.

Mather, J.D. (1975) Development of the groundwater resources of small limestone islands. *Quarterly Journal of Engineering Geology*, 8, 141–150.

Mather, J.D. & Buckley, D.K. (1973) Tidal fluctuations and groundwater conditions in the Bahamian Archipelago. *Transactions of the Second International Symposium on Groundwater, Palermo, Italy*. pp. 165–171.

McGuinnes, C.L. (1948) *Ground water resources of Puerto Rico*. Report Puerto Rico Aqueducts and Sewer Service, San Juan.

Nieuwolt, S. (1977) *Tropical Climatology, An Introduction to the Climates of Low Altitudes*. London: John Wiley & Sons.

Riddel, J.O. (1933) Excluding salt water from island wells, a theory of the occurrence of groundwater based on experience at Nassau, Bahama Islands. *Civil Engineering*, 3, 383–385.

Robins, N.S., Lawrence, A.R. & Cripps, A.C. (1990) Problems of ground-water development in small volcanic islands in the Eastern Caribbean. *Proceedings of the International Symposium on Tropical Hydrology and Fourth Caribbean Islands Water Resources Congress, San Juan.* pp. 257–267.

Serin, A. (1946) *Geological investigation of the ground water resources of Barbados, BWI.* Report British Union Oil Company.

Stanley Associates. (1978) *A joint study by the Government of Barbados.* Report Stanley Associates Limited.

Stubbs, G.C. & Langlois, A.C. (1954) Water supply of Nassau, Bahamas. *Journal of the American Water Works Association*, 46, 220–230.

Tahal (1971) *Hidrogeologia del Valle de Azua.* Report, Tahal Consulting Engineers.

Tempany, H.A. (1914) Memorandum on the geology of the groundwater of the island of Antigua, BWI. *West Indian Bulletin*, 14, 280–303.

United Nations (1976) *Groundwater in the Western Hemisphere.* United Nations, New York, Department of Economic and Social Affairs Natural Resources/Water Series No. 4.

UNDP (1977) *The present status and future potential of government wellfields, groundwater studies in New Providence, Bahamas, Technical Report 1.* Report UNDP, New York, BHA/74/004.

Vacher, H.L. (1974) *Ground water hydrology of Bermuda.* Report Public Works Department, Bermuda.

Vacher, H.L. (1978) Hydrogeology of Bermuda: significance of an across-the-island variation in permeability. *Journal of Hydrology*, 39 (3–4), 207–226.

Versey, H.R. (1959) The hydrologic character of the White Limestone Formation of Jamaica. *Transactions of the Second Caribbean Geological Conference, Puerto Rico.* pp. 59–68.

Watson, D. (1984) *Implications of low recharge to New Providence wellfields.* Water & Sewerage Corporation, Nassau/UNDTCD Report, Project BHA 150-2-001.

White, M.E. (1985) Groundwater movement and storage in karstic limestone aquifers in Jamaica. *Journal of the Geological Society of Jamaica*, 23, 1–16.

History of hydrogeology in China

Han Zaisheng[1] & Chen Mengxiong[2]
[1]*China Geological Survey, Xicheng District, Beijing, P.R. China*
[2]*Ministry of Land and Resources, Beijing, China*

ABSTRACT

The history of Hydrogeology of China can be divided into four stages: the Primitive stage, Preliminary stage, Foundation stage and Development stage. Hydrogeology as a branch of geological sciences was actually started and established in the 1950s. The regional hydrogeology, agricultural hydrogeology, environmental hydrogeology, quantitative hydrogeology and information hydrogeology have developed.

THE PRIMITIVE STAGE (PRIOR TO THE 20TH CENTURY)

Knowledge on the exploitation and utilisation of groundwater had a long history in ancient China. The oldest well discovered in Zhejiang Province was built before 5700–3710 years BC. As early as 2000 years ago, the "karez", also known as the "kanr-well", for irrigation in desert areas and the exploitation of deep brine-wells for salt mines in the Sichuan red basin are known to have been used (Figure 1). A great deal of information about springs may be found in various encyclopedic and geographic writings. The book *Erya* is the earlies record whiich was written in 200 BC. A good description of the springs occurs in the early 6th century in the book entitled "Records of Strange Things". Xu Xiake (1586–1641) was an explorer of the regions of China. He described the characters of karst cave and physiography in the south-western part of China in his travel notes. His discoveries on karst geology were advanced for that time.

PRELIMINARY STAGE (1900–1949)

In China, the geologic sciences used for studying underground water were begun only in the first half of the twentieth century. The first deep well in Shanghai city was dug in 1860. There were only 22 deep wells for water supply until 1921 in Shanghai, and the yield of groundwater was more than 300 000 m^3 per year. Geologist Xie Jiarong published a report about groundwater for urban water supply in Nanjing city in 1929. The preliminary groundwater investigation in Tianjin city was begun in the 1930s. There were also reports about groundwater for urban water supply for cities including Peking, Xian, and Lanzhou during 1930–1940. Several geologists surveyed the natural springs in the northern part of China, and a paper about the hot springs in China was

Figure 1   Deep well for saline groundwater in Sichuan, China in the 12th century.

published by geologist Zhang Hongzhao in 1934. There were also other publications on groundwater during that period.

FOUNDATION STAGE (1950–1970)

Hydrogeology as a branch of geological science was actually started and established in the 1950s after the founding of P.R. China. The decade of the 1950s was the period characterised by the development of regional hydrogeology; while the 1960s was the period of agricultural hydrogeology. In both periods, the development of hydrogeology was much influenced by the basic conceptions or theories of the scientists of the former Soviet Union. Hydrogeology science achieved great progress in parallel with rapid economic development.

Since the mid-1950s, regional hydrogeological surveying has been carried out all across China, stimulating the development of regional hydrogeology. In 1958, the first 1:3 000 000 scale Chinese hydrogeology maps and the first monograph on the "Regional Hydrogeology Outline of China" were published. In 1959, the first book on "Practical Hydrogeology", compiled on the basis of domestic material, was published for the 10th anniversary of the foundation of China. During the 1970s and 1980s, the regional hydrogeological surveys of China were completed, except for a few difficult/complex regions. Subsequently, China began to carry out surveys to evaluate regional groundwater resources and environmental hydrogeology in key economic development regions.

The Bureau of Hydrogeology and Engineering Geology was set up under the Ministry of Geology in 1955, and was responsible for the investigation and research of groundwater resources across the whole country. It co-operated with related sections of the water conservancy, urban reconstruction and industrial sectors in groundwater exploitation for agriculture and industry. Provincial bureaus set up their own hydrogeology teams to undertake their tasks. To coordinate the prospecting works, some units were set up to engage in scientific research. In respect of education, five geological colleges were established with departments of hydrogeology and engineering geology to train specialists in those fields. The first Chinese language university teaching material- "*Basic Hydrogeology*", written by Prof. Wang Dachun, was published in 1956.

In the early 1960s, North China suffered a severe drought which lasted for several years. In order to overcome the drought, extensive well drilling and drought-defying campaigns were conducted in the North China plain. The main focus of Chinese hydrogeology changed from regional hydrogeology to agricultural hydrogeology including hydrogeology for developing well-irrigation. In order to realise water supply for cropland and the reclamation of alkaline land, substantive investigation work was implemented and many maps were compiled. These activities made an important contribution to the development of well irrigation and water conservancy over farmlands.

Much specialised prospecting works had been accomplished for the exploitation of groundwater and the improvement of saline soils as well as the control of swamps, especially in North China. Experiences were also accumulated in water exploration in mountainous regions and pasture lands. A great number of hydrogeologic maps at scales of 1:50 000–1:100 000 were completed for agricultural water supply or other purposes, which provided a scientific basis for the rational arrangement of groundwater use.

DEVELOPMENT STAGE (AFTER 1971)

From the 1970s onwards, the main feature was the development of environmental hydrogeology and quantitative hydrogeology (water-resources hydrogeology). During this period, environmental hydrogeology became the main task to meet the needs of rapid urban development and expansion. Enormous investigation works were undertaken to facilitate water supplies for large and medium-sized cities and industrial bases. Many important cities, including Beijing, Xi'an, Taiyuan, Tianjin and Shenyang are largely dependent upon groundwater supply. With the growth of the urban population and the rapid development of the industrial sector, water demand increased greatly. In the urban areas, environmental problems related to groundwater abstraction frequently arose, such as excessive exploitation and water pollution, leading to seawater intrusion and land subsidence. Many monitoring stations were set up in important cities. Research investigations studied the origins, mechanisms, and basic rules of water pollution and seawater intrusion to assist with their control or elimination. Intensified experimental tests and research on artificial recharge were undertaken in Beijing and other cities in order to expand groundwater resources.

Since the 1970s, surveys and investigations have been carried out in many cities in China to assess the current status of groundwater pollution, including pollution sources, pollution means, composition, degree, distribution and trends. Such investigations improved continuously, moving from single to multiple pollutant evaluations; from single factor to comprehensive evaluations, and from assessing current status to evaluating future trends; from statistical analysis to the construction of mathematical models of groundwater quality. These cities conducted studies of groundwater pollution processes including pollutant movement, accumulation, transformation and self-remediation, especially filtering, physical chemical sorption, ion exchange, enrichment or diluting effect of pollutant, and extenuative effect of radioactive elements. Studies on water quality modelling have made great progress, allowing the trends in water pollution to be predicted and controlled. As to calculation methods, most models adopted finite element or finite difference methods to improve the accuracy of parameters and the creditability and reliability of results.

Various side effects caused by groundwater over exploitation became important components of environmental hydrogeology, such as groundwater depletion, water quality deterioration, seawater intrusion, land subsidence, karst collapse, and ecological environmental degradation. In addition, the emphasis within environmental hydro-geochemistry has shifted from regional hydrogeochemisty to the relationship between environmental hydro-geochemistry and human health and diseases, forming a new discipline of medical environmental geochemistry.

From the 1970s to the 1980s, studies on soil water movement in the unsaturated zone made substantive progress with the foundation of many new experimental approaches within water balance studies and the introduction of new measurment and testing techniques such as tensimeters and neutron probes. The zero-flux plane method was a newly developed field method which was proposed by foreign experts to study the recharge, loss and balance of soil water, and became of great practical value for soil water studies.

The introduction of groundwater system theory and transient flow theory, and the widespread application of computers and modern applied mathematical methods such as numerical simulation and analytical solutionls enabled the study of groundwater to undergo radical changes in the 1980s. The emphasis of groundwater studies moved towards modelling rather than traditional field-based research methods, and great changes have been made with calculation methods. Moreover, the domain of groundwater studies has been expanded to include the relationship between groundwater and social economic systems as well as of that between groundwater and natural environmental systems.

An important emphasis of groundwater resource studies in the late 1980s was on groundwater management. In other words, the aim of research was to find out how to exploit, utilize, regulate and protect groundwater resources so that it was maintained in a state favourable for human life and production. Therefore, the research dealt with problems within the natural environment related to human activities, social environment and the technological economic environment. The research used mathematical models and optimisation methods to develop groundwater management models to realise the objectives of groundwater management.

Groundwater dynamics deals mainly with the flow theory of groundwater. In terms of water-bearing media, flow theory can be classified into three branches corresponding to pore water, fracture water and karst water respectively. In past ten to twenty years, groundwater dynamics has absorbed the contents of modern applied mathematics and evolved into a combination of mathematical hydrogeology and groundwater resource assessment. Mathematical hydrogeology mainly includes groundwater flow simulation, hydrogeologic statistics, stochastic simulation, analytical methods of transient flow, and so on, forming different independent branches. The introduction of groundwater system theory has had a great influence on studies of water resources. Groundwater modelling has become the main content of water resources research. With the guidance of system engineering, a series of groundwater models have been developed, from conceptual models, mathematical methods, and optimisation methods to management models. It has covered all of the processes from water resources evaluation to water resources management, and evolved into water resource hydrogeology.

On the basis of information systems, inference systems and knowledge acquisition, an expert decision system for urban water resource-environment management was developed to realise the automated management of groundwater resources and the water-dependent environment. It offers a technical tool to urban water resources institutions, providing them with real time analysis, process simulation and information output to realise optimum decision making for water resources management. Information systems have become important components of water resources research. They include the development of data management systems, regime monitoring information systems, remote sensing information systems, expert decision systems, and their application to modelling studies in three-dimensional geographic information systems.

Regional hydrogeological mapping, mainly at the 1:200 000 scale, was basically completed for the whole country except for some high mountainous and desert regions. A great number of hydrogeological maps in separate sheets and with relevant monographs have been published. Maps or atlases at various scales have also been compiled for provinces and basins according to the needs of national economic construction. One of the highlights was the publication of the Hydrogeologic Atlas of the PR China in 1979, which was compiled by the Institute of Hydrogeology and Engineering Geology in cooperation with the provincial agencies.

Since the middle of the 1970s, international scientific exchange and corporation have been resumed and advanced, especially with the western countries. It has caused the rapid advancement of hydrogeological sciences in China. Meanwhile, China joined some important international organisations in the 1980s and became a country member of the International Association of Hydrogeologists (IAH) and the International Association of Hydrological Science (IAHS), establishing the China National Committee for IAH and of IAHS. Chinese hydrogeologists have visited many foreign countries and attended international symposiums including the IAH Congress and the Scientific General Assembly of IAHS. In addition, Chinese hydrogeologists have also joined some working groups of the scientific projects of the International Hydrological Programme (IHP) of UNESCO.

CONCLUSION

The history of groundwater investigation and utilization in China may be subdivided into four stages, including ancient China (Table 1). The latter two stages in the recent

Table 1 Summary of the stages in the development of hydrogeology in China.

I	Primitive stage		ancient China (prior to the 20th century)		hydrogeology as pre-science	
II	Preliminary stage		beginning to use geology in studies of groundwater (1901–1950)			
III	Foundation stage (mainly under the influence of USSR and on the basis of geology)	1	1950s	period of regional hydrogeology	traditional hydrogeology	modern hydrogeology
		2	1960s	period of agricultural hydrogeology		
IV	Development stage (mainly in cooperation with the western countries and with application of cross sciences)	3	1970s	period of environmental hydrogeology	current hydrogeology	
		4	1980s	period of quantitative hydrogeology		

60 years represent the development of modern hydrogeological sciences within China. The 1950s–1970s represents the stage of traditional hydrogeology, whilst after 1970 is the stage of current hydrogeology.

REFERENCES

Shen Shurong, Wang Yangzhi, Li Erong. (1985) *History of Hydrogeology in Ancient China*, Geological Publishing House, 1985. 4.

Chen Mengxiong. (1986) The four periods on the development of hydrogeological science in New China. *Proceedings of the XVth International Symposium of the INHIGEO, Beijing*, 1986.

Chen Mengxiong, Cai Zuhuang. (2000) A historical review of the development of hydrogeologic science in China, *Groundwater resources and the related environ-hydrogeologic problems* in China, Seismological Press, Beijing, 2000.

Hydrogeology in the Czech Republic

Radomir Muzikar
Slamova, Czech Republic

The present territory of the Czech Republic consists of two parts, Bohemia and Moravia which were formerly called the Crown Lands of Bohemia and which had fallen under the rule of the House of Austria in 1620. This influenced all development in the country. The German language was used in all official documents and also in research. Only when the independent Czechoslovakia was formed in 1918 did Czech or Slovak become used in all documents. Czechoslovakia then became separated into two countries, the Czech Republic and Slovakia in 1993.

As in many countries, use of groundwater preceded an adequate understanding of the physical principles governing its origin, occurrence and movement. Any knowledge that existed was mainly connected with the use of groundwater for supply. Although the time span covered by this chapter could be from around 1800, it would not be correct to neglect some very important works which appeared in the Czech countries before that time. Attention was given in earlier times both to groundwater in mining and to mineral water rather than to groundwater for supply. Nevertheless, the knowledge gained on groundwater in mines and mineral water occurrence was useful for the future developing discipline – hydrogeology.

Georgius Agricola, whose proper name was Georg Bauer (1494–1555, Fig. 1), worked in the Czech countries. He was a doctor in the well known mining town of Jachymov in the Czech Republic. He was probably the first to describe the occupational disease suffered by the miners working in a radioactive environment, although the principles of radioactivity were unknown at that time. Besides his medical activities, he engaged in mining and mineralogy. He used to visit the Jachymov mines, where he made many observations and investigations. The knowledge obtained from his observations was set out in 12 books "De re metallica libri XII" written in Latin (Agricola, 1553).

The ideas presented were illustrated by means of pictures using wood engraving. In the Europe of that time the hypothesis of Aristotle that air enters into cold dark caverns under the mountains where it condenses into water and contributes to springs was recognized. Agricola developed the hypothesis of the Greek and Roman philosophers on water circulation based upon his own observations. He showed that in dry regions the quantity of groundwater is less than in those with high rainfall and that spring discharges are lower in dry years. He had the correct idea on the equality of groundwater levels according to the principle of communicating vessels

Figure 1   Georgius Agricola (1494–1555).

and on the movement of groundwater from the higher lying points to lower lying points. He also described dewatering using pumps manufactured mostly from wood (Fig. 2).

Thomas Jordan of Klausenburg (1539–1586) was a doctor active in Brno from 1569–1586. He described the mineral springs in the Moravian region. The manuscript was written in Latin and was issued in a Czech translation (Jordan, 1581). The first part of the book contained his ideas on the genesis of mineral springs, formation of chemical characteristics and "investigation" in which he summarised the hydrochemical and analytical knowledge of the second part of 16th century. The second part was dedicated to the use of mineral water and the third part to local descriptions of mineral springs. Like Agricola Jordan was a doctor, as was Erasmus Darwin in England, showing that the science of those times was very universal. In the 17th and 18th centuries, scientific interest continued to be predominantly centred on mineral springs or on groundwater in mines rather than on groundwater for water supply and more general research of the circulation of fresh water.

A more profound interest in the origin, occurrence, movement, use and protection of groundwater in respect to its possible use for drinking water supply began only in the

Figure 2 Dewatering of a mine by means of wooden pumps (Agricola, 1553).

19th century. Groundwater studies were undertaken in three ways. The first considered the geological conditions of the origin, occurrence and movement of groundwater while the second applied the physical laws for the movement and evaluation of groundwater resources and the third was dedicated to the technology of groundwater abstraction.

The first accessible written data on groundwater referred to terminology. Jan Krejci (1825–1887), professor at the Technological University and later at Charles University in Prague is considered as the founder of modern Czech geology and was the author of the first text book of geology written in Czech (Krejci, 1877). The final version issued in 1877 completed a series of text books on geology and mineralogy which had been appearing since 1850. The book explained the occurrence of groundwater and differentiated between shallow water flowing in fluvial sediments and called it "underground water" and fissured or karst water which was defined as "water in the system of sandstone or limestone layers". Krejci used this terminology in the assessments used for the prospecting of groundwater sources for water supply in Prague. He can be considered the first Czech hydrogeologist even though the term "hydrogeology" was not used in the Czech countries at that time. As a member of Austro–Hungarian Imperial Assembly he gained the recognition that reports of research in the natural sciences by Czech scientists could be published in the Czech language.

Frantisek Posepny (1836–1895) comprehended the difference between the circulation of shallow and deep groundwater and was the founder of a new discipline – the geology of mineral deposits. He explained the flow of groundwater related to the genesis of ore deposits at the international congress in Berlin (Posepny, 1889). Posepny defined the descending shallow groundwater flow to the lowest point of the landscape (the so called base level of erosion) as vadose water. Franz Eduard Suess (1867–1942), who was active in the mapping and investigation of the Bohemian Massive, extended the description of the term. According to Suess, vadose water comprised all the waters penetrating to the geological environment or water condensed there (Suess, 1903). This is used in Czech terminology until the present but is different to the English terminology in which vadose water is used for the water in the unsaturated zone.

The dewatering of mines influenced groundwater levels in the surroundings. The catastrophic break through of mining water in the Doellinger brown coal mine in Teplice in 1879 caused by the break of a cavity considerably influenced the discharge of a mineral water spring at the 7 km distant Teplice spas. The piezometric groundwater level decreased below the surface. Such mine water breaks and other problems with mine dewatering called for investigations and monitoring. Groundwater level monitoring in the brown coal mine of Duchcov (Dux in German) belonging to the same basin as that mentioned above detected groundwater level fluctuations caused by earth tides (Klonne, 1880).

Johann Georg Mendel (1822–1884) prelate and abbot of the Augustinian convent in Brno established the first systematic monitoring of groundwater levels. Mendel was a researcher with profound interests. His main international fame comes from his genetic research and he is considered the founder of the science of genetics. Besides genetic investigations, he carried out meteorological monitoring, including observing the sunspots, probably in connection with their eleven-year periods, and the monitoring of groundwater levels in the well located in the convent in 1879 (Vesely, 1965). The results of the monitoring of groundwater levels in Brno and in Prerov were published (Anonymous, 1882).

At the end of the 19th century, serious problems occurred with drinking water quality in Prague and other towns. Water was drawn from wells located near the River Vltava to supply Prague. These collected both water which had infiltrated from the river and groundwater contaminated with domestic wastewater. The contaminated water caused epidemic outbreaks of cholera, typhoid and dysentery. Over pumping often occurred. For that reason, a commission for the water supply of Prague was appointed in 1875 with the aim of solving the problems of water supply and providing the city with water of satisfactory quantity and quality. It was decided to abstract groundwater from the fluvial aquifers both of the River Jizera before the confluence with the River Elbe and of the River Elbe itself. Hydrogeological investigations were carried out in the following years. The law governing the abstraction wells with the water supply network and other necessary infrastructure was confirmed only in 1899 (Jasek et al., 2000). The abstraction wells were located near the small town of Karany.

The design of the abstraction works made Adolf Thiem (1836–1908) a very famous German hydrogeologist. He was the winner in a competition for the work in 1902 and construction works started in 1906. After Thiem's death his pupil and successor Emil Prinz (1891–1938) continued working. He was another famous German hydrogeologist born in Brandys nad Labem (Bohemia). The aquifer is of high conductivity (4×10^{-4} m/s). A total of 680 abstraction wells were drilled to depths of 8–12 m. These were distributed in lines of 26 km length and were joined by siphons. A long-term pumping test of the whole multiple well system was carried out in 1912–1913 and operation for supply started in 1914. The discharge reached of 900–1100 l/s. The Karany abstraction area is considered a supreme work at that time due to its importance, size and technical level. Water supply from the Karany abstraction area decreased the outbreaks of infectious diseases in Prague. Abstraction from the aquifer here has been providing groundwater for Prague until the present time.

Hydrogeological investigation for groundwater supply for Brno, the second town of the Czech Republic, also began at the end of the 19th century. The investigations covered the south–eastern part of the Bohemian Cretaceous strata. These investigations established the monitoring of springs in 1884–1885. Initial one-time discharge monitoring led in 1898 to the long – term monitoring of selected springs and to the drilling of wells and groundwater level monitoring in the wells in 1899. This is the longest continuous monitoring of groundwater levels and spring discharges in the Czech Republic and continues to the present time. The results of long – term monitoring enabled a more detailed assessment of groundwater regime and eventually to the identification of a "secular" periodicity with a period of 26 years. In an initial abstraction test and during operational abstraction, both quantity and quality monitoring have been carried out since the very beginning. The quality monitoring included both chemical and microbiological analysis. The abstraction area at Brezova has been abstracting groundwater since 1906 at a discharge of about 700 l/s.

Jan Vlastimil Hrasky (1857–1939) produced a technical point of view of hydrogeological evaluation. He had been a water engineer and was appointed professor of waterworks engineering, sewerage and land improvement at the Technological University in Prague in 1897. He determined the direction of water management for coming generations. One of his publications was entitled "On quantitative research of groundwater" (Hrasky, 1904). He consulted for and assessed many projects for water supply and sewerage, among them Thiem's project at the Karany abstraction area with its infrastructure.

Parallel with the understanding of groundwater flow, were formed the conditions for the evaluation of groundwater quality due to the development of analytical chemistry and bacteriology. In 1877 six associations of doctors, natural scientists and engineers issued the publication "On water problem" which stated that water quality should be examined by both chemical and bacteriological analysis. On the basis of this suggestion, Frantisek Vejnovsky undertook the first bacterial analysis of water in the Czech lands in 1877.

Max Pettenkofer founded the institute of hygiene in Munich in 1873. Though he did not recognise the discovery by Robert Koch that the decisive factor in the origin and spread of cholera was a bacillus discovered in 1883, his merit in the development of experimental hygiene is indisputable. He stated that the basis of municipal hygiene is care of potable water, water supply and the public sewer system. The institute became a consulting and "training" centre. Two scientists from Prague University, Isidor Soyka and Gustav Kabrhel consulted there on the problems of municipal hygiene in 1877 (Jasek *et al.*, 2000). Both were later acquainted with the methods of Robert Koch in Berlin. The knowledge gained was used in the practice of Czech municipal hygiene and also for groundwater protection. Gustav Kabrhel (1857–1939) participated in the construction of the abstraction wells in the above mentioned Karany abstraction area (Jasek *et al.*, 2000) and developed the instruction for the source protection zone (Kabrhel, 1910). He suggested a diameter of 50 m around the well in which protective measures were imposed. In connection with microbial pathogens, which may plausibly contaminate groundwater, it is worth mentioning the protozoan *Giardia lamblia* which is associated with chronic diarrhoea. It was described by the Czech doctor Dusan Vilem Lambl (1824–1895) after whom it was named. Its importance has increased in hydrogeology only in more recent years.

With the end of the First World War, the understanding of groundwater as a part of other disciplines ceased and hydrogeology gradually became an independent discipline. This was also the time of the beginning of the independent state when new authorities and institutes originated. So the State Geological Institute, the State Institute of Hydrology and Hydraulic Engineering and the Hydrometeorogical Institute were established. All were engaged to some degree with groundwater. Nevertheless the previously mentioned courses remained: the geological view and the technical view.

Otta Hynie (1899–1968) is considered to be the founder of modern Czech hydrogeology. He started his professional carrier both in the State Geological Institute and at the Mine Academy in 1924. Assessment of geological and hydrogeological conditions prior to groundwater abstraction is his very important contribution to the development of groundwater supply. The prospecting undertaken before did not usually take into consideration the geological structure. He likewise established the practice of close co-operation between geologists and water engineers and designers. From the very beginning of his career he solved problems relevant to hydrogeology, applied geology and engineering geology in the territory of the whole country. His assessments of projects were constructive, easy to survey and very clearly formulated. His ideas on the hydrogeological structure were completed in the field with detailed sketches. He used to collect data on the geological and hydrogeological conditions of the country, which later served as the basis for published geological maps. He surveyed and designed groundwater abstraction at large scales even before the Second World War.

Hynie founded the first department of hydrogeology in the country at the Charles University in Prague in 1952. He summarised his knowledge of the geological and hydrogeological conditions of the country in the two volume book "Hydrogeology of ČSSR". The first volume was dedicated to fresh water (Hynie, 1961). The first part of the volume comprises general hydrogeology with theories of the origin and occurrence of groundwater, assessment of quality, well hydraulics, groundwater resource estimation and groundwater protection. The second part consists of a description of the regional hydrogeology of the country. The second volume was dedicated to mineral water (Hynie, 1963). It was also divided into two parts, explaining in the first part the theory of the origin, outflow and classification of mineral water and in the second part the regional hydrogeological conditions of mineral springs in the country.

The second volume was considered to be a unique work in the world at that time. Despite the considerable advances in the development of hydrogeology, both volumes remain the basic source of information for the hydrogeological solutions for water supply and mineral water use. The two book set on hydrogeology was completed with a third book dedicated to the hydrogeology of mineral deposits. Hynie was unable to finish the book because he died in 1968 so it was finished by Vladimir Homola and Stanislav Klir (Homola and Klir, 1975). The two-fold division of the text was the same as in the previous books.

In the 1930s, among its other tasks connected with groundwater studies, the State Institute of Hydrology and Hydraulic Engineering also developed an understanding of the Bohemian Cretaceous where the largest groundwater resources in Czech Republic are found. The results were described in several sheets of hydrogeological maps at a scale of 1:75 000 issued in a publication by Podvolecky (1935).

The monitoring of groundwater levels was established especially for planned structures. A network of monitoring wells was established after 1933 along the proposed Oder–Danube channel. Although the channel was not constructed, the monitoring network was partly incorporated into the state monitoring network from 1951. Monitoring of 250 monitoring wells and springs had also been established in 1937–1940 during hydrogeological investigations for groundwater abstraction in the Bohemian Cretaceous rocks. A proportion of these monitoring wells were also included in the state network from 1951. The monitoring results played a very important role in all hydrogeological investigations for large groundwater abstractions. The monitoring also resulted in further development of the evaluation of groundwater regimes and groundwater level forecasting. Technical solutions for the impacts of construction, especially hydraulic structures, on groundwater were also developed. Zdenek Bazant used flow nets to assess the impacts (Bazant, 1938).

The main development of hydrogeology began after 1945. This was supported by the establishment of new companies with drilling machines, pumping technology and laboratories, from which grew other companies engaged in hydrogeology and engineering geology. Methods for unifying geological documentation were among the first steps in the development of hydrogeology. Investigations for groundwater abstraction with the main aim of drilling the abstraction wells predominated at the very beginning. Hydrogeological investigation for water supply changed gradually into phases of investigation consisting of the following: prospecting, preliminary investigation and detailed investigation.

For hydrogeological investigation and groundwater use and management the main approach involved hydrogeological zoning. At the end of the 1950s preparation of the principles of hydrogeological zoning began (Zima and Vrba, 1959). This had the objective of sub-dividing the country on mainly geological considerations, and provided the information basis for systematic groundwater investigations. In 1965 a new zoning scheme was proposed and approved as a territorial planning basis for the preparation of hydrogeological investigations and evaluation of exploitable groundwater resources (Vrba, 1965). The principles of zoning proposed by Vrba (1965) were retained in essence in further versions. These included the hierarchy of successive division of territorial units into hydrogeological zones, sub zones and units. A hydrogeological zone was defined as a tectonically and geologically consistent area with similar hydrogeological conditions on whose territory a certain type of groundwater circulation predominates. Considerations of structural geology, stratigraphy and lithology were the main starting points for the zoning of the territory. Until that time, zoning had not been applied to hydrogeological concepts due to lack of knowledge and data. The zoning was published in survey maps at a scale of 1:500 000 (Franko, Kulmann and Vrba, 1965) and later re-edited to a scale of 1:200 000. Both hydrogeological and hydrochemical conditions were represented on the hydrogeological maps, and cross sections and the legend were an integral part of the map. The legend – a book – contained descriptions of geological and hydrogeological conditions, hydrogeological zones, hydraulic properties of the soils and rocks, circulation of groundwater, chemical properties of groundwater, data on mineral water and groundwater use. Jan Jetel and Jiri Krasny introduced a new comparative regional parameter of permeability, producing regional assessments of hydraulic properties (Jetel, 1968; Jetel and Krasny, 1968).

In 1967 the hydrogeological sub commission for the classification of resources was established subordinate to the cabinet. The sub commission approved all groundwater resources supporting abstractions of 30 l/s or more. These were registered and included in the balance of exploitable groundwater resources of the country, which forms a component part of the overall water balance of the country. The groundwater balance of the country consisting of a groundwater hydrological balance, a balance of exploitable groundwater resources and a groundwater management balance has been in use since the 1970s. The groundwater balance investigates and evaluates the origin, circulation and regime of groundwater. Groundwater flow is evaluated by hydrograph separation. The balance of exploitable groundwater resources was used for the water management balance. This compares the data on exploitable groundwater resources with the actual groundwater abstraction and with the hydrological balance. Such a system of control for the balance formed the preconditions for admissible disturbance of the natural groundwater flow and also provided the data for planning the investments for water supply and administrative decisions and control of water management. In the second half of the 1960s were seen large projects for prospecting for groundwater resources. These were so called regional hydrogeological investigations which led to an estimation of the exploitable groundwater resources approved by the hydrogeological sub commission of the classification of resources which identified hopeful and less hopeful areas for groundwater abstraction.

This system of investigation and registration in the state water management balance was accompanied by considerable development of hydrogeological methods. It is impossible to mention all the contributions, but some of the most important are briefly introduced here. Firstly came methodology manuals for aquifer tests by means of unsteady flow and its interpretation, and the determination of hydraulic properties of water bearing soils and rocks and the evaluation of boundary conditions. The computer greatly facilitated the computations. The first determination of hydraulic properties using the computer was probably carried out in 1964 in the company Geotest Brno. Estimation of exploitable groundwater resources was based on the analytical solution of the equations for well hydraulics with the principle of superposition. This also began before 1970. The method of analogy was also developed, based on the analogy of groundwater flow with flow though media other than porous media, such as electrical models, spheroid models or slot models (Halek, 1965). The principles of the theory of groundwater flow and analytical solutions for the tasks of water management were described by Halek and Svec in the book "Groundwater hydraulics" in 1973 which was translated into English (Halek and Svec, 1979).

Hydrological methods were developed in parallel with the analytical solutions. Kliner and Knezek worked out a method of groundwater flow separation from streamflow discharge in unconfined aquifers which are in the direct contact with a stream channel. The method using stochastic and genetic approaches is based on the relationship between fluctuations of groundwater levels and stream discharges (Kliner and Knezek, 1974). The systematic long- term monitoring of groundwater levels and spring discharges by the state began in 1951 and was gradually developed over the following years, especially from 1965 to 1974 during the International Hydrological Decade. Jaroslav Vrba and Milan Vrana proposed the concept of a deep aquifer monitoring network (Vrba and Vrana, 1967). They suggested a network to monitor both shallow and deep groundwater circulation. The overall density of the network reached about one monitoring point per 42 km^2, comprising one monitoring well for 66 km^2 and one spring per 113 km^2. This long-term monitoring programme facilitated the development of methods for groundwater level forecasting (Muzikar, 1976).

Investigation of groundwater chemical quality forms an inseparable part of hydrogeological investigation. Besides investigation of the suitability of groundwater for water supply and identification of contamination, evaluations of chemical equilibrium in the natural system of water – rock – atmosphere were introduced (Paces, 1969; Paces, 1972). Environmental isotope techniques were introduced to help determine the origin of groundwater and to follow the hydrological cycle in the long-term and to examine residence times and movement of groundwater, especially in cases where this had not been possible by using conventional hydrogeological methods (Silar, 1976).

The protection of groundwater resources was a subject of interest to hydrogeologists and water management authorities in parallel with the development of groundwater abstraction and hydrogeology. The instruction for groundwater protection was developed in the second part of the 1960s. Miroslav Olmer made maps of potential contamination, locating the high risk installations and areas with pollution hazards. Milan Vrana produced a map of the protection of fresh groundwater for the Czech Republic at a scale of 1:500 000 (Vrana, 1968). This map indicated areas with different grades of necessary protection related to the hydrogeological zone and the protected area. The explanatory text evaluated the exploitable groundwater resources, the present abstraction and outlook for future abstraction and the characteristics of pollution sources. The more detailed tools of groundwater protection were the vulnerability maps at a scale of 1:200 000 issued at the beginning of the 1970s (Olmer and Rezac, 1974). These complimented the existing sets of geological and hydrogeological maps. Indicated on them are the vulnerability, grade of protection, hydrogeological data from the hydrogeological maps such as direction of groundwater flow, groundwater divides and present or future groundwater abstraction. The maps defined seven classes of vulnerability and five grades of groundwater protection (e.g. protection of the whole area, partial protection surrounding the abstraction area, etc). Definition of grade of necessary groundwater protection differs the approach used in other countries.

Remediation of contaminated groundwater began in about 1970. Groundwater contamination by petroleum hydrocarbons predominated. In addition to the usual hydrogeological approaches, site investigations to delineate the contamination plume also used soil gas sampling because of the high volatility of these compounds. Mapping of volatile organics in this way minimised the number of conventional monitoring wells which had to be drilled (Pelikan, 1973). The gas sampling approach was not entirely new in hydrogeological investigation as Czech hydrogeologists had used it in prospecting for mineral water at the beginning of the 1960s. The remediation approach adopted consisted first of removing the source of contamination, followed by withdrawal and treatment of contaminated groundwater with separation of any nonaqueous phase liquid. Hydrodynamic control of the contaminant plume was applied in some cases to manipulate the local hydraulic gradient through withdrawal of groundwater or injection or both. The largest groundwater contamination problem at the Slovnaft Refinery at Bratislava in Slovakia had been worked on by the Czech company Geotest Brno since 1971. When the remediation began, the plume of nonaqueous phase liquid covered an area of $4.2\,km^2$ with an average thickness of about 24 cm (Pelikan, 1984). The approach of hydrodynamic control with "pump and treat" was used at this site.

In the 1970s investigations for sanitary municipal landfills and industrial hazardous waste landfills also began. Hydrogeological investigations for groundwater source protection zones were also initiated.

Hydrogeological investigations were also applied to the complicated tasks connected with underground constructions and dewatering of foundation excavations. Three large integrate hydrogeological investigations aimed at the interference of mining and mineral water resources should be noted: the Sokolov Basin (protection of mineral springs at Karlovy Vary Spa), the Cheb Basin (protection of mineral springs at Františkovy Lázně Spa) and the Teplice Basin (protection of thermal and mineral springs). Within these investigations extensive drilling, pumping tests, groundwater sample collection and monitoring were undertaken. During the evaluation and interpretation of the investigations new evaluation approaches were developed such as hydrogeochemical assessment as for example for the Sokolov Basin (Jetel, 1972).

The approaches of evaluation of exploitable groundwater resources, groundwater contamination by all human activities and remediation were developed after 1975. After 1990 groundwater remediation predominates.

REFERENCES

Agricola, G. (1553) *De re metallica libri XII (Twelve books on mining and metallurgy)*. Basel, Jer. Froben.

Anonymous (1882) *Bericht der meteorologischen Commission des naturforschenden Vereines in Bruenn ueber die Ergebnisse der meteorologischen Beobachtungen im Jahre 1881 (Report of the meteorology Commision of the Society of nature research in Brno on the results of meteorology monitoring in 1881)*. Bruenn, W. Burkart – Verlag des Vereines.

Bazant, Z. (1938) *Groundwater Flow and its Influence on Structure Foundations Especially Weirs* (In Czech). Praha, Masarykova akademie prace.

Franko, O., Kulman, E. & Vrba, J. (1965) *Zoning of Groundwater in CSSR 1:500 000* (In Czech). Praha, Reditelstvi vodohospodarskeho rozvoje.

Halek, V. (1965) *Hydraulic Engineering Research 3. Analogy Methods* (In Czech). Praha, SNTL.

Halek, V. & Svec, J. (1979) *Groundwater hydraulics*. Praha, Academia.

Homola, V. & Klir, S. (1975) *Hydrogeology of CSSR III. Hydrogeology of Mineral Deposits*. (In Czech). Praha, Academia. 428 p.

Hrasky, J.V. (1904) *On Quantitative Research of Groundwater* (In Czech). Praha.

Hynie, O. (1961) *Hydrogeology of CSSR I. Hydrogeology of Fresh Water* (In Czech). Praha, Academia. 564 p.

Hynie, O. (1963) *Hydrogeology of CSSR II. Hydrogeology of Mineral Water* (In Czech). Praha, Academia. 800 p.

Jasek, J. et al. (2000) *Water Supply in Bohemia, Moravia and Silesia*. (In Czech). Praha, Milpo Media. 240 pp.

Jetel, J. (1968) A new comparative regional parameter of permeability for hydrogeology maps. In: *Mém. Ass. Int. Hydrogeol., 8, Congr. Istanbul*. pp. 101–107.

Jetel, J. & Krasny, J. (1968) Approximate aquifer characteristics in regional hydrogeological study. In: *Vest. Ustr. Ust. Geol.*, 51, 1, Praha. pp. 47–57.

Jetel, J. (1972) Hydrogeology of the Sokolov Basin (function of rocks, hydrochemistry, mineralwaters). In: *Sbornik geologickych ved, rada HIG*, vol. 9. Praha, Ustredni ustav geologicky. pp. 8–146.

Jordan, T. (1581) *The Book on Medicinal Waters and Hot Waters* (In Czech). Olomouc, Fridrich Milichtaler.

Kabrhel, G. (1910) *Paper on Protection Rayon for Establishment of Groundwater Abstraction*. (In Czech). Praha, Cas. Lék. Ces.

Kliner, K. & Knezek, M. (1974) *Method of Groundwater Flow Separatin Using the Groundwater Level Monitoring* (In Czech). Bratislava, Vodohosp. Cas. 5.

Klonne, F.W. (1880) Die periodische Schwankungen in den inundierten Kohlen Schaechten von Dux (The periodical fluctuations in the flooded coal shafts in Duchcov). *Akad. Wiss. Wien, Sitzungsbericht*, 81, Wien, 101 p.

Krejci, J. (1877) *Geologie*. (In Czech). Praha.

Muzikar, R. (1976) *Groundwater Level Forecasting* (In Czech). Bratislava, Vodohosp. Cas. 24 (1), pp. 71–78.

Olmer, M. & Rezac, B. (1974) Principles of maps for groundwater protection in Bohemia and Moravia 1:200 000. In: *Mém. Ass. Int. Hydrogeol., X, Congr*. Montpelier.

Paces, T. (1969) Chemical equilibrium and zoning of subsurface water from Jachymov ore deposit. *Geochinica et Cosmochimica Acta* 33, 591–609.

Paces, T. (1972) Chemical equilibrium in naturak system water–rock–atmosphere (In Czech). *Knihovna Ustredniho ustavu geologického*, 43, 194 p.

Pelikan, V. (1973) Remediation of groundwater contaminated by petroleum hydrocarbons (In Czech). *Vodni hospodarstvi*, B, 23, 36–40.

Pelikan, V. (1984) Groundwater protection of Zitny ostrov (area limited with the River Danube and the River Little Danube) against petroleum hydrocarbons (In Czech). *Prace a studie CSAV 16*. Praha, Academia. 108 p.

Podvolecky, F. (1935) Systematic research of groundwater and springs in Bohemian Cretaceous formation and its results in the period of 1928–1935 (In Czech). *Prace a studie c. 17*. Statni ustav hydrologicky a hydrotechnicky T.G. Masaryka v Praze.

Posepny, F.A. (1889) Ueber die Bewegungsrichtunge der unterirdischen Fluessigkeiten. (On the movement of underground liquids). In: *Compes rendu du Congres geol. Internat*., Berlin.

Silar, J. (1976) Radioactive groundwater dating in Czechoslovakia – first results. In: *Vestnik Ustredniho ustavu geologickeho*, 51. Praha. pp. 209–220.

Suess, F.E. (1903) Ueber heissen Quellen (On hot springs). *Verh. Ges. Dtsch. Naturfoescher u. Aertzte in Karlsbad*, Leipzig.

Vesely, E. (1965) In memory of Gregor Mendel in archive of Hydrometeorogical Institute (In Czech). *Meteorologické zprávy*, 18 (2), 28–30. Praha.

Vrana, M. (1968) Map of groundwater protection in Bohemia and Moravia 1:500 000 (In Czech). *Reditelstvi vodnich toku*, Praha. 30 p.

Vrba, J. (1965) Principles of hydrogeological zoning (In Czech). *Vodni hodpodarstvi*, Nr. 6, Praha.

Vrba, J. & Vrana, M. (1967) Principles of establishment of monitoring network of deep aquifers. (In Czech). *Vodni hospodarstvi*, Nr. 12, Praha.

Zima, K. & Vrba, J. (1959) Zoning of the area in the River Elbe Basin (In Czech). *Reditelstvi vodnich toku*, Praha.

History of French hydrogeology

Jean Margat, Didier Pennequin & Jean-Claude Roux
BRGM, Orléans, France

THE 19TH CENTURY: FROM ORIGINS TO THE CREATION OF AN EXACT NATURAL SCIENCE

When he created the word "hydrogeology" in 1802, Jean-Baptiste Lamarck defined it as the "study of the influence that water has on the surface of the earth . . .". This is a meaning that is, in fact, different from its present-day meaning, but which nevertheless foresaw the advent of the "Science of groundwater", which it has come to mean and as it has been adopted by all languages (Figure 1).

The science of groundwater in France was founded in the 19th century based on fluid mechanics in porous media and was rapidly rooted in the expanding field of Earth Sciences, along with the increasing knowledge and mapping of the geology of France, instigated by the Mining Corps. From the start, groundwater science combined the concepts and methods of both the physical and natural sciences, of hydraulics and the earth sciences. Moreover, its development would closely associate knowledge and application to understand the phenomena governing the origin and circulation of water in the subsurface and to promote its exploration and capture, by coupling the knowledge and technology of engineers with the spirit of observation, deduction and experimentation of naturalist-geologists.

Engineers take the lead

The first theoreticians and practitioners of hydrogeology in France were engineers, most of whom graduated from the *"Ecole Polytechnique"* and were knowledgeable about hydraulics. In general their practical objectives were related to developing urban drinking water supplies (Paris, Dijon, etc.) and sometimes, more rarely, protection of groundwater quality. Their approach was based on a general theory. These men, Louis-Etienne Héricart de Thury, François Arago, Eugène Belgrand, Jules Dupuit, and then Henri Darcy, introduced a "hydraulic interpretation and conversion" of the first geological descriptions of shallow and deep underground formations, notably in

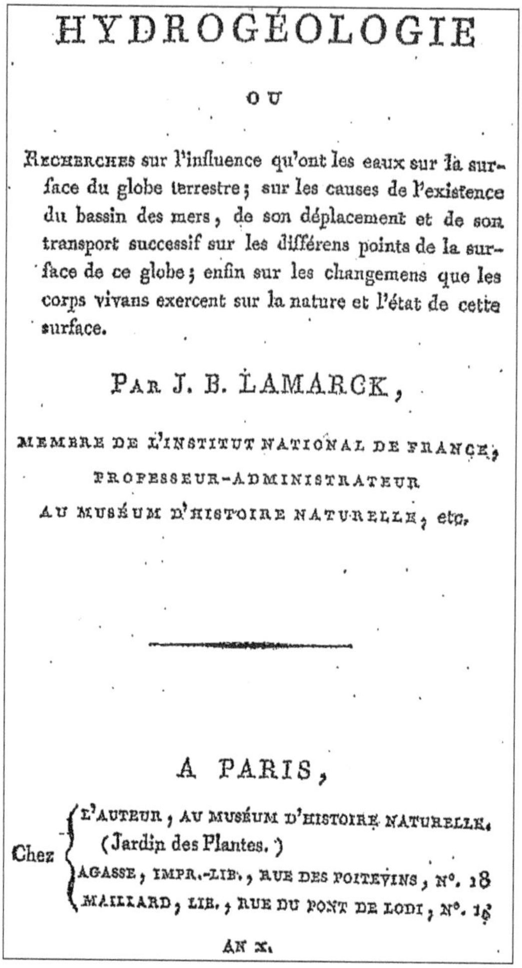

HYDROGÉOLOGIE

O U

Recherches sur l'influence qu'ont les eaux sur la surface du globe terrestre; sur les causes de l'existence
du bassin des mers, de son déplacement et de son
transport successif sur les différens points de la surface de ce globe; enfin sur les changemens que les
corps vivans exercent sur la nature et l'état de cette
surface.

Par J. B. LAMARCK,

MEMBRE DE L'INSTITUT NATIONAL DE FRANCE,

PROFESSEUR-ADMINISTRATEUR

AU MUSÉUM D'HISTOIRE NATURELLE, etc,

A PARIS,

Chez
L'AUTEUR, AU MUSÉUM D'HISTOIRE NATURELLE,
(Jardin des Plantes.)
AGASSE, IMPR.-LIB., RUE DES POITEVINS, N°. 18
MAILLARD, LIB., RUE DU PONT DE LODI, N°. 15

AN X.

Figure 1   J.B. Lamarck's original definition of the term "Hydrogeology".

the large sedimentary basins. The first observations of groundwater regimes and their behaviour began and the basics of their dynamic factors were understood. In 1842, for example, the engineer Dausse showed the inefficiency of summer rains for recharging shallow aquifers.

The major defining act was Henri Darcy's experimentation in 1856 on water flow through sand. His equation for the linear relationship between the flux and the hydraulic head gradient, named "Darcy's Law" by Dupuit in 1857, became one of the pillars of quantitative hydrogeology and groundwater hydraulics (Figures 2 and 3 and Box 1).

Box 1 – Darcy's Law

Henri Darcy was born in Dijon in 1803 and graduated from the *Ecole Polytechnique* and was a member of the Corps of the "Ponts et Chaussées" (Bridges and Roads) engineering. In 1840, he was a pioneer of urban drinking water supply in France when he conceived and supervised the work to divert the Rosoir Spring in the Val Suzon east of Dijon, and to pipe the water by means of a 12 km long aqueduct to the Porte Guillaume reservoir next to Dijon.

Internationally, Darcy is known primarily for the law that he discovered governing groundwater flow in porous media. In 1856, in his paper, "*Les fontaines publiques de la ville de Dijon*" (The Public Fountains of the City of Dijon), he was the first to propose a law for the hydrodynamics of flow in porous media, which is applied today by hydrogeologists and hydraulic engineers throughout the world.

In fact, the objective of his experiments was to calculate "the filtering of water" derived for the public drinking water supply. His experiments were carried out in 1854–56 in the courtyard of Dijon hospital where he had installed a 2.5 m high and 0.35 m diameter column, filled with "Saône River sand" held by a 2 mm grid metal screen placed 0.2 m above the bottom. He was able to vary the grain size of the sand and the water height (pressure).

He arrived at an equation that expresses flow discharge as a function of permeability. The initial equation was:

$$q = \frac{ks}{e[h+e]}$$

with:
q = flow rate or discharge
s = cross-sectional area through which water flow occurs = surface of the top of the sand filter
k = coefficient depending on the permeability of the sand layer
e = thickness of the sand filter
h = water elevation above the sand filter.

Today, his equation is written more simply as follows:

$$V = Ki$$

with:
V = discharge velocity
K = coefficient of permeability or hydraulic conductivity
i = hydraulic gradient.

Based on the experimental results, Darcy concluded that "it seems therefore that for a given sand, we can admit that the volume discharged is proportional to the head and inversely so, to the thickness of the layer crossed."

Later Dupuit applied this equation to flow in aquifers.

Darcy's mathematical expression also applies in a very general fashion to fluid flow (oil or gas) in porous media, and later American petroleum engineers (Wyckoff *et al.*, 1933) called the unit of measure of intrinsic permeability the "Darcy".

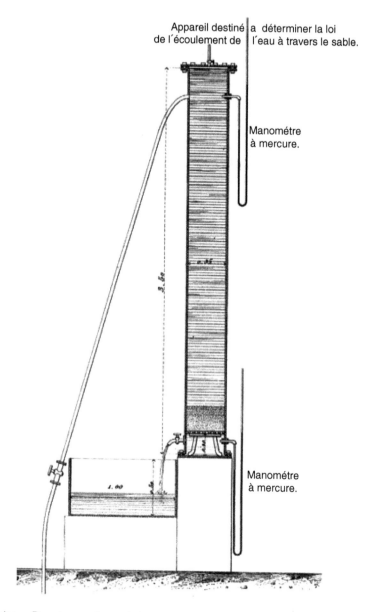

Figure 2 Henry Darcy's experimental device used for his experiment on water flow through sand published in 1956 (to calculate the "filtering of water" for the Dijon water supply).

Although most of the engineers' work understandably focused on quantitative aspects, some attention was also paid to water quality. In 1846 Belgrand, for example, was the first to study the relationship between the composition of groundwater and the nature of the rocks through which it flows; hydrogeochemical concepts taken up later by Edouard Imbeaux in 1897.

Figure 3 Henry Darcy. Portrait by F. Perrodin from the collection of the bibliothèque Municipale de Dijon.

Geologists come into play

Early in the 19th century, well-educated "amateur" naturalists, who were good at combining observation and deduction, took an interest in rationalising groundwater exploration and water well siting. One of the most active and well-known after 1827 was Father Jean-Baptiste Paramelle who, as J. Tixeront so interestingly observed in 1956, published "The art of discovering springs" in 1856, the same year Darcy published his paper (Figure 4). Several others also contributed to this effort, and the great naturalist Etienne Geoffroy Saint-Hilaire pointed out in a paper to the French Academy of Sciences in 1836 that the success of "the Lot Department's hydroscope" was in fact based "on science and observation".

During the second half of the century, eminent university geologists laid down the foundations of the interpretation of geological, lithostratigraphic and structural conditions, which aimed at understanding the formation and functioning of aquifers. This work led to the first major publications by Jules Gosselet in 1886–88, Gabriel Auguste Daubrée in 1887, who introduced the concept of *"surface piézométrique"* (water table), and Albert-Auguste de Lapparent in 1882–98. Hydrogeological mapping was also invented, and the "Hydrological map of the Seine Department" by Achille Delesse (1862) was the first time "piezometric" contours had ever been represented on a map (Figure 5).

Figure 4 L'Abbé (Father) Paramelle 1790–1875.

From capturing springs to drilling deep wells

While the art of capturing springs and piping water by gravity flow to towns had a long history in France – as attested by Roman structures like the famous "Pont du Gard" aqueduct – increasing numbers of projects were undertaken during the 19th century. From the reign of Napoleon III to the Third Republic, modernizing the thousand-year-old Roman aqueducts, engineers built large water supply networks to provide the major cities with drinking water captured from minutely studied springs. After Dijon and the work of Darcy carried out at the end of the 1830s and in the 1840s (Box 1), Paris was supplied at the instance of Baron Georges Haussman and Eugène Belgrand (Figure 6) with 600 km of aqueducts bringing water from the springs of la Dhuys in 1865, la Vanne in 1874 and l'Avre in 1893. Thus groundwater was captured at a distance of more than 100 km from the capital city and is still used today to supply 50% of the city of Paris (Figures 7, 8 and 9).

Tapping shallow groundwater with shallow wells in France is an ancient story – for example drawing from the *nappe des puits* that G.A. Daubrée suggested in 1887 should be called more academically the "phreatic" groundwater. However, in the 19th century advances in drilling techniques that allowed engineers to reach increasing depths, together with more reliable geological predictions, enabled the prospecting for and exploitation of groundwater in deep aquifers in the major sedimentary basins,

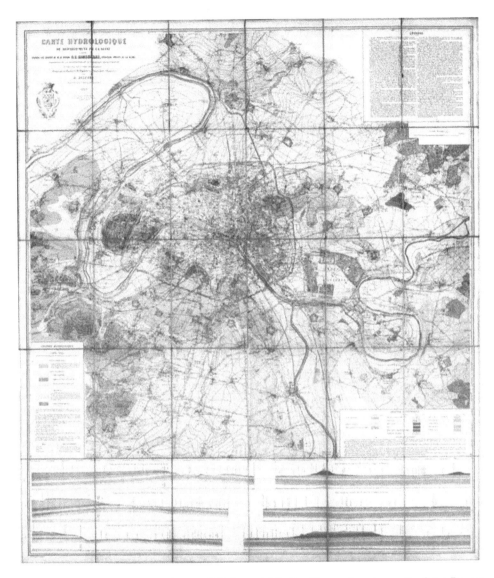

Figure 5 The first hydrogeological map – F. Delesse, 1862. On this map focusing on the Seine Department, the engineer Delesse represented among other features, the local geology, the water courses, shallow aquifers and piezometric contours. The map had been requested by the Prefect Hausmann (*Atlas de l'Architecture et du Patrimoine de la Seine-Saint Denis*).

and more specifically in the Paris, the Nord and the Aquitaine basins. These drilling techniques were described and codified very early in detailed technological publications by F. Garnier (1821) and Héricart de Thury (1829). The most spectacular results were the first deep artesian wells in Tours, Lille and later in Paris (Box 2).

Belgrand

Figure 6 Eugène Belgrand 1810–1870.

Figure 7 La Vanne springs: captage d'Armentières (built in 1874).

Figure 8 Syphon of the Villemarche aqueduc (Yonne – France).

Figure 9 Montsouris Reservoir (Paris area).

Box 2 – Artesian wells

One tradition, although not documented before the 19th century, claims that the province of Artois, in the north of France, is the type area of wells from which confined groundwater flows without pumping. The earliest mention of this is by F. Garnier (1821): "The first research on upwelling springs seems to have been undertaken in the area included in the Pas-de-Calais department, made up of the former provinces of Artois, Boulonnais, Calaisis, Ardresis and a small part of Picardie. At least this is the general opinion, and what seems to confirm it, is the fact that the name "artesian wells" is given to the same type of springs in other countries".

Later, in 1834, François Arago wrote that "the oldest known artesian well dates back, it is said, to 1126. It is in Lilliers, in Artois ..." This is, however, a shallow well with wooden catchment works, tapping the confined fringe of the chalk aquifer.

The adjective "artesian" thus designated both (i) the method and the profit gained from the groundwater pressure, eliminating the need to draw up or pump, and (ii) the hydrodynamic conditions that make it possible, hence the expressions "artesian aquifer" and "artesian spring", then "artesian basin", universalized thereafter in many languages in different countries. It was only in the 19th century that deep artesian wells were drilled in France. The most renowned was the "Grenelle Well", planned by Arago and drilled in Paris between 1833 and 1841 to a depth of 548 m, to reach the Greensand aquifer from which the water initially shot up 33 m above ground level, with a discharge of 160 m³/h (Figure 10).

Figure 10a Drilling works for the Grenelle deep artesian well – 1933–1941.

Figure 10b The Grenelle deep artesian well completed and flowing.

At that time, although interest in the deep aquifers was mostly to ensure proper drinking water supply, the idea of using them for absorbing wastewater also emerged in some cases. In 1833, for example, a report by the Public Health Council (Conseil de Salubrité) to the Seine Police Prefect (head of the Seine Police Department) proposed using "drilled or artesian wells" to "evacuate dirty or infected water" in the area around Saint-Denis and Bondy in the Paris region (Public Health Annals, X, 1833).

Groundwater abstraction in the XIX century ...

Except for exploitation linked to the discovery of these deep aquifers (and that on-going from the shallow wells), groundwater abstraction in France overall remained minor and increased at only a moderate rate during the 19th century. During this period, the rapidly growing mining industry (e.g. coal in the Nord, iron in Lorraine) certainly entailed the largest abstraction of groundwater, as demonstrated in the Nord by the work of Jules Gosselet (1886–88).

First specializations

The end of this century also saw the birth of more specialised fields of research in hydrogeology, such as the exploration and study of karst formations with Edouard Alfred Martel (1894) the "father" of modern speleology, or the scientific study of thermomineral springs, which remained mainly the prerogative of mining engineers (Louis De Launay, 1899). Scientists also began to understand the metallogenic and "mineralising" role of groundwater (Gabriel-Auguste Daubrée, 1887). In fact, advances in general knowledge that were at the basis of groundwater science in France were inexorably linked to the progress of regional and local investigations, to specialised studies and to increasing groundwater abstraction. By the end of the 19th century, the outlines of the hydrogeological conditions in France were understood.

THE 20TH CENTURY – FROM 1900 TO WORLD WAR II: CONSOLIDATION AND EXTENSION OF THE QUANTITATIVE APPROACH AND BIRTH OF THE QUALITATIVE APPROACH

The two hydrogeological currents founded during the 19th century continued during the first half of the 20th century. Early in the 20th century, several works and new scientific advances allowed for significant progress in groundwater hydraulics and in the understanding of aquifers. In 1905, for example, Edmond Maillet formulated the equation (subsequently named after him) describing the drying up of springs, and the work of Joseph Boussinesq in 1904 contributed to clarifying the relationship between groundwater and surface water. In 1923 Marcel Porchet developed a method for calculating aquifer characteristics by carrying out a pumping test under transient state conditions. Later A. Vibert led several studies and further developed the understanding of the dynamics of groundwater in aquifers and of catchment works (1937–49). In 1917, during the Battle of the Somme, the British Army Corps of Engineers backed up by French geologists drilled almost 200 wells in the Chalk aquifer to supply its troops, thereby providing substantial new knowledge about chalk aquifer behaviour. These are only a few examples.

However, hydrogeological investigations were now being carried out throughout the whole of France. Many geologists set out to explore, describe and explain the hydrogeological conditions of most of the regions of France. Among the most active were:

- J. Gosselet and L. Dollé (who, in 1923, wrote and defended the first thesis on a subject concerning regional hydrogeology) in the North,
- R. Abrard and F. Dienert in the Paris Basin,
- A. Bigot and G. F. Dollfus in Normandy,
- J. Welsh in the Poitou,
- L. Armand and Ph. Glangeaud in the Massif Central (particularly on thermal and mineral springs),
- E. A. Martel, H. Fournier and B. Gèze on the karst regions in the Alps, Jura, Pyrenees and the Causses.

In 1933, A. Bonte drew the first water level contour map of the chalk aquifer west of the Somme River.

The first inventories and the call of Africa

By 1909, research had advanced to the point that Edouard Imbeaux was able to present a first synthesis of "French aquifers" to the French Geological Society and later to publish the first set of regional hydrogeological monographs of France based on stratigraphy, in his "Hydrogeology Essay" of 1930. "Deep drilling in the Paris Basin", published in 1939 by P. Lemoine, R. Humery and R. Soyer, was a model of the methodical inventory that heralded future subsurface databases.

Although hydrogeological work proceeded over the whole of France, some French hydrogeologists were already looking beyond the French borders, particularly towards North Africa. The first efforts were focused on Tunisia, for which J. Archambault published a first hydrogeological synthesis in 1947 (The Tunisian hydrogeology). This work resulted from the investigations carried out by the Hydraulic Resources Inventory Bureau (BIRH – Bureau d'Inventaire des Ressources Hydrauliques) which was created in 1930 mainly for groundwater exploration and development. The BIRH was the first of many specialised institutions which rapidly became a model to be followed elsewhere.

After Tunisia, similar institutions were created in Morocco and Algeria to enhance groundwater exploration and undertake better groundwater inventories. In Morocco for instance, the Groundwater Study Committee created in 1931 played an instigating role and several of their hydrogeological missions launched the first groundwater studies in this country. Later, in 1946, the "Centre des Etudes Hydrogéologiques" – CEH – was established to further enhance groundwater inventories and studies in Morocco.

Some of the major achievements in terms of groundwater surveys carried out in Tunisia, Morocco and Algeria prior to or during the Second World War are summarised in the works of J. Archambault, P. Bellair, G. Castany, M. Gosselin, J. Margat, M. Samsoen, J. Savornin, H. Schoeller, J. Tixeront and others.

Beginning of tracer experiments, water geochemistry, water geothermics

Prior to the Second World War, field investigations mostly remained individual actions like the map surveys which they often accompanied. The methods used included for a large part field observation, backed up by drilling data, hydrometric data from springs, water analyses and, in some cases, tracer experiments in karst areas associated with speleological exploration. Tracer experiments were already regarded as a powerful tool for investigating karst systems. In 1931, for example, Norbert Casteret discovered the source of the Garonne River by pouring tracers into the Forau des Aigualluts at the foot of the Maladetta massif in the Spanish Pyrenees.

The first part of the 20th century also saw the first true developments of geochemistry with, in the early 1930s, Henri Schoeller who laid the groundwork for groundwater geochemistry in Aquitaine and in the Maghreb region, before later developing the groundwork for groundwater geothermics.

Increased abstraction, first alarm call and first regulations

Simultaneously and interactively, groundwater abstraction increased as knowledge grew, particularly in the deep "artesian" aquifers discovered during the previous century in the Paris and Aquitaine Basins, but also in the Nord (Carboniferous limestone) and in the Algerian Sahara (the artesian "Continental Intercalaire" – or Albian aquifer –, and the "Continental Terminal" aquifer). The number of wells increased. In the Paris region, about 40 deep wells with an initial total discharge of 8 000 m³/hr (a rate that dropped quite rapidly thereafter) were added to the four artesian wells that had been drilled before 1900 to tap the Albian Greensand aquifer.

The total withdrawal from deep aquifers in France increased from 2 to 10 m³/s during the first half of the 20th century. In the Algerian Sahara, withdrawals from the Continental Terminal aquifer doubled between 1900 and 1930, reaching 6 m³/s in the Oued Rhir. In all of the Saharan aquifers in both Algeria and Tunisia, withdrawals increased from 284 Mm³/year in 1900 to 557 Mm³/year in 1950.

The first signs of overexploitation soon began to appear, showing the limits of these aquifers. The Greensand aquifer at the heart of the Paris basin was particularly affected: total abstraction reached a maximum of 34 Mm³/year in 1935 and alarming drops in head and subsequent decreases in production started to appear (Figure 11a). This motivated the first public conservation regulations for groundwater in France: the decree-law of 1935 "concerning groundwater protection" in the Paris region, which was extended later to other departments. This amounted, in fact, to the creation of a "drilling permit" for wells deeper than 80 m rather than a pumping permit. Nevertheless, it is interesting to observe the reasoning behind this major political act:

> "This issue, imperfectly dealt with by articles 552 and 641 of the Civil Code, which made the use of groundwater dependent, by right of accession, on ownership of the land, was, for a long time, merely theoretical. However, recent circumstances, that is to say new technical developments and the considerable increase these have made possible in the number of very deep wells drilled without any rules, have

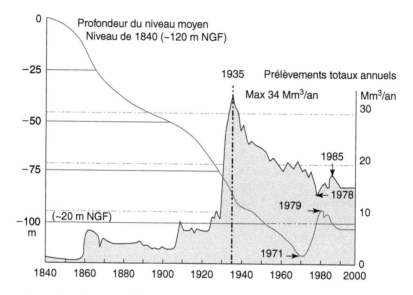

Figure 11a Evolution of the water level in the Albian aquifer in the Paris Basin.

shown the need for regulations similar to those that already exist for surface water and, for over a century, for mines. The aim is to prevent, mainly in the Paris region where the issue is more pressing than elsewhere, the squandering of this water— both quantitatively by depleting deep aquifers, and qualitatively by using it for inappropriate purposes—which threatens, over the very short term, to seriously compromise the preservation of one of our most valuable resources, which must be protected" (Exposé des motifs – JO of August 8th, 1935).

The evolution of hydraulic heads in these deep aquifers is, in fact, a dependable record of both the evolution of regulation and industrial development including the breaks caused by wars: the effects of the two world wars, are for instance, clearly visible on groundwater levels recorded in the Nord and Paris regions (Figure 11b).

Wellhead protection for drinking water purpose

The idea of protecting aquifers around water wells also arrived with the 20th century. Indeed, early in the century, guaranteeing potable water by implementing preventive measures around water wells became a major preoccupation: the notion of "protection zones" around wellheads and well fields used for drinking water purposes appeared for the first time in the Law of 15 February 1902. Initially, only the bacteriological risk was taken into account. "Official geologists" were appointed by the Ministry of the Interior from the ranks of mining engineers to define these zones. The law required the owners of some sensitive wells to acquire full ownership of the land on which their wells were drilled and enabled the administration to forbid activities entailing risk on the site. These measures were later reinforced by a regulation text in 1924 and, more specifically, by the Decree-law of 1935 mentioned earlier.

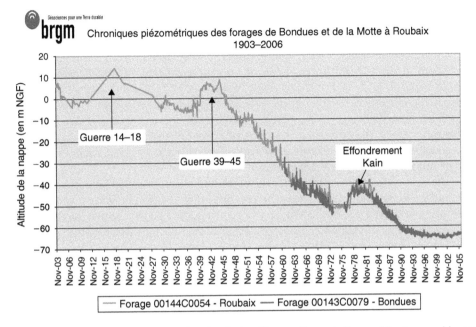

Figure 11b   Evolution of the water level in the Carbonifère aquifer and effects of the two world wars (Roubaix area, northern France).

THE 20TH CENTURY – FROM THE END OF THE SECOND WORLD WAR TO THE BEGINNING OF THE 1960S: ORGANIZING, STRUCTURING, TRAINING

The period between the end of the 1940s and the beginning of the 1960s was marked by several events that had a major impact on the development of professional hydrogeology in France.

Awareness of the stakes involved, structuring training and inventories, the first piezometric networks

The first consulting company working exclusively in the field of groundwater, BURGEAP, was founded in 1947 by Jean Archambault. The growing awareness of the importance of water problems, and in particular those linked to groundwater, encouraged the creation of structured courses. The first degree programme was created in 1945 at the Ecole Nationale Supérieure de Géologie in Nancy, after which degrees in hydrogeology were created in 1958 at the universities of Bordeaux (H. Schoeller), Montpellier (J. Avias), and Paris (L. Glangeaud and G. Castany). The first professional hydrogeologists obtained their diplomas in 1946 from the Ecole Nationale Supérieure de Géologie of Nancy and in 1959 from the universities mentioned. Later, in the early 1960s, several PhD theses were presented in hydrogeology. The Paris School of Mines also started delivering training in hydrogeology, especially through research programmes in hydrogeology, including mathematical hydrogeology.

The development of water supply systems after the Second World War required the drilling of many new wells. Many of these replaced traditional spring wells. At that time, wells were still sited by university geologists.

In 1955 a presentation by P. Lafitte and J. Ricour at the centennial meeting of the French Mineral Industry Society had considerable repercussions in the field of water. It resulted in the publication in 1956 of an article entitled *"La recherche minérale la plus importante de la France métropolitaine: l'eau"* (The most important mineral prospecting in France – water) in the *"Revue de l'Industrie Minérale"* (Journal of the Mineral Industry Society). The article discussed the increasing water needs of large industrial regions, the poor understanding of water problems, the lack of investments, and concluded with the vital role of hydrogeologists in the fields of water exploration and groundwater resource protection.

The first consequence of this paper, which had a significant impact in administrative and governmental circles, was the creation of a "National Water Commission". Soon after, at the request of the Ministry of Industry, the BRGM (Bureau de Recherches Géologiques, Géophysiques et Minières) created the first unit to assemble inventories of hydraulic resources (IRH) in the Nord – Pas-de-Calais region in 1956. This was followed later by the Lorraine, Normandy and the Gironde, all highly industrialized regions where problems with water resources and decreasing water levels were particularly marked.

The work carried out by these units mainly consisted of:

- identifying all wells and boreholes used for drinking water supply, industrial activity and agriculture – still rare at that time – and assessing their withdrawal;
- gathering geological, hydrogeological and technical documents relating to these wells which were scattered in the archives of drilling companies, administrations and towns;
- drawing maps locating the wells and indicating groundwater levels.

In the 1960s the Geological Mapping Service of Alsace and Lorraine (SGAL) undertook a similar study for the Alsatian aquifer instigated by L. Simler. From 1962 onwards this minutely detailed collecting of documents and making field inventories furnished a large quantity of valuable data for BRGM's Subsurface Databank (B.S.S.) which was already supplied with prospecting and mining data from declarations required by the Mining Code.

In 1963 BRGM created Regional Geological Surveys and regionalized some of its activities. The inventory, study and exploration of and for groundwater resources which now officially fell within its jurisdiction (previously BRGM's jurisdiction was limited to mineral exploration and subsurface data collection) is one of such activities, which progressively spread throughout France. With the hiring of university-trained hydrogeologists, groundwater specialists were now posted to all regions of France. This trend went even beyond the French borders, especially in West Africa where inventories and surveys for groundwater required an increasing number of hydrogeologists.

Two fundamental works on hydrogeology were published during this period: *"Les Eaux Souterraines"* (Groundwater) by H. Schoeller in 1962, and the *"Traité*

Pratique des Eaux Souterraines" (Practical Treatise on Groundwater) by G. Castany in 1963.

The first piezometers had been installed very early on, usually in areas where over-pumping had been observed. However, the first true network to monitor groundwater levels was installed in 1945 near Bordeaux to monitor the deep Tertiary aquifers. After the early 1960s, other piezometric networks were developed, notably in the North of France (Carboniferous limestone and Chalk aquifers), and later in the Paris Basin.

Africa – the other training and playground for French hydrogeologists

During the post-war period, many hydrogeologists continued to be drawn to French-speaking Africa, in particular to the Maghreb where it was increasingly difficult to find sufficient water supplies and where groundwater represented a major resource for both urban and rural areas. After 1945, public hydrogeological services were set up and/or developed in North Africa (BIRH in Tunisia, the Scientific Study Service of the Colonization and Hydraulics Department in Algeria, the Hydrogeological Studies Centre in Morocco), long before any similar types of institutions were created in France itself.

For several decades, the work carried out by these pioneers in hydrogeology in arid and tropical environments consisted of making inventories and synthesising hydroge-ological data, as well as testing and developing groundwater resources. Both these activities were closely linked (Figures 12, 13 and 14). This work greatly furthered both French and international natural or field hydrogeology and laid down the basis for modern hydrogeology. Furthermore, it was also in North Africa at this time that the idea emerged for gathering hydrogeologists together to organise their profession.

Figure 12a Pumping test for a well drilled in the Beni-Amir area during the 1950s (Tadla region in Morroco).

Figure 12b Pumping test in the Rharb region during the 1950s (Morroco).

Figure 12c Tazzarine's underground dam (Anti-Atlas) built around 1955.

Some of the most prominent of the French hydrogeologists who worked in Africa–Ambroggi, Schoeller, Castany, Robaux, Margat, Tixeront – indeed made a proposal at the International Geological Congress of Algiers in 1952 to create an International Association of Hydrogeologists. Four years later in 1956 at the next such congress in Mexico, IAH was created.

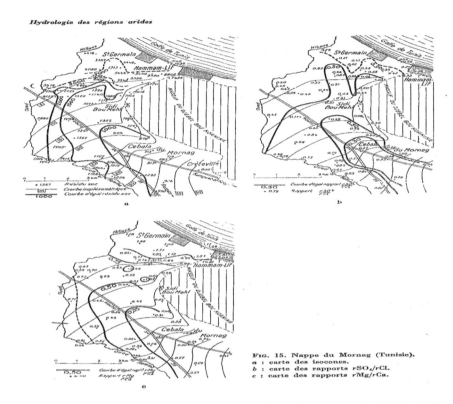

Figure 13 Example of an early hydrogeochemical map for the Mornag aquifer (Tunisia, 1940's).

With decolonisation, and in particular between 1956 and 1962, many of these famous hydrogeologists returned to France. These included G. Castany and H. Schoeller returning from Tunisia, J. Margat and L. Monition from Morocco, and P. Gevin from Algeria. Most of them joined French universities and public research institutes where they trained future hydrogeologists and participated actively in structuring and organising French hydrogeology. Others, like R. Ambrozzi, F. Mortier, R. Dijon and E. Stretta became international experts working for the United Nations agencies.

THE 20TH CENTURY – FROM 1964 TO THE END OF THE 1990S: MODERN HYDROGEOLOGY AND WATER RESOURCES MANAGEMENT

A new concept, new means, a new institutional organisation ... to boost hydrogeology

In the early 1960s water resource management took on a new dimension with the concept of management by catchment units. For the first time legislators recognised how complex the functioning of water resources systems actually was and the need for

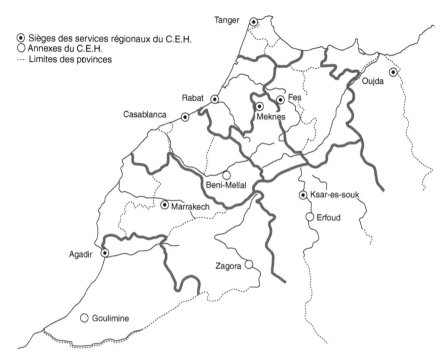

Figure 14a Centre for Hydrogeological Studies in Morroco ("Centre des Etudes Hydrogéologiques – CEH") – situation in 1960: one central office and 10 regional offices.

Figure 14b Centre for Hydrogeological Studies in Morroco ("Centre des Etudes Hydrogéologiques – CEH") – Inside the Rabat Office.

a broad vision to be able to manage them properly. It was decided that water resources should be managed at the catchment level.

In 1964, the first water law was enacted, creating six Agences Financières de Bassin, or Basin Authorities, (now known as the Agences de l'Eau or Water Agencies).

These were financed by taxes on water consumed and wastewater discharged. Among other things, the law also made the declaration of withdrawal of groundwater mandatory, regulated discharges into the subsoil, and required that wellhead protection zones be set up around all drinking water wells. Later, in 1972, the creation of a Ministry of the Environment endowed with a service for the study of issues concerning water, derived from DATAR (Délégation à l'Aménagement du Territoire et à l'Action Régionale – Territorial Development and Regional Action Delegation), would, for the most part, complete the establishment of the major institutional water structures in France.

Spurred on by this new law and within this new institutional context, up to the beginning of the 1970s, hydrogeology in France was given a considerable boost with the creation of hydrogeology laboratories in more universities (Marseille, Besancon, Lille and Strasbourg). It also led to the establishment of hydrogeologist positions in the French departments, in the River Authorities and in the administration (SRAE/Service Régional d'Amenagement des Eaux, CETE/Centre d'étude Technique de l'Equipement, etc.). In addition, this also prompted the development of "groundwater" activities at BRGM, which created a hydrogeology department in 1964 and increased the number of hydrogeologists in the regions. The private sector followed this trend and several consulting companies that dealt with hydraulic engineering and water technology, including large companies such as SOGREAH and SAFEGE, started to turn to hydrogeology.

From descriptive hydrogeology to the study of how aquifers function

During the early decades of the 20th century, hydrogeologists' work remained very "naturalist" and involved mainly descriptive hydrogeology. It consisted for the most part in (1) compiling inventories of public or private wells which, depending on the region, were extensive or descriptive, (2) drawing groundwater level maps and (3) synthesising data. This culminated in the publication of the first detailed 1:50 000-scale hydrogeological maps (e.g. Douai – 1964; Amiens – 1964) and regional syntheses (e.g. the Atlas of Aquifers in the Paris Region (1970), the Quercy Causses – 1976, North-Montpellier Causses – 1965, Istres et Eygières – 1964).

The first 1:1 000 000 scale map of groundwater in France was published in 1965 and was followed by the 1:500 000 scale Hydrogeological Map of the Paris Basin in 1966 and several maps at 1:200 000 scale (Nord – Pas-de-Calais – 1965, Picardy – 1966). The first "Atlas of Groundwater in France" showing the state of achievement of the mapping and the bibliographic knowledge by administrative region was published in 1970 by BRGM and DATAR.

Many detailed hydrogeological monographs for individual aquifers or hydrological basins were prepared, either within the framework of doctoral theses or of public service projects carried out at BRGM for the Water Boards (mainly in the Nord and in the Paris Basin). Similar monographs were also produced by BURGEAP and others at the request of the Rural Engineering Services of the Ministry of Agriculture. In 1970 a first attempt was made by the Basin Authorities to estimate groundwater withdrawals in France; the estimated total withdrawal at the time was 5 km^3/year.

In addition to essential descriptive and inventory work, hydrogeologists also increasingly participated in regional and local prospecting for domestic and industrial water supplies and in the siting of wells and supervision of well drilling. They contributed to increasing the water supply in large cities such as Lyon, Paris, Bordeaux, Lille, Rouen, Le Havre and Strasbourg, notably by studying and designing large well fields, and in numerous rural towns. These various interventions led to the further development of geophysical exploration, well design and drilling techniques.

As an example, the reconnaissance of potential groundwater resources in the Bassée aquifer in the Seine River Valley upstream from Montereau carried out in 1967 for the Paris water supply was one of the largest studies ever undertaken in France in terms of number of wells, piezometers and pumping tests constructed and was achieved within a year. The results showed that 50 wells could tap about 260 000 m^3/d to cover a significant portion of Paris' water needs.

Moreover, hydrogeologists also recognized the importance of understanding the functioning of groundwater resources to be able to manage them correctly. To this effect, funds allocated by the DGRST (Direction Générale de la Recherche Scientifique et Technique – General Direction for Scientific and Technical Research) within the framework of the International Hydrological Decade (1965–74) made it possible to launch research programmes and equip experimental basins to better understand their functioning and their recharge mechanisms (i.e; Hallue in the Somme Département; Baget in Ariège; Lez in Hérault; Orgeval in Seine-et-Marne; Fontaine de Vaucluse near Avignon).

The first analogue simulation models (rheo-electric tank, resistance-capacity network, conductive paper) were also developed as early as the 1960s to improve the understanding of aquifer behaviour in several areas (Géohydraulique, SCGAL, then BRGM).

Significant progress was thus made in knowledge of aquifer recharge and functioning, as well as in the measurement of hydrodynamic parameters. Aquifer-river interactions were also recognised through hydrometric gauging of rivers, piezometric monitoring and later in the 1970s by mathematical modelling. This period also saw a rise in the development of structured methods for resources evaluation – or Water Resources Assessment (WRA). This represented the final step toward quantitative analysis. The methodology of water resources management was now being clearly outlined too.

Development of powerful tools for synthesising and managing groundwater resources – mathematical models

From the 1960s onwards, predictive groundwater management intensified with the development of quantitative hydrogeology, based on hydrogeological data processing and computer sciences. Numerical models replaced analogue models and, with the advent of microcomputers, progress was increasingly rapid, with mathematical approaches being applied to hydrodynamic structures and the behaviour of aquifer systems at all scales.

The first mathematical models were developed during the 1960s and 1970s under the direction and guidance of several pioneers of mathematical hydrogeology, including

G. de Marsilly and Y. Emssellem. In 1965, the Greensand aquifer (Albian) in the Paris Basin was the first large regional aquifer to be modelled (DICA/Direction des Carburants, Ministry of Industry). All of France's major aquifers were subsequently and progressively modelled: the Albian, Champigny and Chalk aquifers of the Paris Basin, the Vosges Sandstone, the Carboniferous Limestone (Nord), the multilayered aquifers of Aquitaine, the Rhine's alluvial aquifer, the Roussillon Plain (Pyrénées Orientales), the Infra-Toarcian of Vendée, the Beauce aquifer, etc. This process allowed a good understanding of the importance of groundwater potential in France and enabled the refined quantification of groundwater resources and groundwater abstraction (Box 3).

Box 3 – Groundwater in France

France is rich in groundwater and ranks first among European Union countries for its water resources. It is estimated that all of its aquifers together hold 2000 billion m^3, 100 billion of which discharges from springs every year. Only about 6 billion are developed – 60% for drinking water, 25% for industries not connected to the public supply network and 15% for irrigation. The most heavily developed aquifers are the Alsace aquifer, the Chalk in the Nord and in the Paris Basin, the Eocene-Oligocene in Aquitaine and the alluvial valley aquifer systems of the Seine, Rhone, Isere and Loire rivers. About 50% of the abstracted groundwater comes from alluvial aquifer systems.

Around 60% of the population of France is supplied from groundwater and for some departments in the Nord and in the Paris Basin, groundwater furnishes up to 100% of the drinking water supply (DWS). Fifty percent of the water supply to the city of Paris comes from several springs captured 100 to 150 km west, east and southeast of the city. This water is piped into Paris by aqueducts built mostly between 1865 (under Napoleon III) and 1893.

There are an estimated 13 500 drinking water supply networks (municipal or groups of towns). Half of these are operated by the towns themselves, while the other half are managed by private companies. There were around 34 000 drilled and dug wells and springs supplying cities and towns in 2005. This number has dropped considerably during the last two decades as many wells have been abandoned due either to high nitrate or pesticide levels exceeding drinking water standards, or to the impossibility of protecting wellheads located in urban or industrial areas.

The first estimates of groundwater abstraction were made in 1970. Between 1981 and 1999 abstraction increased moderately, from 5.7 billion to 6.1 billion m^3. It is assumed that during the 1970s there was a sharp increase in pumping, especially for drinking water. Total groundwater abstraction seems to have varied little in the last 15 years.

After the 1990s, as coal (Nord-Pas-de-Calais) and iron (Lorraine) mines closed down, the pumping of mine water decreased considerably.

Considering groundwater withdrawal according to usage since the 1970s, different trends appear. For drinking water, abstraction increased sharply during the 1970s, then slowed down during the 1980s, and stabilised during the 1990s, in spite of a 20% increase in the population and an improvement of living conditions. This was due to a reduction of leaks in the networks, the installations of water meters in all homes and a dissuasive increase in the cost of water.

Groundwater abstraction for industrial use has decreased steadily since 1970 in the north of France and in the Seine-Normandy Basin due to the closure of factories and to pressure from the administration to recycle water or to pump from rivers. It remained stable in the Loire-Bretagne Basin, which is not heavily industrialised. In the Rhine-Meuse Basin, abstraction increased until 1985 and then stabilised.

The evolution of the abstraction rate for irrigation is still poorly known because the installation of water meters on agricultural wells is too recent. It is suspected to have significantly increased in the recent past.

The first numerical models, which were excellent data assessment tools, also enabled great advances in the understanding of how groundwater resources functioned. Interactions between aquifers, and also between surface water and groundwater, and their evolution through time, were evidenced, quantified and simulated, further reinforcing the need to use a broad and extensive approach to water resource issues. These models also helped lead to new concepts such as that of "aquifer systems" which J. Margat inventoried over the whole French territory ("Margat Codes"). These aquifer systems were mapped and published in 1980 in the "Hydrogeological Map of France – Aquifer Systems" at 1:1 500 000 scale. This project was later taken up again in 1995 and completed.

The numerical models were also used to assist in the management of large drinking water well fields, starting in the 1970's (Croissy, Aubergenville, Vallées de l'Oise, de l'Yerres ... in the Paris area, Illzach near Mulhouse, Sorgues and Bédoin close to Avignon, Moulle in the Dunkerque area and Les Ansereuilles in the Lille area). They were also used to ensure their protection with the design of efficient anti-pollution protection systems – hydraulic traps (Aubergenville, Ansereuilles, Illzach, Avignon, Lyon ...).

The ensuing results from these models were numerous and varied including, for the quantitative aspects, aquifer potential, aquifer yield, recharge and discharge rates, flow rates, flow fields, water exchange areas and rates between adjacent aquifers or between surface and groundwater. They also included water resources management schemes or master plans and optimization scenarios for water resources exploitation, including artificial recharge. For qualitative aspects, the model results demonstrated the likely evolution of water quality in space and time, well field protection schemes, plume migration in 1D, 2D or 3D, and pollution removal and remediation scenarios ...

Numerical models and simple mathematical or statistical tools based, for example, on the correlations between groundwater levels and spring discharges and/or global pumping rates, made it possible to determine useful limits for water resources management at the aquifer or aquifer system scale. Proposed limitations, such as groundwater withdrawal thresholds that must not be exceeded in given climatic conditions, started to serve as a basis for regulations (i.e., for the Beauce, the Champigny, the Chalk or the Greensand aquifers in the Paris Basin, for the La Rochefoucault Karst system and the Touvre springs, for the Eocene and Oligocene aquifers in the Bordeaux area).

Mathematics opened up new horizons towards increasingly sophisticated tools. Several prototype models were developed and implemented such as the integrated optimal management model of the Croissy aquifer system in 1988, which enabled the operator to optimise abstraction, artificial recharge and induced recharge from the Seine river in a large well field area, while minimising operational and maintenance costs, and ensuring proper water supply to satisfy the evolving water demand (Pennequin et al., 1991).

Droughts – an impetus for turning to unconventional resources and setting up prediction support tools

In 1976, a prolonged drought in the northern half of France caused a water shortage in the Armorican Massif, a low-permeability crystalline bedrock area where surface water is the principal source for the municipal water supply. This situation allowed attention to be focused on groundwater resources to look for possible back-up supplies, and notably on formations that had been erroneously thought to "contain no groundwater". Techniques developed in West Africa were applied to the hydrogeology of fractured crystalline rocks in these ancient massifs. This prompted the elaboration of conceptual models that made prospecting for new sources of water more efficient. Productive layers were identified and numerous wells were drilled in the bedrock with the "down-hole hammer" technique to supply significant quantities of back-up water. Fracture identification and interpretation techniques were used to facilitate their siting (i.e., satellite images, image analyser, data processing, radon surveys).

Many studies were also carried out on carbonate aquifers in limestone regions and massifs (Chalk in the Nord and in the Paris Basin, limestone in the Jura, Pre-alps, Causses, Provence and Pyrenees). The karst systems of the Fontaine de Vaucluse and of Baget in the Pyrenees are among the best-known and were the focus for continuous monitoring and for many fundamental research programmes.

The Lez karst system near Montpellier was also studied for about ten years by the University of Montpellier as a possible source for the city's municipal water supply, although local officials preferred at that time a solution involving pumping from the Lower Rhone Canal. In the late sixties J.Avias showed that, during the summer, it was possible to abstract the reserves in the karst system that were recharged during the winter, and in the early 1980s, an underground pumping station was built (and inaugurated in 1982) to supply 300 000 inhabitants. This is also a good example of "active aquifer management" that has been operating now for the last 27 years. An attempt was also made to capture the Port-Miou submarine spring near Marseilles. In 1977, an experimental dam was placed in the karst conduit, unfortunately without success because the water in the karst aquifer turned out to be too salty for human consumption.

Aside from pushing engineers to turn to unconventional water resources, the drought periods, and particularly the one that lasted several months in 1976, also made the government aware of the risk of drying up of wells in sedimentary basins. As a result, hydrogeologists were asked to make medium-term predictions of groundwater levels and participate in the "Drought Committees" that were being set up in the departments. Predictive models were developed for several aquifers in the regions most affected by drought. In 1977, BRGM began publishing a quarterly "Low-Water-Level Prediction Bulletin" that provided information concerning groundwater quantitative status in the country's main aquifers and the water levels in many reservoirs.

Water quality becomes a major issue – the growing importance of hydrogeochemistry

As the number of studies of groundwater resources increased, the negative impacts of industrialisation and intensive agriculture became increasingly apparent. Beginning in the late 1960s and early 1970s, there was a growing awareness of the risk of groundwater contamination and an increasing preoccupation with determining and preserving groundwater quality. Qualitative aspects of hydrogeology were added to quantitative aspects. In France, Gil Michard (with the help of Jacques Faucherre) started teaching and developing water geochemistry in 1966. A Masters degree was created at the University of Paris 7 in 1970.

The notion of groundwater vulnerability to contamination was introduced by J. Margat and the first map of groundwater vulnerability in France at 1:1 000 000 scale was published in 1970. It was followed by many regional maps of individual departments (e.g., Seine-Maritime, Eure), of geographic units (e.g. the Var Plain), or by quadrangle at the 1:50,000 scale (e.g., Rhone River Valley, Allier River Valley ...). The methodology for vulnerability assessment was established.

Groundwater quality monitoring networks were set up in several regions (Normandy, Nord, centre of the Paris Basin) to monitor the general evolution of groundwater hydrogeochemistry. Specialised monitoring networks were also set up in some major industrial areas in order to detect specific contaminants (Lyon, Rouen, mining basin of the Nord ...). Data from these regional networks made it possible to publish a first water quality map of France in 1977 at the 1:1 000 000 scale. Unfortunately, this project was not continued in such a comprehensive way and was set aside until much later, partly because the attention of hydrogeologists and of the government was rapidly focusing on the nitrate problem.

Indeed, data from the groundwater quality monitoring networks, added to those of previous desktop studies, rapidly revealed that nitrate concentrations had been steadily increasing since 1955–1960 in most of France's unconfined aquifers. In 1980, the publication of a map of nitrate concentrations of groundwater in France at 1:1 500 000 scale showed their geographic distribution, and proved that there was a direct correlation between high nitrate concentrations, on the one hand, and the major livestock and intensive agricultural regions, on the other hand. Natural denitrification in semi-confined aquifers was also observed in 1980 in the Douai region (Nord). The same phenomena were observed later in other regions (Calvados, Poitou), and then in granitic areas in the Armorican Massif, where denitrification was associated with the presence of pyrite (Coët-Dan Basin, Morbihan). The first predictive models of nitrate concentrations in springs were developed during this period – the early eighties (Provins springs, Seine-et-Marne).

Although nitrate retained much of the attention of hydrogeologists due to the numerous instances of exceedance of drinking water limits in many regions, other substances or elements, most of which were included in the list of these same standards, were analysed and studied, such as, pesticides, heavy metals, PAH and VOC. Much attention was paid to their transfer processes in soils, in the unsaturated zone and in aquifers, involving many university, BRGM and INRA research groups, as well as private consultants. Conceptual and mathematical models were developed and these underscored the importance of having reliable and representative data.

In 1985 the National Groundwater Quality Observatory (ONQUES) was created, and managed by BRGM under the auspices of the Ministry of the Environment and the Ministry of Public Health. This was the first national groundwater database and its objective was to consolidate the results of all water analyses referenced by drinking water supply wells for the whole of France. Studies carried out on water quality led to an improvement of sampling and analytical techniques. As a result, new molecules were discovered and added to the list of groundwater pollutants. New tools also were developed such as radon, CFCs and isotopes. René Letolle and Jean Charles Fontes, for example, were two of the pioneers in France who developed stable isotopes for groundwater studies, starting in the 1970's.

Isotopes, including Carbon 14, were used to date groundwater, notably in confined aquifers. Ages of 20,000 to 40,000 years were identified, for example, in the deep aquifers of the Paris Basin (Albian), in Touraine (Cenomanian), in Aquitaine (Eocene-Oligocene) and in Charente (Infra-Toarcian), and up to 100,000 years in the deepest part of the Alsatian aquifer. In 1999 an isotope database was created, which included rainfall data and data from various aquifers.

The 1992 Water Law – Heralding the future European Water Framework Directive

The second French Water Law was published in 1992. It notably reiterated that ground-water and surface water are one and the same resource and that water is a common heritage of mankind. It introduced the obligation to establish a SDAGE (Schéma Directeur d'Aménagement des Eaux – Water Management and Development Master Plan) for each major river basin, and a SAGE (Schéma d'Aménagement et de Gestion des Eaux – Water Management and Development Scheme) for each catchment and sub-catchment area. The polluter pays principle (PPP) was officially endorsed. In addition, it reinforced and completed the concepts and principles already introduced by the 1964 Water Law, and the organisation of the water scene in France was strengthened, notably with the creation of a Water Division within the Ministry of the Environment and of Local Water Commissions (CLE – Commissions Locales de l'Eau) in the regions, whose role were to develop, revise and monitor the application of the SAGE.

In 1996, the General Council of Mines published a report entitled "Sustainable Groundwater Management" (Y. Martin), which notably supported the implementation of the 1992 Water Law. The monitoring networks were reorganised and coordinated, with global management at the national level and a harmonisation of existing water resources databases took place.

Wellhead protection gets organized, but bogs down ...

The Water Law of 1992 emphasised the necessity of completing the protection of all wells supplying drinking water (except for some naturally well-protected deep wells) during the next five years (i.e., targeted to be completed by 1997). Hydrogeologists played an essential role in the creation of drinking water wellhead protection zones. Indeed, in France, this mission was given to "certified hydrogeologists" who were appointed by regional prefects, on the basis recommendations made by the regional divisions of the Ministry of Public Health, according to their competence and regional experience.

In 1989, BRGM published "Methodological guidelines for the creation of drinking water wellhead protection zones" for these hydrogeologists (the document was later updated in 1998). After 1995, the definition of protection zones had to be based on hydrogeological and environmental studies recommended by a certified hydrogeologist. However, although the obligation to create protection zones dated back to 1935 (for some wells), and in spite of reiteration in the 1964 and 1992 Water Laws, only 12% of wellheads had protection zones in compliance with regulations in 1986. The percentage had risen to 25% by 1992 (and 55% in 2007). The problem was due, for the most part, to the complexity of the administrative procedure and the absence of any strong incentives or sanctions for non-compliance.

The importance of economic criteria

The introduction of economic concepts into the management of water resources dates back to the beginning of the 1970s. Initially, this involved provisional and comparative estimations of the cost of developing groundwater, illustrated by specific regional mapping works (France, BRGM and BURGEAP-1971, and West Africa). Analyses later involved evaluations of external phenomena and environmental impacts (cost of protecting wellheads, cost of pollution, etc.). This work led to the publication in 1983 of a textbook entitled *Introduction à l'Economie Générale de l'Eau* (Introduction to General Water Economics), then to the national and commercial accounting of water (French Water Accounts—INSEE, 1986). This useful approach notably showed that groundwater, often a bit forgotten by government officials because it is invisible, was important and that it had to be taken into account in resource planning, as well as in the SDAGE and in the SAGE. However, these concepts, which were somewhat beyond the normal scope of the hydrogeologist, began to really develop only later in the management of water resources, and especially with the arrival of the European Water Framework Directive (WFD).

And the ever-wandering French hydrogeologist ...

While hydrogeology was taking on a new face in France, work in Africa continued along more traditional lines with other teams. Hydrogeological inventories and syntheses continued in West African countries like Senegal, Mauritania and the Ivory Coast. Growing water needs also led to what was called for many decades "hydraulique villageoise" (village water supply). In Mali, during the 1972–73 drought, use of the down-hole hammer drilling technique, structural geology and photogeology made it possible to drill wells rapidly and successfully in unweathered crystalline rock. The drilling production rate increased from three boreholes a year to 12 boreholes a month. Vast "village water supply programmes" were carried out to provide drinking water to villages. At least 12 000 villages were equipped with wells in Chad, Gabon, Niger, Senegal, Morocco, Mauritania, Burkina Faso, Togo and Cameroon, financed by Co-operation Funds from the French Government, the European Union and, in some cases, Saudi Arabia. These wells also made it possible to create "small irrigated perimeters" to grow vegetable crops.

Hydrogeological studies of major aquifers like those of the Northern Sahara (the ERESS project) carried out for UNESCO showed, among other things, how the artesian aquifer in the Continental Intercalaire functions in all of the countries where it is present (Libya, Tunisia, Algeria, etc.). French hydrogeologists had discovered this aquifer during the 1950s in the Hassi-Messaoud region (Algeria) when prospecting for groundwater to supply oil well fields. Models were also developed to determine the groundwater potential of large aquifers, including the Maestrichtian aquifer.

Beginning in the 1970s and 1980s, French hydrogeology took another step forward and was exported to Asia, Latin America and even Australia. In the Middle-East, for example, the Saq aquifer system in Arabia with a surface area of 370 000 km^2 underwent a detailed study between 1981 and 1986 requested by the Ministry of land and water. Non-renewable groundwater resources in deep aquifers down to the Cambrian Sandstone were studied and their potential estimated. Boreholes 2 500 m deep were drilled to prospect for fresh water and predictive modelling of abstraction for drinking and agricultural water was carried out for a period of 20 years to help set the framework for economic development in the region. Extension of this work to Kuwait and Oman took place later up until and beyond the 1990s.

Similar studies using regional mathematical models, geophysical prospecting tools and remote sensing were carried out in China, India, Indonesia, Argentina, Brazil, Iran, Peru, and in many other countries in Asia and in the Americas. In 1999–2000, for example, French hydrogeologists headed the "groundwater" part of a major study with the objective of demonstrating the advantages of using SPOT satellite imagery for the development of irrigated surfaces in the Shandong province in China. In India, between 1997 and 2000, two large studies of the impact of mining activities on water resources were carried out – the first in the iron mining area in the State of Goa, the second in the region of the Sukunda chromite mines in eastern India. Likewise, in Australia during the 1970s, French hydrogeologists contributed significantly to the study, estimation and development of water resources on the western slope of coastal hills and of the Great Artesian Basin.

Several hydrogeologists were internationally recognised experts and were invited to participate in projects carried out by international bodies like UNESCO (e.g., international hydrogeolgical map of Africa at 1:5 000 000 scale, the hydrogeological world map at 1:25 000 000 scale), the FAO, the World Bank and the European Union.

Hydrogeologists extend their scope of action

In addition to activities such as mapping, exploration, resource estimation, resource protection or quantitative and qualitative management of groundwater resources – work that might be called "traditional hydrogeology", French hydrogeologists also started to contribute more and more during this period to many specialised areas such as mineral and thermal waters (Box 4), geothermal energy (Box 5), aquifer storage and waste storage site safety (Box 6).

Box 4 – Thermal-mineral water in France

The development of thermal and mineral water in France (Figure 15) has been regulated through time by several laws and decrees including the Law of July 1956 and the Decree of 1957 regarding their exploitation, the Water Law of 1964 concerning bottling, the Decree of June 1989 concerning bottled mineral water and the Water Law of 1992. Many of these regulations aimed at ensuring the preservation of the special character of these waters.

Figure 15   Source Eugénie in Royat (Auvergne – France).

Today, thermal-mineral water in France tends to follow the same trend as that for drinking and normal bottled spring water so that it is now regulated by the 2007-49 decree from 11 January 2007 which includes the prescription of all relevant EU directives, including 80/777/CEE of July 1980 regarding the exploitation and selling of natural bottled mineral waters. The protection of thermal-mineral waters is ensured by "emergence sanitary safety zones" (périmètres sanitaires d'émergence) established around the abstraction wells. It can be reinforced by protection zones at the scale of the resource when "declared to be in the public interest" – (DIP).

A national inventory of mineral and thermal springs combining physico-chemical and geological characteristics has been progressively assembled up to 1998, in association with the ministries concerned. There are in France over 700 springs, 400 of which are captured for spas or bottling plants, and 158 of which are declared to be in the public interest, but less than 40 springs have fully established DIP protection zones. Their development and exploitation used to be controlled by the Ministry of Health and the Ministry of Industry and today is under the control of the Prefects of the Departments.

Since the 1950s, hydrogeologists have been requested to identify geological reservoirs of mineral water, to characterize the properties of both the embedding rocks and of the mineral water) and to optimise and preserve their exploitation. Since 1980, several projects supervised by hydrogeologists have been undertaken to capture springs at greater depth, either to increase yield or to protect resources from surface pollution, as for example in Boussanges (Vichy Basin), Chaudes-Aigues, Bagnols-les-Bains, Watwiller, Aix-les-Bains, Dax, Luchon and Gréoux-les-Bains.

Research has been carried out to determine the origin and age of the water and of the CO_2, notably using isotopes and geothermometers (i.e. Cauterets, Luchon, Evian, Perrier, Balaruc-les-Bains). The circulation times of deep geothermal fluids can be very long (e.g., 5,000 years in Cauterets, 14,000 years in Dax). These studies have supported the creation of protection zones for many sites such as Evian, Vittel, Aix-les-Bains and others.

Since the 1970's, large industrial bottled water groups (Danone, Nestlé ...) have constituted teams of hydrogeologists to be in charge of the development, protection and exploitation of their mineral water springs in France and abroad. Many of them also went to search for new mineral springs in foreign countries, like in California for Vittel in 1978-1979. The picture is different for thermal water operators who tend to use more the expertise of Regional Hydrogeologists. Several programmes are launched by the thermal regions too in order to assess the local hydrogeological risks and better secure the supply of thermal water.

The first map of mineral waters in France at a 1:1 000 000 scale was published in 1973. It was updated in 1994 and then in 2005, this time integrating spring waters in addition to thermal and mineral waters. A guide to thermal-mineral waters was published in 2005 by BRGM (Vigouroux, 2005).

Box 5 – Geothermal energy

The first geothermal well in France was drilled in 1961 for the *Maison de la Radio* in Paris which tapped the Albian aquifer for heating and cooling. However, it was the first energy crisis in 1973 that triggered a major research programme on geothermal energy starting in 1976. This created new types of missions for hydro-geologists, including inventorying geothermal potential in each basin, region and department, preparing geothermal guidelines and guidebooks, siting boreholes and supervising drilling operations. Between 1976 and 1994, 73 low enthalpy wells (50–100°C) were drilled for urban or greenhouse heating (and sometimes cooling), 52 of which were in the Paris region and 12 in Aquitaine.

The largest geothermal project in France was developed by BRGM geologists and hydrogeologists in Bouillante (Guadeloupe). Two exploration boreholes (338 m and 2500 m deep) were drilled between 1969 and 1977 and yielded water at 242°C, and an experimental 4.4 MW pilot power plant was built in 1975. In 1996, an operational power plant was connected to the national grid. In 2000, three new production wells were drilled and electricity from geothermal energy is now intended to supply 8% of Guadeloupe's needs.

Hydrogeologists have also been and are still participating in the large Franco-German "Hot Fractured Rock" research project in the granite in Soultz-sous-Forêts (north of Strasbourg) where researchers hope to use the earth's heat to produce energy. A network of natural fractures has been identified down to a depth of around 3900 m where temperatures exceed 165°C and reconnaissance work is continuing down to a depth of 5000 m where the temperature is expected to reach 200°C. A 6 MW pilot plant is planned with one 5000 m deep injection well and two extraction wells.

At the same time, and up until the present, the use of very low enthalpy geothermal energy (10–15°C) with water/water heat pumps has enhanced the development of the hydrogeology market by the drilling of several thousands of boreholes, mainly in alluvial aquifers. Areas favourable for the siting of these heat pumps have been mapped for many regions.

Box 6 – Underground storage

Underground storage projects have involved many hydrogeologists in the last three or four decades. The storage of natural gas in aquifers is widespread in France with 12 sites, mostly located in the Jurassic and Triassic formations of the Paris and Aquitaine Basins. These sites began operating between 1956 and 1993. Hydrogeologists participated extensively in their development and monitoring, notably to look for potential reservoirs, calculate their permeabilities and available volumes, ensure proper piezometric surveillance and undertake the necessary modelling work to design a safe and sustainable operation mode. GDF (*Gaz de France*) for example employs its own hydrogeologists who manage the hydrogeological monitoring of the sites in operation and assess the risk of interference with pumped aquifers or thermal springs.

Projects to store hydrocarbons in natural or artificial cavities have also required the participation of many hydrogeologists (e.g. at May-sur-Ornes, Manosques and at salt mines in the Lorraine). Likewise, CO_2 storage in saline aquifers in order to reduce greenhouse gas emissions and protect the ozone layer has been the objective of many studies and much research over the last decade or so. Hydrogeologists were and are directly involved in these studies, and this trend grew significantly in recent years with an increasing number of sites being studied.

It is the development and safety of domestic, industrial and radioactive waste disposal sites which has probably required the greatest number of hydrogeologists. Studies began in the early 1970s for non-radioactive waste disposal, with regional inventories of very low-permeability sites which were "a priori" suitable for siting domestic landfills, industrial waste disposal, sludge and mine tailing storage or burial grounds for animal carcasses. Later the studies were reoriented toward the characterisation of the waste and of the sites (aquifer vulnerability, measurement of very low permeabilities with development of measuring equipments, role of the overlying formations). The results of these studies serve the fine tuning of French regulations for waste storage conditions – active and passive barriers.

After 1976, hydrogeologists also became highly involved in radioactive waste storage. Between 1976 and 1980, an inventory of formations that might be used to store radioactive waste was launched by the European Community, at the instance of the IAEA. This work called in geologists in the nine then Member states to select suitable formations according to geological and hydrogeological criteria (rock type, structure, permeability, hydrodynamic gradient). As a result of this work, the Soulaines site (Aube) was chosen and equipped for the storage of low- and medium-level radioactive waste. The waste was stored in drums placed in concrete silos, themselves set in a clay butte whose outflow to the water table aquifer was (and is still) "monitored" at the boundaries. The installation became operational in 1990.

Hydrogeologists also worked on the "Manche" storage site in La Hague, on the sites of EDF nuclear power plants and on the inventory of formations to study for the deep burial of high-level radioactive waste (granite, shale, clay and salt). After 1992, they also participated in studies of various sites in the north, centre and south of France (geological and lithological characterisation, in situ hydrodynamic tests, permeability measurements and global hydrogeological approaches). In 1999 the Bure site (Meuse and Haute Marne) in the Callovo-Oxfordian clay at a depth of 500 m was chosen for the creation of an underground laboratory.

Even more so, hydrogeologists became increasingly involved in the field of environment, and particularly in soil and groundwater decontamination. Indeed, at the end of the 1980s, new environmental issues arose in connection with contaminated sites and soils. Very naturally, hydrogeologists and hydrogeochemists became involved in this field and particularly in the decontamination of groundwater polluted by old industrial storage sites or accidental spills (Alsace, Rhone Valley, Grand Stade de France in Paris, etc.). Involvement in diffuse pollution issues grew also.

Many groundwater specialists also participated in large civil engineering projects such as the Canal du Nord (1961), the Rhone Canal (CNR), the Lille and Lyon metro lines, the Halles Châtelet in Paris, the RER train stations on the EOLE and Météor lines in Paris, several EDF nuclear power plants (Blayais, Nogent-sur-Seine, Fessenheim), the Channel Tunnel, the Lyon-Turin railway tunnel, and the stabilisation of the cliffs along the Basque coast in Biarritz, just to give a few examples. Their important contribution during the post-mining era, notably evaluating the consequences of the cessation of mine water pumping, studying the recovery of the water table and evolution of the quality of overlying water resources (Lorraine mining basin, Nord mining basin) needs to be mentioned too.

THE BEGINNING OF THE 21ST CENTURY: THE PROFESSION OPENS UP TO OTHERS AND OTHER FIELDS

The Water Framework Directive: organisation at the European scale

In France, as elsewhere in Europe, the Water Framework Directive (WFD) is fashioning the landscape of current and future hydrogeology. It has, in particular, introduced the concept of water bodies as management units and the notion of "good water status". The Directive sets objectives (reaching good water status) and a working framework common to all of Europe, and prescribes penalties if deadlines are exceeded. It promotes transparency, public participation in decisions and schemes and requires not only the economic viability of "resource management-water supply" systems (water pays for water), but also the use of economic tools (cost-effectiveness analysis, cost-benefit analysis, etc.) to ensure sustainable management of water resources, and more specifically for the quantification of the various elements to be taken into account in the decision making process. Economic analysis is taken a step further and places value on environmental elements.

The WFD therefore requires a revision of hydrogeological units. In France two options are chosen based on past work: a division into management units – the water bodies (>500 water bodies) and a refined division into physical units, the aquifers, which generally serve as the basis for the former.

Consolidation of national piezometric and water quality monitoring networks is in progress, by gathering, harmonising, rationalising and extending existing networks. Many new tools have been developed based on databases and GIS, such as the "SEQ EAUX souterraines", a fine groundwater quality assessment grid, and ADES, the national groundwater database designed and managed by BRGM for the Ministry of Ecology and Sustainable Development, which is available on the Internet, and which significantly improves and broadens the field of action of the former ONQUES.

Faced with the requirements imposed by the WFD, numerous decision support tools are being developed, everywhere emphasising the informative visualisation of results, such as the integrated model of the Herault River Valley, which incorporates hydraulic, economic and behavioural aspects of both the context and the actors, and which enables viewers to actually see the consequences of each management policy in the basin.

The WFD has also resulted in the development of (1) original methods such as one that enables the reconstruction of the natural quality of groundwater (natural geochemical background or baseline quality), and (2) innovative technical tools, like nitrogen and boron isotopes, which, when linked with inverse models, make it possible to trace back to the source of nitrate pollution. These increasingly lead hydrogeologists to study non-point source contamination (one of the major challenges of the 21st century) and the interactions between surface water and groundwater which, they recognise better than anyone else, are fundamental and need to be understood to ensure proper water resources management and to control contaminant transfer.

The WFD is not unrelated also to the acceleration of the protection of wellheads in France (almost 50% of which were protected in 2006), notably due to the simplification of the administrative procedure and to the implementation of the National Health and Environment Plan (Plan National de Santé et Environement – PNSE, the objective of which is to have 80% of wells protected by 2008 and 100% by 2010).

Global changes, climate change and greenhouse-effect gases

The so-called "global" changes and the media coverage of climate change and greenhouse-effect gas problems have intensified the development of a new generation of mathematical models, the complex "reactive transport models" which couple chemistry and transport. These were initially developed for studying contaminants that react with the environment and are now refined and used increasingly in the field of CO_2 storage on which a growing number of hydrogeologists are working.

Hydrogeologists are now studying high water and flooding phenomena too. Indeed, the 2001 Somme River floods demonstrated how the inflow of groundwater could be one of the factors that cause flooding in many sedimentary environments (rising water table), challenging long-held beliefs concerning flood-triggering mechanisms. BRGM research teams have developed a new high-water and flood risk prediction tool that integrates the entire water cycle according to a rain-transfer-water level-discharge principle. This type of tool has been in use since the winter of 2001–02 in the Somme river basin.

Hydrogeologists have also resumed their study of problems associated with droughts and their impact on water resources, and are developing new palliative measures that are based notably on the concept of active water resource management, going as far as recommending multiple water use and the reuse of waste water. On the international scene, the Millennium Development Goals and global changes have opened up important perspectives for young hydrogeologists.

The grand opening ...

This new century therefore continues and amplifies trends that began in the late 1980s. Water resource management requires a truly multidisciplinary approach. Hydrogeologists are opening up to other professions, and must acquire new skills in various fields. These include the "soft sciences" – water resource management must take into consideration both the social dimensions (the wishes and needs of all stakeholders) and the economic dimensions incorporated into a global analysis of the interests and constraints of the system studied. Today, hydrogeologists are often dual-skilled, something that was rarely encountered 10 or 15 years ago. He or she must, however, not lose touch with field work because good water resource management must be based on models that are correctly structured and calibrated, with representative and reliable data knowledgably collected in the field.

CONCLUSION

France was one of the cradles of hydrogeology from the 19th century. French hydrogeology sprang on the one hand from 19th century hydraulic engineers and their encounter with geologists and, on the other, from studies and research carried out in the arid and semi-arid zones of Africa, where groundwater is a major component of usable water resources, to be developed and preserved. Its evolution resulted from a constant conjunction and interaction between the physical and mathematical sciences, water science and technology, and the earth sciences.

Originally a branch of applied geology, hydrogeology became not only a multi-disciplinary science and a recognized profession, but also a field of education and research with many concrete and multiple applications. French hydrogeology has contributed greatly to the advancement of knowledge and use of groundwater, both in France and in other countries where French hydrogeologists were (and are still) called in as experts. It also contributed much to the awareness of government officials concerning the position of groundwater in water policy and land development, and has for many years recognized and prescribed the necessity to fully integrate groundwater in water resources management schemes (integrated water resources management).

Today, there are an estimated one thousand groundwater specialists in France (hydrogeologists, hydraulic engineers, hydrogeochemists) working in research institutes, universities and industry. Most of them are members of one or more of the five major professional associations which gather hydrogeologists and geologists in France: the International Association of Hydrogeologists (IAH), the French Hydrogeology Committee (CFH), the Association of Public Service Hydrogeologists (AHSP), the French Independent Geologists (GIF), and the French Union of Geologists (UFG).

MILESTONES

1802 – Jean-Baptiste Lamarck invents the word "Hydrogéologie" (Hydrogeology).

1821 – F. Garnier publishes the first technical guidebook on drilling and artesian wells. The paper was awarded a prize by the French Society for the Promotion of National Industry.

1830 – First artesian well drilled in Tours (Loire river region).

1833–41 – First artesian well drilled in Paris: the Grenelle well designed and decided by François Arago, who was both an engineer and the Mayor of Paris.

1840 – Artesian well drilled in the Carboniferous limestone near Lille.

1846 – Eugène Belgrand – First monograph published on the upper Paris basin hydrology and hydrogeology – data gathered from the works done for the water supply of Paris.

1856 – Henri Darcy publishes his famous law in *"Les fontaines publiques de la ville de Dijon"* (The public fountains of the city of Dijon).

1856 – Father Jean-Baptiste Paramelle publishes a book on the art of finding groundwater.

1857 – Jules Dupuit introduces the term "Darcy's Law".

1862 – Achille Delesse draws the first hydrogeological map using piezometric contours: the hydrogeological map of the Seine department, Paris.

1870 – Beginning of the oldest known piezometric time series (Toury-en-Brie, Sugar refinery).

1886–88 – Jules Gosselet gives the first lessons on hydrogeological science in Lille in the North of France, published by the Geological Society of Northern France.

1887 – A. Daubrée invents the first scientific terms regarding hydrogeology and groundwater: piezometric level, the term "phreatic"...

1901 – An informative accident: the fire in the Pernod absinthe factory in Pontarlier shows the link between the losses in the Doubs River and the source of the Loue River (Jura).

1902 – A law provides for the creation of wellhead protection zones for drinking water wells, according to expert advice of "official geologists".

1904–05 – Edmond Maillet and Joseph Boussinesq propose mathematical expressions to clarify and reproduce the relationship between surface and groundwater.

1909 – E. Imbeau presents the first systhesis of French aquifers.

1923 – L. Dollé completes the first doctoral thesis on hydrogeology (University of Lille).

1923 – Marcel Porchet develops a method for computing aquifer characteristics using pumping test under transient state.

1924 – Regulations requiring wellhead protection zones are broadened to include all type of drinking water wells.

1933 – First water level contour map of the chalk aquifer west of the Somme River produced by A. Bonte.

1935 – Decree-law of 8 August 1935 "concerning groundwater protection".

1939 – M. Gosselin and H. Schoeller publish the hydrogeological map of Tunisia at a 1:50 000 scale together with an explanatory document.

1947 – Creation of BURGEAP, the first engineering firm specialized in hydrogeology.

1955 – Pierre Lafitte and Jean Ricour establish groundwater as the most important mineral mined in France: this started the large scale inventory of groundwater resources in France.

1956 – First inventory of hydraulic resources launched in the Nord—Pas-de-Calais region by BRGM.

1957 – First IAH conference held in Paris.

1958 – Creation of university degree programmes in Hydrogeology in Bordeaux, Montpellier and Paris.

1960 – The SGAL launches its hydrogeological activities.

1961 – First geothermal well drilled for the Maison de la Radio in Paris.

1963 – Launching of the journal *Chronique d'Hydrogéologie* (Hydrogeological Chronicles) by BRGM.

1964 – First Water Law: Creation of the Agences Financières de Bassin (Water Boards).

1965 – First mathematical model of a major regional aquifer – the Albian aquifer, Paris Basin.

1965–66 – Equipping of the first experimental hydrogeological basin – the Hallue-Somme Basin.

1966 – Creation of the "Centre d'Informatique Géologique" (Centre of Geological Computer Sciences" at the Paris School of Mines.

1966 – BRGM creates a hydrogeology department.

1969 – Beginning of reconnaissance work for a major geothermal site in Bouillante, Guadeloupe.

1972 – Creation of a Ministry of the Environment.

1973 – Creation of the French Hydrogeological Committee (Comité Français d'Hydrogéologie).

1976 – Drought in the north of France.

1976 – Beginning of major geothermal programmes (Paris Basin and Aquitaine).

1976 – Identification of aquifer systems; creation of "Margat Codes".

1980 – Evidencing of natural denitrification phenomena in groundwater in the north of France.

1980 – Obligation to carry out hydrogeological studies prior to the creation of domestic and industrial waste landfills.

1980-84 – G. Castany is appointed president of IAH.

1982 – First "active aquifer management" operation on the Lez Spring karst aquifer (Montpellier) by Prof. J. Avias.

1987 – Launching of the geothermal energy programme on "Hot Fractured Rock (HFR)" at Soulz-sous-Forêt in Alsace.

1992 – Second Water Law: creation and launching of the SDAGEs and SAGEs concept.

1993 – Enactment of the "Nitrate Directive"; definition of vulnerable zones.

2000 – Passing of the European Water Framework Directive.

2003 – Creation of the ADES database (Accès aux Données sur les Eaux Souterraines – Access to Groundwater Data) managed by BRGM.

2006 – 3rd Water Law.

Selected Congresses which impacted the history of French hydrogeology (non exhaustive list)

1974 – 10th International Congres AIH/IAH, Montpellier, September 1974.

1977 – Protection de eaux souterraines captées pour l'alimentation humaine – Colloque National, Orléans, March 1977.

1977 – Eaux souterraines et approvisionnement en eau de la France – Colloque National, Nice, October 1977.

1978 – Hydrogéologie de la craie du Bassin de Paris – Colloque Régional, Rouen, May 1978.

1988 – Hydrogeology and safety of radioactive and industrial hazardous waste disposal – International Congrès AIH/IAH, Orléans.

1990 – L'eau souterraine, un patrimoine à gérer en commun – Colloque National, Paris.

2006 – "150ème anniversaire de la Loi de Darcy" – Managing Aquifer Systems, International Symposium AIH/IAH – CFH, Dijon, May–June 2006.

REFERENCES

Adams, U.von & Gert, M. (1987) Der Wassersucher Abbé Paramelle (1790–1875) – Hexer oder Heiliger? (Le chercheur d'eau Abbé Paramelle (1790–1875) – Sorcier ou saint?), Brunnenbau, Bau von Wasserwerken, Rohrleitungsbau, n° 4, avril, 38ème année, pp. 149–158.

Albinet, M. et al. (1967). Carte hydrogéologique du bassin de Paris. Éch. 1/500,000. éd. BRGM.

Albinet, M. (1970) Carte de vulnérabilité des aquifères aux pollutions. Éch. 1/1,000,000. éd. BRGM.

Ambroggi, R. & Margat J. (1961) Les recherches hydrogéologiques au Maroc. Le Centre des études hydrogéologiques (1946–1960). (Notes et Mémoires du Service géologique du Maroc, 153, 55 p., Rabat).

Arago, F. (1835) Sur les puits forés connus sous le nom de puits artésiens, de fontaines artésiennes ou de fontaines jaillissantes. Annuaire 1835 du Bureau des Longitudes, Paris.

Archambault, J. (1947) Hydrogéologie Tunisienne, Direction des Travaux Publics de Tunisie.

Atkinson, G. et al. (1998) Eugène Belgrand (1810–1878): Civil Engineer, Geologist and Pioneer Hydrologist. (The Newcomen Society for the Study of the History of Engineering and Technology, Transactions, 69(1), 1997–1998).

Banton, O. et al. (1997) Hydrogéologie – multiscience environnementale des eaux souterraines. éd. Presses de l'Université du Québec/AUPELF.

Belgrand, E. (1846) Études hydrologiques dans les granites et les terrains jurassiques formant la zone supérieure du bassin de la Seine, Paris.

Bodelle, J. & Margat J. (1980) L'eau souterraine en France. (éd. Masson).

Boussinesq, J. (1904) Recherches théoriques sur l'écoulement des nappes d'eau infiltrées dans le sol et sur le débit des sources. Journal Mathématiques Pures et Appliquées, France, 1 (5), 10.

BRGM (1970) Atlas des eaux souterraines de la France. éd. BRGM/DATAR.

BRGM (1977) Création du bulletin hydrologique de prévision des basses eaux.

Castany, G. (1950) L'œuvre hydrogéologique française en Tunisie, Cahiers Charles de Foucauld, vol. 18.

Castany, G. (1963) Traité pratique des eaux souterraines, éd. Dunod, Paris.

Castany, G. (1968) Prospection et exploitation des eaux souterraines. éd. Dunod, Paris.

Castany, G. & Margat, J. (1977) Dictionnaire français d'hydrogéologie. éd. BRGM.

Castany, G. (1980) Principes et méthodes de l'hydrogéologie. éd. Dunod.

Castany, G., Margat, J. & Roux, J.-C. (1986) Origine, évolution et applications de l'hydrogéologie. Géologues, 76–77, pp. 19–26, Paris.

Chapoutier, P. (1960) L'hydraulique souterraine dans son cadre historique, Soc. hydrotechnique de France, 6èmes journées de l'Hydraulique, Nancy, juin, T. 1, pp. XXXII–XL, La Houille Blanche.

Chery, L. & Thouin, C. (coordonateurs – 2005) Guide technique: Qualité naturelle des eaux souterraines – Méthodologie de caractérisation des états de référence des aquifères français, BRGM édition.

Collectif (1970) Atlas des Eaux Souterraines de la France. éd. BRGM-DATAR.

Collin, J.-J. (2004) Les eaux souterraines; Connaissance et gestion. éd. BRGM/HERMANN.

Comité Français d'Hydrogéologie (Direction J.-C. Roux & de T. Pointet) Aquifères et eaux souterraines en France. éd. BRGM.

Conseil Général des Mines & Martin, Y. (1996) Gestion durable des eaux souterraines.

Crampon, N., Roux, J.C. & Bracq, P. (1993) Hydrogéologie de la craie en France. Hydrogéologie, N02.

Daubrée, A. (1887) Les eaux souterraines à l'époque actuelle, Revue des deux Mondes, librairie Dunod, Paris.

Darcy, H. (1856) Les fontaines publiques de la ville de Dijon. Paris.

Dausse (1842) De la pluie et de l'influence des forêts sur les cours d'eau. *Annales des Ponts et Chaussées*, 1842, 198–201.

De la Rozière Bouillin, O. (1965) Carte des eaux souterraines de la France. Éch. 1/1,000,000. éd. BRGM.

Delesse, A. (1862) Atlas de l'Architecture et du patrimoine de la Seine-Saint Denis.

De Marsily, G. (1981) *Hydrogéologie quantitative.* éd. Masson.

Detay, M. (1977) *La gestion active des aquifères.* éd. Masson.

Dollé, L. (1923) Étude sur les eaux souterraines du département du Nord – first doctoral thesis on hydrogeology, University of Lille.

Dupuis, J. (1848) Études théoriques et pratiques sur le mouvement des eaux dans les canaux découverts et à travers les terrains perméables, Paris, 1st edition.

Dupuis, J. (1857) Mémoire sur le mouvement de l'eau à travers les terrains perméables, Acad. Sciences, Paris.

Erhard-Cassegrain, A. & Margat, J. (1983) Introduction à l'économie générale de l'eau. éd. Masson.

Freeze, A. (1994) Henry Darcy and the Fountains of Dijon. *Ground Water*, 32 (1).

Garnier, F. (1821) De l'art du fontenier-sondeur et des puits artésiens, Imprimerie de Madame Huzard, Rue de l'Eperon St. André des Art, Paris.

Gilli, E. *et al.* (2004) Hydrogéologie. Objets, méthodes, applications. éd. Dunod.

Guillemin, C. & Roux, J.-C. (1992) Pollution des eaux souterraines en France; rapport n° 29 de l'Académie des Sciences. éd. BRGM. Manuels et méthodes, n° 29.

Girard & Parent-Duchatelet (1833) Des puits forés ou artésiens. *Annales d'Hygiène Publiques*, tome X.

Gosselet, J. (1886–88) Leçons sur les nappes aquifères du Nord de la France. Ann. Soc. Géol. Nord, Lille.

Gosselet, J. (1906) Étude de la nappe du Calcaire carbonifère, Roubaix – Tourcoing.

Gosselin, M. and Schoeller, H. (1939) Notice générale de la Carte hydrogéologique de la Tunisie au 50,000e (Tunis, Tunisie).

Héricart de Thury, L.E (1829) Considérations géologiques et physiques sur la cause du jaillissement des eaux des puits forés ou fontaines artificielles, Paris.

Imbeaux, E. (1909) Les Nappes Aquifères de France – Essai d'Hydrogéologie – proceedings of the conference of the Geological Society of France – 6-12-1909.

Imbeaux, E. (1930) Essai d'Hydrogéologie. éd. Dunod.

Lafitte, P. & Ricour, J. (1956) La recherche minérale la plus importante de la France métropolitaine: l'eau. Rôle de l'hydrogéologie. Revue de l'industrie minérale.

Lallemand-Barrès A. & Roux J.-C. (1989) Guide méthodologique d'établissement des périmètres de protection des captages d'eau souterraine destinée à la consommation humaine. (Manuels et Méthodes n° 19, éd. BRGM, réédité en 1998, Manuels et Méthodes n° 33).

Lallemand-Barrès A. & Roux, J.-C. (1989) Cartes des teneurs en nitrates dans les eaux souterraines de la France. éd. BRGM.

Lamarck, J.B. (1802) Hydrogéologie, Agasse, imprimeur-Lib, Rue des Poitevins, n° 18.

Landreau, A. & Lemoine, B. (1977) Carte de la qualité chimique des eaux souterraines de la France, 1/1,000,000, Ministère de l'Environnement et Cadre de Vie, éd. BRGM.

Lanini, S., Courtois, N., Giraud, F., Petit, V. & Rinaudo, J.D. (2004) Socio-hydrosystem modelling for integrated water resources management: the Hérault catchment case study. *Environmental Modelling & Software*, 19, 1011–1019.

Lapparent (de), A.A. (1882–98) Traité de Géologie.

Launay (de), L. (1899) Recherche, captage et aménagement des sources thermo-minérales, Paris.

Lemoine, P., Humery, R. & Soyer, R. (1939) Les forages profonds du bassin de Paris, éd. Mémoire du Muséum National d'Histoire Naturelle, n° 11.

Maillet, E. (1905) Essais d'hydraulique souterraine et fluviale. éd. Librairie Scientifique A. Hermann.

Margat, J. (1960) Carte Hydrogéologique de la plaine du Tafilalt. éd. Service Géologique du Maroc.

Margat, J. (1980) Carte hydrogéologique de la France. Systèmes aquifères à 1/1,500,000. éd. BRGM.

Margat, J. (1986) Hydrogéologie: passé, présent et avenir, Ingénieurs-géologues, n° 47–48, Av., 40ème anniv. ENSG, Nov. 1985, pp. 59–61, Nancy.

Margat, J. (1986) Les comptes des eaux continentales, Chapitre 5 dans les comptes du Patrimoine Naturel, Les Collections de l'INSEE, n° 137–138.

Margat, J. (2001) Histoire de l'hydrogéologie, Colloque international OH2 – Origines et histoire de l'hydrologie, Dijon, 9–11 mai, 8 p.

Margat, J. (2003) Le grand siècle de l'hydrogéologie, Géochronique, 88, Décembre, pp. 36–37, Paris.

Margat, J. (2007) L'hydrogéologie. Géochronique, 101, De la Géologie aux Géosciences, pp. 45–46, Paris.

Margat, J. (2008) Les Eaux Souterraines dans le Monde éd. UNESCO/BRGM.

Martel, E.A. (1894) Les abîmes. éd. Delagrave, Paris.

Martel, E.A. (1921) Nouveau traité des eaux souterraines. éd. G. Doin.

Martin, Y. et al. (1996) Rapport sur la gestion durable des eaux souterraines. Ministère de l'Industrie, Conseil Général des Mines, Paris.

Mégnien, C. et al. (1970) Atlas des Aquifères de la Région Parisienne.

Paramelle, J.B. (1856) L'art de découvrir les sources. éd. Bailly, Divry et Co., Paris.

Pennequin, D., Suzanne, P. and D'arras, D. (1991) SOPHOS: Modèle de gestion optimale de la nappe de Croissy – XXIème Journées de l'Hydraulique – SHF, 29–31 janvier 1991, Sophia-Antipolis, Question III, rapport n° 3.

Pennequin, D., Vernoux, J.F., Saout, C., Château, G. and Pillebout, A. (2007) Protection des captages d'eau souterraine destinée à la consommation humaine. Géosciences, 5 (February).

Pointet, T., Amraoui, N., Golaz, C., Mardhel, V., Pennequin, D. & Pinault, J.L. (2003) La contribution des eaux souterraines aux crues de la Somme: observations, hypothèses, modélisation – La Houille Blanche, n° 6.

Pomerol, Ch. & Ricour, J. (1992) Terroirs et Thermalisme en France. éd. BRGM.

Porchet, M. (1923) Étude sur l'écoulement souterrain des eaux. Annales of the Ministry of Agriculture, France.

Porchet, M. (1930) Étude des eaux souterraines de la Crau, Congrès de l'eau de Marseille, 29 juin–1er juillet 1930, pp. 81–173.

Risler, J.-J. (1973) Carte des eaux minérales et thermales en France. éd. BRGM – rééditée en 1994 et 2005.

Roux, J.C. et al. (1980) Carte des Concentrations en Nitrates dans les Eaux Souterraines de France. éd. BRGM

Saint-Hilaire, E.G. (1836) Comptes Rendus de l'Académie des Sciences.

Schoeller, H. (1934) Les échanges de base dans les eaux souterraines, trois exemples en Tunisie. Bul. Société Géologique de France.

Schoeller, H. (1962) Les eaux souterraines. Masson.

Tixeront, J. (1956) Note sur les rôles respectifs de Darcy et Paramelle dans la fondation de l'hydrologie moderne. Assoc. intern. Hydrol. Scient., Symposia Darcy, Dijon, Sept., Publ. AIHS n° 41, T. 11, pp. 7–9.

Vibert, A. (1937) Hydrogéologie et Hydrodynamique des nappes Albiennes du Bassin de Paris, Le Génie Civil, vol. 111 (other articles followed this one).

Vigouroux, P. (2005) Guide qualité pour la ressource en eau minérale et thermale. édition BRGM, collection scientifique et technique.

Wychoff, R.D., Botset, H.G., Muskat, M., & Reed, D.W. (1933) Rev. Sciences Instruments, 4, 394–405.

History of hydrogeology in Central Europe, particularly relating to Germany

Eckehard Paul Loehnert
Formerly Dept. of Geology & Paleontology, University of Münster, Germany

INTRODUCTION

Treatment of historical aspects of hydrogeology in relation to German speaking nations in Central Europe appears rather neglected in the professional literature. This refers particularly to textbooks although most of them contain short remarks on early researchers and on historical evolution of hydrogeological sciences in this area. Only Matthess & Ubell (1983) have devoted a short paper to the "History of Hydrogeology" (literally). From this it is derived that ideas and contributions by German or German speaking authors emerged first in the 15th/16th Century, for example by F. Faber, G. Bauer (Agricola), C. von Megenberg and others.

Under the heading "Ways to modern groundwater exploration" Hölting (1984) dealt with specific trends in the area under discussion. Important developments since 1856 are tabled in this paper, i.e. since the publication of Darcy's law.

The author of this publication benefited also greatly from a draft paper by Jacob de Vries, kindly provided in June 1996 (de Vries, 1996). This paper deals with specific concepts and trends in neighbouring The Netherlands. The same author had earlier paid tribute to two eminent Dutch groundwater hydrologists of the late 19th/early 20th Century, namely W.D. Badon Ghijben and J.M.K. Pennink, both "Pioneers of coastal dune Hydrology" (de Vries, 1989; 1994). The first-named introduced what is now called the Ghijben-Herzberg principle since the German engineer A. Herzberg (1901) arrived independently at the same conclusions on the saline/fresh water interface in coastal areas.

In this context it makes sense to introduce briefly German groundwater domains in which ideas on hydrogeological sciences developed.

Roughly half of the area of the Federal Republic of Germany (FRG) is made up of unconsolidated, porous aquifers, predominantly of Quaternary age, underlain by Neogene (Miocene and Pliocene). These aquifers are widespread in the North German Plain, furthermore in the Lower and Upper Rhine Basin and in the Molasse Basin, to the north of the Alps. The systems are crossed by large river plains, remnants of the last glaciation, such as rivers Elbe, Weser, Rhine, Oder, Isar and Danube, the first three flowing into the Atlantic (North Sea) and the latter into the Black Sea. Hence the Main European Water Divide is crossing southern Germany.

It is of major note that big German towns were founded and developed in river plains making use of surface water first and later of bank filtrate and/or groundwater.

Examples are: Berlin, the federal capital, Hamburg, Munich, Cologne, Dresden, Magdeburg and others (Loehnert, 2002).

Less prospective with respect to groundwater resources is the central-southern part of Germany consisting of hard rocks, few of them with transmissive properties, especially karstified limestones in the Swabian-Frankonian Alb (Jurassic), within the Hercynian Rhenish Shield (Devonian) and in the surrounding of the Münster Basin (Upper Cretaceous). It is mainly these areas where concepts of karst groundwater hydrology and hydraulics grew rapidly in the 18th/19th Century in Germany, e.g. in the Harz Mountains and the Swabian Jura (Pfeiffer, 1963).

Vast karst reservoirs have been exploited by springs to supply towns in neighboring Austria, for example the capital Vienna since 1875 through long-distance pipelines from the eastern Alps at the suggestion of Eduard Suess (1831–1914), the renowned Austrian geologist who among other contributions to hydrogeology introduced the concept of "juvenile water" (Pinneker, 1989).

On the other hand, techniques of artificial groundwater recharge were introduced as early as 1902 in water deficiency areas of the Hercynian Shield, for instance to supply the highly industrialized Ruhr district with drinking water from the Ruhr valley (König & Bruns, 1930; Frank & Schwarz, 1966).

EARLY CONCEPTS (19TH CENTURY TILL WORLD WAR I)

As pointed out by Matthess & Ubell (1983) first hydrogeological publications by German authors arose in the middle of the 19th Century and are related to chemical groundwater composition. Early investigators are G. Bischof (1863/71) and B.M. Lersch (1864, 1870), the latter author likely to present some sort of first textbook on hydrogeochemistry.

As early as 1870 the German government building surveyor Adolf Thiem modified Dupuit's equation which requires, however, stationary conditions to calculate groundwater flow and yield as derived from pumping tests (Hölting, 1984). Few years later the Austrian hydraulic specialist Philipp Forchheimer (1886) applied extensively modern methods including the Laplace equation, to groundwater flow (Matthess & Ubell, 1983; de Vries, 1989). The same author (Forchheimer) published in 1914 his text book "Hydraulics" which became rather popular even abroad, for example has this book "widely been used in the Netherlands" (de Vries, 1996).

The origin of groundwater remained point at issue toward the end of the 19th Century. G.H.O. Volger (1877), a German mineralogist and geologist from Frankfurt, insisted that "no water beneath ground surface is derived from rain". "He even argues that the concern for contamination of groundwater is a fear for phantoms, which would lead to unnecessary costs by the construction of abstraction wells for public water supply outside the villages and cities (!)" (de Vries, 1996). Indeed, these ideas originated from old Greeks (Aristoteles) and can hardly serve as an example of modern times. Konrad Keilhack (1858–1944) as early as 1902, also opposed Volger's "Theory of Condensation" and was an advocate of the "Infiltration Theory" as "the today generally predominant doctrine of the connection between groundwater and atmospheric precipitation".

Reliable hydrogeological concepts developed in the 2nd half of the 19th Century with the introduction of central water supply systems to urban and later to rural areas in Germany. Early approaches are closely associated with the German civil engineer Otto Lueger (1843–1911), famous for his book "Water Supply of Towns" (1895) in which he advocated spring and groundwater instead of surface water, particularly in towns of Southwest Germany. He remains a personality of merit in his time although de Vries (1996) criticizes sharply "Lueger's doctrine" as well as his "authoritative book" and his wrong groundwater flow conception which did not fit Amsterdam dune water conditions.

A striking example of town supply history is the city of Hamburg, briefly presented by the author (Loehnert, 1985) and extensively dealt with by Meng (1993). In 1842, after the "great fire", the English engineer William Lindley (born 1808 in London) was commissioned to set up a central water supply. He strongly recommended sand filtration of Elbe river water upstream the city border but the plant which opened in 1848 was established with sedimentation basins only. Consequently, the disastrous cholera epidemic occured in 1892 which claimed about 7600 lives. Valuable background information on social and hygienic conditions which prevailed during that time period was provided recently by the Englishman Richard J. Evans (1987).

Consisting search for groundwater in Hamburg began in the 1830s but abstraction from deep confined aquifers commenced in the 1870s with spectacular artesian outflow in the low-lying Elbe valley, the piezometric surface reaching 20 m above ground (Figure 1). Darapsky (1903) advertised both the capability of Deseniss & Jacobi's drilling company and deep-seated groundwater as against filtered surface (Elbe) water. The company's borehole of 377 m depth drilled 1901 in the neighbouring Prussian town Altona used to be the deepest well of that time in Germany (Loehnert, 1967).

Two practical aspects are worth emphasizing:

(i)   Propagation of ocean tides in deep artesian wells was recognized and described as early as 1904 by Olshausen. Tidal fluctuation rates decrease according to this author with inverted square of distance to the Elbe river, and Olshausen discussed two hypotheses, namely the barrier effect by confining layers and the flood impact on the sequence of strata (the tidal fluctuation amounts to approx. 2.5 m in the Elbe river at Hamburg's harbour). Decades later the first-mentioned effect was rejected by Hermann Keller (1928) in his PhD thesis, submitted to the Technical University of Dresden. H. Keller referred to E. Koch's findings that the confining layer differs in geological age and is not continuously deposited (Figure 2).

(ii)   Emil Koch (1886–1960) was the outstanding geologist to lay the foundation for the complicated aquifer system of Tertiary/Quaternary age. His results were summarized in 1928. He used to be a grammar school teacher but was commissioned in 1911 to set up bore archives at the Mineralogical-Geological Institute (which later got the designation "State Institute" and became part of the Hamburg University). Thoroughly collecting and working on borehole samples Koch gained his intimate knowlege of Hamburg's underground (Figure 3). In 1948 he was appointed the first director of the small but efficient Geological Survey.

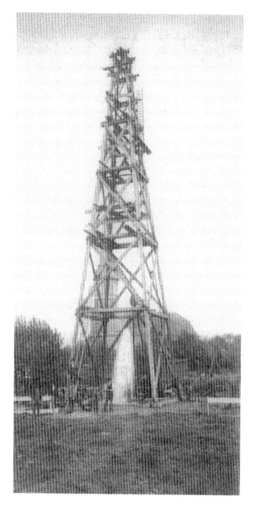

Figure I Artesian flow hit in the Elbe valley early in the 20th Century (Fig. taken from Meng, 1993).

In mountainous areas research was focused on spring utilisation. Thus Hippolyt J. Haas, professor at the University of Kiel, published in 1895 his book "Quellenkunde" (Spring Science) in which springs were grouped taking into account hydrological and geological criteria (Zetinigg, 1998). Haas (1895) did already sum up terms like "Schichtquelle" (boundary spring) or "Schuttquelle" (talus spring) and treated also thermal-mineral springs.

The first ever textbook on hydrogeology (sensu lato) written in German was – to the best of the author's knowledge – the one by J. Gustav Richert "*Die Grund-wasser*" (1911). Richert used to be professor at the Royal Technical University of Stockholm and consulting engineer regarding himself as "layman in geology". He called himself student und friend of Adolf Thiem and subtitled his book addressed to

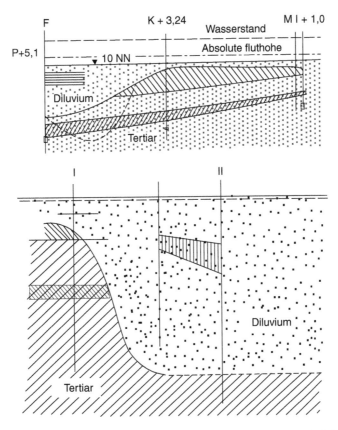

Figure 2 Sketch of Hamburg's hydrogeology after E. Koch, interpreted by H. Keller (1928).

engineers "with particular consideration of Sweden's groundwater". Indeed, Thiem's school of thought is to be felt throughout the calculations of groundwater velocity, quantity and artificial recharge. Although not trained as geologist Richert presents in his book nice sketches of water table configurations (as commented on by Prof. John M. Sharp Jr., The University of Texas at Austin, in a letter of 1999). This refers in particular to chapter I (General Hydrology), while chapter II is dealing with Sweden's geology including "hydrological" implications and III with the author's investigations in this country, among others to supply water to Gothenborg, Malmö and Upsala. Of the 33 Swedish towns hydro(geo)logically surveyed by the author, 26 of them got supplies from groundwater out of Quaternary deposits. Some of the bores showed artesian flow in those early days, whereas few aquifers were artificially recharged. The latter technique included the so-called Groundwater Factory which was established north of Gothenborg in 1897–98 after lowering artesian pressure and overexploiting the natural resource. The situation is sketched in Figures 3 and 4.

Major disadvantage of Richert's stimulating book is, however, the total lack of bibliographical references.

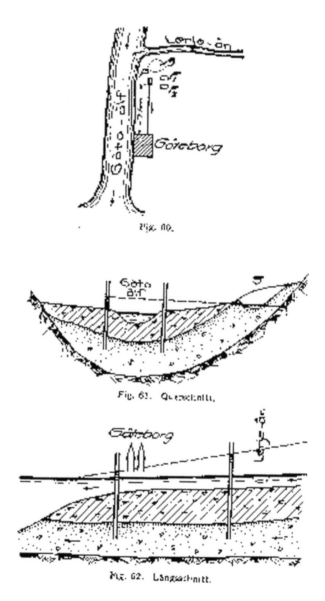

Figure 3 Hydrogeology north of Gothenborg (after Richert, 1911).

Two major German language textbooks were published simultaneously (1912), namely Konrad Keilhack's "*Lehrbuch der Grundwasser- und Quellenkunde*" (Textbook on groundwater and springs science) and Höfer von Heimhalt's "*Grundwasser und Quellen*" (Groundwater and Springs). Both authors were the leading figures in groundwater science of that time in their respective countries Germany and Austria. However, only Höfer did use the term "Hydrogeology" (Figure 5) as quoted by Zetinigg (1990). Keilhack had earlier published his textbook on applied

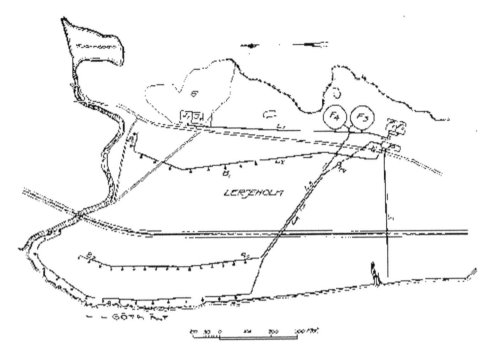

Figure 4 "Water Factory" to supply Gotheborg after Richert (1911). 5 = sand pit like Fig. 4. J1 + J2 = Infiltration Basins and F1 + F2 = Filtration Basins for river water from Götaälf, B1 = 20 wells. Additional basins F3 + F4 and wells B2 + B3 as suggested by Richert.

geology (*"Lehrbuch der praktischen Geologie"*) in 1896 which included already two groundwater-related chapters. He was a prominent member of the Prussian Geological Survey, with its headquarters in Berlin (see section 5.) and at the same time teaching professor at the Mining Academy (which became later the Technical University Berlin) while Höfer used to be professor at the Montanistic University of Leoben. Keilhack dedicated his book of 1912 to "Geologists, Hydrologists, Drilling Entrepreneurs, Well Constructors, Miners, Mining Engineers and Hygienics", thus covering a broad field and stressing the practical orientation of his publication.

DEVELOPMENTS AFTER WORLD WAR I

Some modest progress has been achieved between the two wars, i.e. between 1918 and 1939. One of the most active researchers, K. Keilhack, retired from the Prussian Geological Survey in 1923 although he continued to work among others on the 3rd edition of his textbook which was published "completely revised and increased" in 1935, now including a proper list of references and an index. It is worth to compare this latest nobreak edition with the 1st one of 1912 in order to check progress over a time period of more than 20 years with World War I in between. Two new chapters, a legal one by Hans Keilhack, Konrad's brother (III), and by the author himself on "Relationship

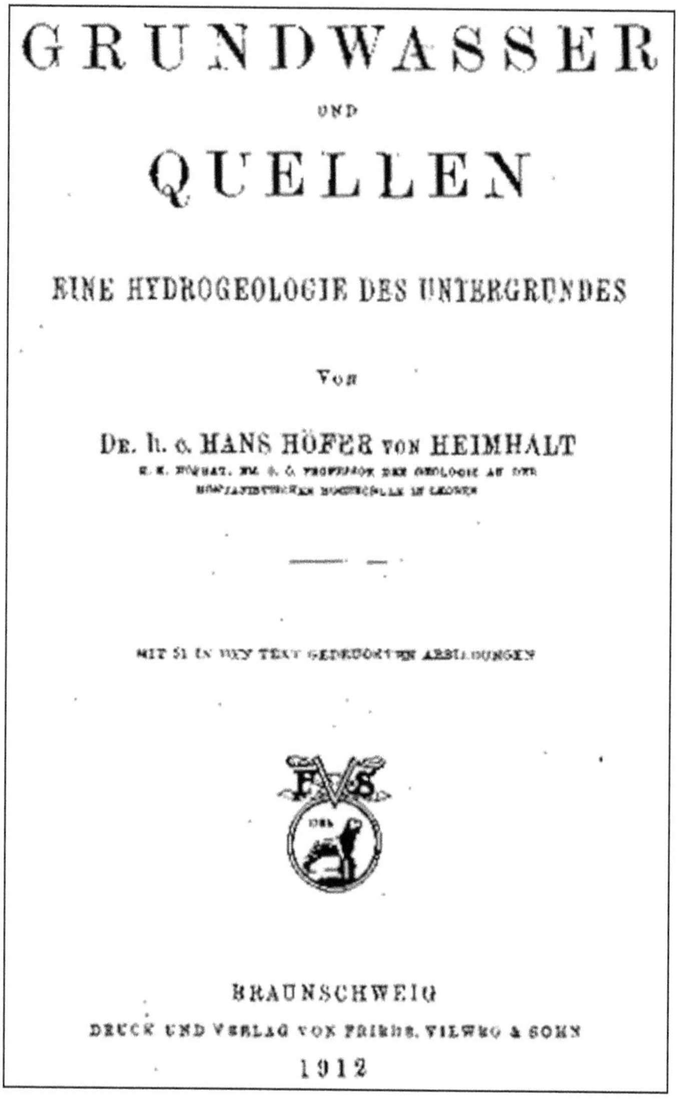

GRUNDWASSER

UND

QUELLEN

EINE HYDROGEOLOGIE DES UNTERGRUNDES

Von

Dr. h. c. HANS HÖFER von HEIMHALT

MIT 51 IN DEN TEXT GEDRUCKTEN ABBILDUNGEN

BRAUNSCHWEIG

DRUCK UND VERLAG VON FRIEDR. VIEWEG & SOHN

1912

Figure 5   Front page of Höfer von Heimhalt's book "*Groundwater and Springs*" published 1912 and subtitled "*A Hydrogeology of the Underground*".

of hydrology to mining, sewage farming, agriculture and forestry" (10.) have been added thus reflecting Keilhack's vast experience in these fields of competence, especially with opencast mining in Lusatia. On the other hand as pointed out by Hölting (1984), specific aspects of hydrochemistry do not appear updated although considerable progress was achieved and published. Indeed, the respective chapter taken over from an early description of spas in Germany (1907), remained unchanged. Keilhack, now at a ripe old age, might have faced difficulties to follow up scientific progress at home and abroad, additionally due to tense political circumstances in the late 1920s and early 1930s.

In 1928, few years prior to the appearance of Keilhack's revised third edition, Werner Koehne (1881–1963) published the 1st edition of his important work "*Grundwasserkunde*" (groundwater science) which focussed essentially on porous aquifers (Zetinigg, 1990). Koehne belonged to the Prussian Survey 1906–08 but joined subsequently the Bavarian Geological Survey (1908–1916) and eventually the State Survey of Hydrology in Berlin where he retired in 1946, right after World War II. Koehne has been regarded as "pioneer of modern groundwater science in Germany"(Johannsen *et al.*, 1972). Indeed, he contributed among others substantially to the standardisation of hydrological terms which were introduced as "DIN 4049" for the first time in 1944.

Surprisingly nothing is known about scientific contacts between Keilhack and Koehne although both resided in Germany's capital for decades.

Two books on springs are hydrogeologically also worth emphasising: "*Die Quellen*" (The Springs) by J. Stiny (1933) and "*Quellen (Süßwasser- und Mineralwasserquellen)*" (Fresh and Mineral Water Springs) by E. Prinz & R. Kampe (1934). Stiny, professor at the Technical University Vienna, used to be the leading "technical geologist" of his time. In his monograph he introduced a new spring classification and nomenclature which deviates to some extent from the one by the German authors Prinz & Kampe. All authors represent the "second phase" of spring research according to Zetinigg (1998), while the "first phase" is associated with names such as B.M. Lersch and H.J. Haas.

ACTIVITIES AFTER WORLD WAR II (UP TO ABOUT 1980)

The term "hydrogeology" seems to have been introduced to Germany as late as in the 1930s (Matthess & Ubell, 1983). Indeed, according to Zetinigg (1990) the term is not to be found in headlines of appropriate textbooks within the German-speaking region (including Austria and Switzerland) prior to World War II.

A considerable number of German language textbooks came out after World War II emphasising different aspects of hydrogeology, most of them edited in West Germany (i.e. the then Federal Republic). These include books by Pfalz (1951), Richter & Lillich (1975), Thurner (1967) and Hölting (1980, 1st edition) on general hydrogeology, by Wundt (1953) on hydrology (including subsurface waters), by Grahmann (1958) on regional hydrogeology, by Schneider, Truelsen & Thiele (1951; these authors preferring the term "geohydrology"), and G. Keller (1969) on applied hydrogeology, by Zötl (1974) on karst hydrogeology, by Carlé (1975) on mineral-thermal waters, by Langguth & Voight (1980) on hydrogeological investigation methods and by Matthess (1973, 1983 translated into English) on hydrogeochemistry. The latter author has been commissioned by Wolfgang Richter in 1965 to edit a series of textbooks on hydrogeology and related sciences of which meanwhile 9 volumes were published (until 1997).

After Germany's reunification in 1990, reorganisation occured in applying, teaching and research of groundwater sciences. Consequences are illustrated in section 6.

SPECIFIC ROLE OF GEOLOGICAL SURVEYS

Geological Surveys have played a decisive part in promoting the science up to the current stage.

The Prussian Geological State Survey (P.G.L.A.) which existed between 1873–1939, deserves special attention in this context (Burre in Udluft 1968). Practical advisory services were provided soon after the foundation of this institution in connection with drilling for groundwater. Two important ministerial decrees issued in 1904 and 1909, paved the road for services which were preferably conducted by mapping geologists. Konrad Keilhack (Figure 6) although not educated in hydrogeology (Loehnert, 1999) and actually head of the lowland mapping section since 1914, became the leading scientist in this field (Loehnert, 1997; Quitzow, 2002). His textbook *"Grundwasser- und Quellenkunde"* (Groundwater and Springs Science) has also been translated into Russian in 1937. Prior to the first edition (1912) Keilhack had started a publication series of what he called *"Groundwater studies Nos. I–VI"* (1908–1913).

Figure 6   Royal Prussian State Geologist Dr. Konrad Keilhack, photo taken 1900 after his appointment as professor at the Berlin Mining Academy.

These six articles constitute the backbone of the textbook mentioned before. They include:

(i)   the occurrence of artesian flow in the lower Ohre valley, the recharge area of which, the Letzlingen Heath, became later an essential source to supply the city of Magdeburg with drinking water (Loehnert, 2002),

(ii)   hydrogeology of a shallow Quaternary aquifer system near Berlin to assess prerequisites of a cemetery,

(iii)   relationship of groundwater to agriculture and forestry; the author strongly recommends systematic groundwater monitoring,

(iv)   the impact of groundwater pumping on water table fluctuation in the Grunewald Lakes area (Berlin) is being considered,

(v)   the influence of dry summer conditions in 1911 is clearly demonstrated by hydrographs recording shallow groundwater in a number of river valleys of east and central Germany (nowadays "New Federal States").

(vi)   The effect of intense groundwater lowering is exemplified by the opencast mining area of Lower Lusatia, where engineer O. Smreker who in his Ph D thesis (1914) became a critic of Darcy's Law, assisted in calculating the vast amount of groundwater discharge.

In summary the above-mentioned contributions by Keilhack reflect the Survey's assignment to provide consultancy services related to practical groundwater issues. Mineral and thermal waters, in particular the protection of springs, also used to be subject of intense research by the Prussian Survey as proved by a number of publications (Burre, 1968). Keilhack (1916 among others) devoted some research to balneology and especially to the therapeutical application of mud and peat.

Another mapping geologist of the Prussian Survey who later became a worldwide recognised structural geoscientist, was Hans Stille (1876–1966) who presented as early as 1903 an investigation report on Upper Cretaceous karst which became trend-setting in karst hydrology in Germany.

The Prussian Geological Survey was replaced by the "Reichsamt für Bodenforschung" (State Authority for Soil Research) in 1939 (Udluft, 1968). Into this institution previously provincial Surveys of Baden, Bavaria, Hamburg, Hesse, Mecklenburg, Saxony, Thuringia, Württemberg and the State Survey of annexed Austria were incorporated. Practically oriented hydrogeological research continued, and results were presented in April 1943 and published in a special Volume "*Hydrogeologische Forschungen*" (1944). This publication is reflecting the state of Hydrogeology science at that time and is, therefore, of particular interest. Twenty six articles are related to water supply of all German provinces including parts now belonging to Poland (Silesia), Czech Republic, France (Lorraine) and Austria. (Ground-) Water was regarded as deposit and hence part of geological sciences. Consequently, authors of the contributions were mainly mapping geologists and excellent local experts in the respective regions. Only few of them became recognized hydrogeologists (in the narrower sense) after World War II.

One of the contributors to the above volume was Rudolf Gramann (1888–1962) who had gained recognition for establishing a groundwater monitoring network in

Saxony (1935). After World War II Grahmann joined the Federal Institute of Hydrology and became editor of the general hydrogeological map of West Germany on scale 1:500 000 (Grahmann, 1958).

After World War II each of the federal states established its own independent geological survey (except the small City State of Bremen which is being looked after by the Lower Saxony Survey). Hydrogeology played continuously an essential part within the Survey's area of responsibility (Johannsen et al., 1972).

An example of hydrogeological research of one of the State Surveys is given by Richter et al. (1968). The federal State of Lower Saxony participates in both lowland with porous aquifers and mountainous country with consolidated aquifers and aquicludes. This subdivision has required different techniques of groundwater exploration and exploitation, among others application of stratigraphy to Tertiary and Quaternary strata, accurate evaluation of natural groundwater recharge rates, and new concepts of hydrogeochemistry with particular respect to the origin of mineral and thermal waters in state-owned spas.

CURRENT SITUATION AND SUSPECTED FUTURE TRENDS

A major break occured due to Germany's reunification in 1990. Nowadays there are sixteen federal states, the Geological Surveys of them being mainly divisions of Environmental Departments.

Teaching of hydrogeology is nowadays being performed at most German universities, with centres at Tübingen, Karlsruhe, Freiberg, Berlin (Free University and Technical University), Bremen and Kiel (selection from the author's subjective point of view). Research subjects are focussing on environmental hydrogeology but problems arise with increasingly reduced number of staff.

To cope with modern requirements of hydrogeology science in Germany, the "Fachsektion Hydrogeologie" (Specialist Section of Hydrogeology, abbreviation: FH-DGG) was established in 1969 as branch of the German Geological Society. This section has nowadays a membership of 1200. It is organizing annual meetings every two years with differing main topics, the latest 2002 in Greifswald with the subject area "Groundwater strained between exploitation and nature conservation". Since 1996 the FH-DGG is editing the national journal Grundwasser (Groundwater) published by Springer, with 27 issues up to the end of 2002.

The smaller German national Comittee of IAH (International Association of Hydrogeologists) comprising currently 200 members was founded in 1967 thanks to the efforts of Prof. Hans-Joachim Martini, the late President of the Federal Institute of Geosciences and Mineral Resources (BGR) in Hannover. Close co-operation between the two hydrogeology groups (FH-DGG and IAH Germany) has been launched both organisationally and scientifically, exemplified by the 31st IAH Congress which took place 2001 in Munich.

CONCLUSIONS

Roots of the science of hydrogeology can be traced back in Central Europe (German speaking nations) to the beginning of the 19th Century, with major developments in the first half of the 20th Century, prior to the First World War. The main impetus came in connection with the establishment of central water supplies to towns following catastrophic epidemics late in the 19th Century. Both engineers such as Adolf Thiem (1836–1908) and his son Günter Thiem as well as geologists, namely Konrad Keilhack (1858–1944) and Werner Koehne (1881–1963) have contributed to hydrogeology, the term proper introduced by the Austrian Hans Höfer von Heimhalt in 1912 but hardly used before World War II. Geological Surveys, especially the Prussian Survey which existed between 1873–1939, have been backing up groundwater sciences by practical research. This tradition was resumed and further developed after the Second World War by surveys of federal states in Germany and the Federal Institute of Geosciences and Mineral Resources (BGR).

Hydrogeology is nowadays a teaching subject at approximately 20 German universities.

ACKNOWLEDGEMENT

The author acknowledges gratefully valuable support and stimulating suggestions by his colleague Prof. Hilmar Zetinigg, Graz, who drew the attention to specific Austrian matters which, however, could only be touched upon marginally in this paper.

Thanks are also due to Dr. Siegfried Lampe, Bangkok, for editorial help

REFERENCES

Bischof, G. (1841) *Lehrbuch der chemischen und physikalischen Geologie I*. Bonn, Adolph Marcus.

Burre, O. (1968) Die Arbeiten auf dem Gebiet der Hydrogeologie. In: Udluft, H. (ed.) Die Preußische Geologische Landesanstalt 1873–1939, 107–114. Beih. geol. Jb. 78, 1–170, Hannover.

Carlé, W. (1975) Die Mineral- und Thermalwässer von Mitteleuropa. Wiss. Verlagsgesellschaft mbH Stuttgart. pp. 643.

Darapski, L. (1903) Die Grundwasserfrage in Hamburg. 2. Aufl., 1–34, Verlag F. Leineweber, Leipzig.

De Vries, J.J. (1989) The historical base of ground-water hydrology in The Netherlands. *Groundwater*, 27 (1), 92–95.

De Vries, J.J. (1994) Willem Badon Ghijben and Johan M.K. Pennink, Pioneers of coastal-dune hydrology.*Applied Hydrogeology*, 4/94, 55–57.

De Vries, J.J. (1996) Historical evolution of hydrogeology in The Netherlands. Draft manuscript. 39 pp., 22 figs.

Evans, R.J. (1987) *Death in Hamburg. Society and Politics in the Cholera-Years 1830–1910*. Oxford, Oxford University Press.

Forchheimer, Ph. (1886) Über die Ergiebigkeit von Brunnen-Anlagen und Sickerschlitzen.- Z.d. Achitekten- u. Ing.-Verein, 32, 539–564.

Forchheimer, Ph. (1914) Hydraulik. 566 p.

Frank, W.H. & Schwarz, D. (1966) Zur Geschichte der künstlichen Grundwasseranreicherung. Veröff. Hydrolog. Forschungsabt. Dortmunder Stadtwerke AG 9, 7–26, Dortmund.

Grahmann, R. (1935) Der Sächsische Landesgrundwasserdienst.- Abh. Sächs. Geol. Landesamt 16, Leipzig.

Grahmann, R. (1958) Die Grundwässer in der Bundesrepublik Deutschland und ihre Nutzung. Forsch. z. deutsch. Landeskunde 104. 198 pp., Remagen.

Haas, H.J. (1895) Quellenkunde. Lehre von der Bildung und vom Vorkommen der Quellen und des Grundwassers. J.J. Weber, Leipzig.

Herzberg, A. (1901) Die Wasserversorgung einiger Nordseebäder. -J. f. Gasbeleuchtung u. Wasserversorgung, 44, 815–819; 842–844.

Höfer von Heimhalt, H. (1912) Grundwasser und Quellen. Eine Hydrogeologie des Untergrundes. Braunschweig (Vieweg).

Hölting, B. (1980) Hydrogeologie. Ferdinand Enke Verlag, Stuttgart.

Hölting. B. (1984) Die Wege zur modernen Grundwassererkundung. gwf Jubiläumsheft 125 Jahre DVGW, H. 6, 165–168.

Hydrogeologische Forschungen (1944). Abh. Reichamt f. Bodenforsch., NF, 209. 344 pp., Berlin.

Johannsen A., Matthess, G. & Richter, W. (1972) Die Tätigkeit der staatlichen geologischen Dienste in der BR Deutschland auf dem Fachgebiet 'Hydrogeologie'. Geol. Jb. C, 3, 3–13, Hannover.

Keilhack, K. (1902) Die geschichtliche Entwickelung der Lehre von der Entstehung der Grundwasser. Jb. Königl. Preuss Geol. Landesanst. u. Bergakademie, XXIII, I–XXI, Berlin.

Keilhack, K. (1908–1913) Grundwasserstudien I–VI. Z. prakt. Geol 1908, 458–464; 1909, 405–412; 1910, 125–130; 1912, 112–118; 1913, 29–41, 362–378, Berlin.

Keilhack, K. (1912) Lehrbuch der Grundwasser- und Quellenkunde. pp. 545, Gebr. Borntraeger Berlin.

Keilhack, K. (1916): Geologie der Mineralquellen und Thermen, der Mineralmoore und Mineralschlämme. Handbuch der Balneologie, Bd. I, 45–116.

Keller, G. (1969) Angewandte Hydrogeologie. Hamburg, Verlag WASSER UND BODEN. pp. 411.

Keller, H. (1928) Gespannte Wässer. 90 S., 50 Abb., W. Knapp Halle/S.

Koehne, W. (1928) Grundwasserkunde. Stuttgart, Schweizerbart'sche Verlagsbuchhandlung. pp. 291.

König, A. & Bruns, H. (1930) Künstliche Grundwasseranreicherung unter Berücksichtigung der Verhältnisse des Ruhrkohlengebietes. GWF, 43, 622 (und Ges.-Ing. 53, 662 u. 740).

Langguth, H.-R. & Voigt, R. (1980) Hydrogeologische Methoden., Berlin/Heidelberg/New York, Springer-Verlag . pp. 486.

Lersch, B.M. (1864): Hydro-Chemie oder Handbuch der Chemie der natürlichen Wässer nach den neuesten Resultaten der Wissenschaft. 2. Aufl., Verlag Hirschwald Berlin.

Lersch, B.M. (1870) Hydro-Physik oder Lehre vom physikalischen Verhalten der natürlichen Wässer, namentlich von der Bildung der kalten und warmen Quellen. 2. Aufl., Verlag A. Henry, Bonn.

Loehnert, E. (1967) Grundwasserversalzungen im Bereich des Salzstockes von Altona-Langenfelde. Abh. Verh. Naturw. Ver. Hamburg, N.F., XI, 29–46, Hamburg.

Loehnert, E.P. (1985):The impact of groundwater and the role of hydrogeology on a city's growth – Case study of Hamburg, Federal Republic of Germany. Mem. 18th IAH Congr., Part 2, 178–186, Cambridge.

Loehnert, E.P. (1997) Konrad Keilhack: Pioneer of German Hydrogeology. Hydrogeology Journal, 5(3), 125.

Loehnert, E.P. (1999): Konrad Keilhack (1858–1944) – seine Promotion im Jahre 1881 in Jena. Geohistor. Blätter, 2(1), 57–63, Berlin.

Loehnert, E.P. (2002) Major Aspects of urban hydrogeology in Central Europe – Examples from Germany. In: Current Problems of Hydrogeology in Urban Areas, Urban Agglomerates and Industrial Centres. The Netherlands, Kluwer Academic Publishers. pp. 243–261.

Lueger, O. (1895, 1908) Die Wasserversorgung der Städte. Darmstadt und Leipzig.

Matthess, G. (1973) Die Beschaffenheit des Grundwassers. Gebrüder Borntraeger Berlin, Stuttgart.

Matthess, G. & Ubell, K. (1983) Allgemeine Hydrogeologie Grundwasserhaushalt.- 214 Abb., 75 Tab., Gebr. Borntraeger, Stuttgart.

Meng, A. (1993) Geschichte der Hamburger Wasserversorgung. Hamburg, Medien-Verlag. 454 pp.

Pfalz, R. (1951) Grundgewässerkunde. Lagerstättenlehre des unterirdischen Wassers. Halle (Saale), Verlag Wilhelm Knapp. pp. 175.

Pfeiffer, D. (1963) Die geschichtliche Entwicklung der Anschauungen über das Karstgrundwasser. *Beih. geol Jb.*, 57, 111 S. Hannover.

Pinneker, E.V. (1989) Eduard Suess als Hydrogeologe. *Steir. Beitr. z. Hydrogeologie*, 40, 165–174, Graz.

Prinz, E. & Kampe, R. (1934) Quellen (Süßwasser- und Mineralquellen). *Handbuch der Hydrologie* 2. Band. Berlin, Verlag J. Springer.

Quitzow, H.W. (2002) KONRAD KEILHACK (16. August 1858–10. März 1944), Leben und Werk.- Nachrichtenblatt Gesch. *Geowiss.*, 12, 57–97, Krefeld u. Freiburg.

Richert, J.G. (1911) Die Grundwasser mit besonderer Berücksichtigung der Grundwasser Schwedens. München u. Berlin, R. Oldenbourg. 106 pp.

Richter, W. & Lillich, W. (1975) Abriß der Hydrogeologie. E. Schweizerbart'sche Verlagsbuchhandlung (Nägele u. Obermiller), Stuttgart. pp. 281.

Richter, W., Preul, F., Dechend, W., Dürbaum, H.-J., Groba, E. & Herrmann, R. (1968) Ein Rückblick auf 20 Jahre hydrogeologischer Arbeiten des Niedersächsischen Landesamtes für Bodenforschung. *Geol. Jb*, 85, 817–840, Hannover.

Schneider, H., Truelsen, H. & Thiele, H. (1951) Die Wassererschließung. Essen, Vulkan-Verlag.

Smreker, O. (1914) Das Grundwasser, seine Erscheinungsformen, Bewegungsgesetze und Mengenbestimmung. Diss. Zürich, 67 S., Wilh. Engelmann, Leipzig.

Stiny, J. (1933) Die Quellen. Die geologischen Grundlagen der Quellenkunde für Ingenieure aller Fachrichtungen sowie Studierende der Naturwissenschaften. Wien, Verlag J. Springer.

Stille, H. (1903) Geologisch-hydrologische Verhältnisse im Ursprungsgebiete der Paderquellen zu Paderborn. *Abh. Kgl. preuß. geol. L.-Anst.*, N.F., 38, 1–129, Berlin.

Thiem, A. (1870): Die Ergiebigkeit artesischer Bohrlöcher, Schachtbrunnen und Filtergallerien. *Journal für Gasbeleuchtung u. Wasserversorgung*, 14, 450–467.

Thurner, A. (1967) Hydrogeologie. Wien/New York, Springer-Verlag. pp. 350.

Udluft, H. (1968) Die Preußische Geologische Landesanstalt 1873–1939. *Beih. geol. Jb.*, 78, 170 S., 2 Tab., 3 Taf. Hannover.

Volger, G.H.O. (1877) Die wissenschaftliche Lösung der Wasser-, insbesondere der Quellenfrage. *Zeitschr. Ver. Dt. Ing.*, XXI (11), 482–502.

Wundt, W. (1953) Gewässerkunde.- 320 S., 185 Abb. Berlin/Göttingen/Heidelberg, Springer-Verlag.

Zetinigg, H. (1990) Bemerkungen zur Entwicklung des Begriffes Hydrogeologie. *Mitt. naturwiss. Ver. Steiermark*, 120, 145–154, Graz.

Zetinigg, H. (1998) Bemerkungen zu Klassifizierungssystemen für Quellen. Mitt. Ref. Geol. u. Paläont. Landesmuseum Joanneum SH 2, 371–386, Graz.

Zötl, J.G. (1974) Karsthydrogeologie. Wien/New York, Springer-Verlag. pp. 291.

History of Hungarian hydrogeology

Irma Dobos, Lajos Marton & Pál Szlabóczky
Hungarian Geological Society

INTRODUCTION

The Hungarian Republic, is situated in central Europe in the Carpathian Basin. Its neighbours are Slovakia to the north, Ukraine to the north-east and Rumania to the east. Serbia and Croatia are south of Hungary, Slovenia is to the south-west and Austria to the west. Hungary dramatically was reduced in area after World War I from an area that covered almost the entire Carpathian Basin ($325\,411\,\mathrm{km^2}$) to the present area of $93\,030\,\mathrm{km^2}$. This history refers to the larger pre-Great War area.

PHYSIOGRAPHY

Physical setting

The Carpathian Basin is the largest intra-mountain basin in Europe, bordering the Alps, Carpathians and Dinarides mountains. It is also called the Pannonian Basin since its extent is determined by the Late Tertiary (Pliocene) sediments of the Pannonian Inland-sea (Figure 1). The Carpathian Basin comprises the Hungarian Plains, the Hungarian Little Plains and the Transylvanian Basin. The Hungarian Plains is separated from the two smaller basins by the Transdanubian Mountains and the Transylvanian Mountains, respectively. The central part of the Pannonian Basin is part of Hungary. The Hungarian Plains is the central area of the Carpathian Basin of which less than half, exactly $45\,000\,\mathrm{km^2}$, is part of Hungary. The basin contains fluvial, limnic and aeolian sediments and lies at an elevation of less than $200\,\mathrm{m}$. The area of Szeged is the lowest part of Hungary and is only $78\,\mathrm{m}$ above sea level. One third of the country lies within the flood risk zones of the rivers and half of its population lives in such areas.

The climate of Hungary is continental but can be affected by oceanic influences from the west, Mediterranean from the south and dry continental from the east. The latter is the strongest influence and results in only $600\,\mathrm{mm}$ long term average annual precipitation. The distribution of the precipitation is uneven, the driest area is the central part of the Hungarian Plains, while the most humid is the Kőszeg Mountains close to the Alps. The average annual mean temperature is $10.8°\mathrm{C}$.

The central part of the Pannonian Basin is filled with poorly consolidated marine, limnic and fluvial Neogene (mainly Pliocene) basin sediments from $100\,\mathrm{m}$ to over

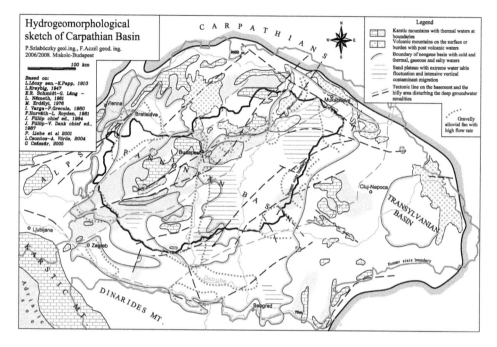

Figure 1 Hydrogeomorphological sketch of the Carpathian Basin.

7000 m thick. This sedimentary series overlies a tectonically influenced basement in which the elevation difference between the horsts and the rifts can reach 5000 m (Tóth & Almási 2001). Beneath the basin the crust has thinned out unevenly, resulting in a high geothermal gradient of 5°C per 100 m, roughly 1.5 times the global average.

Hydrogeology

The storativity of the crystalline basement is negligible. Boreholes located on tectonic features yield minor amounts of generally poor quality water.

The main aquifers are the Triassic basement, the Tertiary limestones and conglomerates and the Late Pannonian (Pliocene) and Quaternary porous sediments. The deep Quaternary fluvial deposits are the most important and 80% of the boreholes penetrate these strata.

Thermal-water occurs in the fractured Devonian dolomite and the Late Triassic dolomitic and limestone units. Thermal-water also occurs in the Late Pannonian (Pliocene) and Quaternary basin sediments, at depths below 500 m. 95% of the thermal-water boreholes penetrate the porous layers of the Late Pannonian (Pliocene) sand.

The groundwater flow regime in the Pannonian Basin is driven by gravitational and lateral compression of the rock (Tóth & Almási, 2001). Gravity flow occurs in the unconfined aquifer and in the compression zone it is confined. Between these two zones the hydraulic potential is controlled by bedding and the structure of the rock. Direct rainfall recharge occurs to the unconfined system although shallow flow paths return much of the recharge to surface (Tóth, 1963; Erdélyi, 1978; Rónai, 1985). The upwelling of groundwater from the deeper, over-pressurized, confined zone is caused by the regional tectonic compression and the resulting loss of pore volume. The hydraulic connection between the two zones is through regional semi-permeable layers, rock pores, highly permeable lenses and conductive tectonic fractures.

HYDROGEOLOGICAL DEVELOPMENT

The early communal water services 1730–1917

The Carpathian Basin offers numerous thermal springs providing mineral and medicinal waters and there are several gas discharges to surface (e.g. Budapest, Hévíz, Harkány, Daruvár, Herkules Spa, Nagyvárad-Félixfürdő, Bártfa, Pöstyén). Large numbers of medicinal and thermal water spas were established, requiring input from various Hungarian hydrogeologists. The thermal spas of Budapest, the mineral waters above the Eastern Transylvanian post-volcanic and evaporate formations, and the sulphuric waters of Upper Hungary are still very popular.

The first classification of the Transylvanian mineral waters is found in the work called "Old and New Diaetetica" which classifies more than fifty types of mineral waters into sulphuric, alkaline, ferric and saline. This work by István Kibédi-Mátyus I. (1787–1793) identified the connection between the composition of the mineral water and rainfall since 'the composition of the water is different in summer and in winter, in rains and in droughts'.

The first detailed analysis of the springs of Buda was given by the physician of the city in the work entitled 'The Thermographia Budensis' (subtitled 'The heat geography of Buda'). The physical and medical aspects of the water are described, the source and location and likely groundwater age are described. The author described how the level and temperature of the sources varied with time, and provided chemical analyses along with descriptions of his investigation methods (Stocker, 1721). 'Hydrographica Hungariae' (Kitaibel, 1829) marked the end of the first era of medicinal water analyses with the publication of analyses of 150 mineral waters. The publication described the geographical and geological setting of the mineral water sources and the physical characteristics (taste, colour, transparency, temperature, density, smell) of the waters.

Water has always had a major role in the mining industry from both mine dewatering and as a source of energy. The establishment of the water management system at Selmecbánya (now Banska Stiavnica) was attributed to Sámuel Mikoviny.

The first borehole in the country was drilled at Ugod in 1825 in search of mineral water. In the 1830s a few artesian boreholes were completed, mainly for drinking water supplies in the Hungarian Plains (Debrecen) and in Transdanubia (Székesfehérvár, Csór, Buda). An exploration programme using scientific approaches started with a 38 m deep thermal water well at Harkány and continued with the 119 m deep well at Margitsziget (Margaret Island, Budapest) and 970 m deep well at the City Park, both at Pest. The first public supply wellfield was at Budapest with work starting in 1868 on the gravel terrace of the River Danube with four large diameter wells. During the late nineteenth century, with the beginning of urbanization and industrialization, the demands for good quality drinking water and the increasing amounts of industrial water could only be met by drilling numerous 100–300 m deep boreholes for artesian water in the plains. Count István Széchenyi (1791–1860) then considered that the basis for water management was river regulation, flood-prevention, inland water extraction and irrigation.

After the civil revolution (1848–1849) conditions were harsh in the country's fastest growing city of Pest (in the eastern part of Budapest), where water borne disease and epidemics begged for improved water services, a problem that was only solved at the turn of the century. In more rural areas similar problems forced Hódmezővásárhely to establish its first communal well in 1880.

From 1876 onwards private and communal wells were drilled by the Béla Zsigmondy company. The introduction of the water circulation type drilling technique in 1890 started a new period in hydrogeological exploration and in the water supply. As the economy grew, increasing amounts of good quality water was needed. At the end of the nineteenth century there were 1500 drilled wells, and after the introduction of the water circulation drilling technique at the beginning of the twentieth century there were 3000. By 1917 almost 5000 drilled wells had been commissioned. In the settlements in the hilly and mountainous areas of Hungary, wells and gravitational spring-fed sources provided most groundwater supplies.

The significance of practical hydrogeology was enhanced by the Water Conservancy Law in 1885 which addressed the legal and administrative issues of water exploration. At the same time the boundaries around mineral water, medicinal water and thermal water sources were described and the development of water treatment works commenced. Halaváts (1896) reported technical and hydrogeological data and the location of water supply system and wells of settlements was reported on maps at the first decade of the 20th century. According to the hydrogeological model accepted at that time for the Hungarian Plains the Pleistocene and Late Pliocene aquifers were lenticular formations which correspond to the former river channels. The high geothermal anomaly of the Hungarian Plains was reported by Lóczy (1912). The first chemical tracer tests for investigating the connectivity of karstic springs (and the velocity of groundwater) were carried out in the Bükk Mountains (Diósgyőr) by Emszt (1913) using the Thiens method. Later, in 1933, a similar investigation was carried out at Tiszafüred by Tihamér Gedeon. The Hydrogeological Department of the Hungarian Geological Society was initiated in 1917.

Broadening of hydrogeological exploration in the inter-war period 1917–1945

In 1925 thermal water exploration began at Hajdúszoboszló where a 1091 m deep borehole exploited thermal water with a temperature of 70°C to supply a thermal spa and bottling plant, and for heating greenhouses. The gas was used for lighting railway carriages and for electricity generation. A number of other thermal water boreholes were drilled subsequently.

By the early 1940s some 24 000 artesian wells were recorded. The monographs and studies dealt with the theoretical and practical aspects of fractured aquifers (Weszelszky, 1922) and the geothermal conditions of groundwater (Sümeghy, 1929).

Several investigators looked at borehole hydraulic tests (Pattantyús-Ábrahám, 1935; Schmidt, 1939; Mazalán, 1939). In 1933 the Hydrogeographical Institute started monitoring the groundwater, and the first monographs in Hungarian on the general hydrogeology and hydrogeochemistry of the capital were published (Horusitzky, 1924; Szádeczky-Kardoss, 1941). Sub-water table mining required extensive hydrogeological exploration and safety measures to prevent mine water ingress were initiated in the coal basins which had karstic water hazards (Vígh, 1944).

Post-World War II to the beginning of the electronic age 1945–1970

Growth of urban areas after World War II required extensive development of groundwater for public supply. A hydrogeological map was prepared for the whole area of Budapest. Reconstruction of the spas and wells of Budapest, damaged during World War II, was helped with advice from the universities.

In 1958 the national artesian well survey was started by János Urbancsek, and the first two reports were published in 1963 by the Hungarian Water-management Scientific Research Institute (Hungarian abbreviation: VITUKI). At that time more than 35 000 artesian wells were recorded. Groundwater recharge and groundwater storage were an issue and subsequent regional hydrogeological investigations by the Hungarian Geological Institute produced the Hydrogeological Atlas (Schmidt, 1961) and an explanatory text (Schmidt, 1962) as well as a potentiomentric map (Rónai, 1961). In 1964 mapping of the Hungarian Plains was started at 1:100 000 scale and several of the maps cover hydrogeology. Hydrogeological investigations became broader, with an increased number of boreholes drilled for thermal water and agricultural supply, and there was innovative use of mine discharge water.

The electronic age

By the start of the 1980s the communal, industrial and agricultural water consumption was close to 5 million m^3 d^{-1}. Groundwater supplied 60% of the total demand, including bank filtration water along the rivers, open-basin artificial recharge, alluvial groundwater, artesian groundwater, karstic water, spring water and mine water. Large water table depressions occurred around the larger cities of the Hungarian Plains. Mine dewatering caused a regional decrease in the potentiometric level in the karstic area of Central Transdanubian; the average decrease was 50 m by the second half of the 1980s, but in certain areas it reached 100 m (Liebe, 2002). Abstraction of karstic water was twice the calculated recharge, but subsequent mine closure karstic water production has now dropped by a third and recovery is taking place. Since the mid-1990s the country's water consumption has decreased significantly.

Isotope-hydrology has been applied since 1969 at VITUKI (Major, 1972). The regional flow regime of the Great Plains and the horizontal and vertical seepage velocity was estimated (average 30–50 mm/yr and 2–3 mm/yr respectively) by Stute & Deák (1989) using isotopic indicators. Based on the high He concentration very slow seepage was modelled in the deeper layers of the Great Plains (Stute et al., 1992). 5 to 15% of mantle component was detected in deep Hungarian groundwaters by $^3He/^4He$ isotope ratios (Martel *et al.*, 1989). During the investigation of the effect of the Gabcikovo-Nagymaros Dam on the Danube River it was shown that the total groundwater storage in the thick gravels derived from the Danube River. The water from the bed of the Danube River flows at 100–500 m/yr (Deák *et al.*, 1996). A new hydrological model was developed which uses ^{14}C concentrations to calculate the recharge (Marton *et al.*, 1990).

TEACHING HYDROGEOLOGY IN HUNGARY

Early days

The first book published in Hungary about hydrogeology was the work of Marcellus Squarcialupus, written in Latin in 1585 at Kolozsvár (Cluj, Romania) about the origin of springs and rivers (*De fontium & fluviorum origine ac fluxu*). Hydrogeological research in Hungary started several hundred years ago, largely because people were compelled to drain deep mines. Mining was the only large-scale industry in Hungary up to the first half of the nineteenth century and technical and natural sciences were then incorporated in higher education.

Mátyás Bél (Mathias Belius in Latin, 1684–1749) was a member of the academies of Berlin and London, as well as of the scientific associations of Olmütz and Jéna. Issue 450, October–November 1738, of the *Philosophical Transactions* published in London included an essay written by him in Latin about the metalliferous mineral waters of Besztercebánya (Neosolium, Banská Bystrica in Slovakian) (De Aquis Neosoliensium aeratis, Ferrum Aere permutantibus). Notitia Hungariae novae historico-geographia (Wien, 1735–42) dealt with the modern history, hydrography and geography of ten Hungarian counties.

Selmecbánya (Banská Štiavnica in Slovak, Schemnitz in German) was a flourishing mining town from the Middles Ages up to the end of the nineteenth century. The industrial sector around Selmecbánya (now Banska Stiavnica) and its environs required the creation of technical higher education. The institute established in 1735 by the 'Hofkammer' (Chamberlain) of the Imperial Court of Vienna for the training of miners, metalworkers, and mining officers (Bergschule) was the first institution of its kind in the world. Sámuel Milkoviny (1700–1750) who was one of the greatest scholars in Hungary in the eighteenth century was appointed the first teacher of the mining officers training school at Selmecbánya. He taught mathematics, mechanics and hydraulics, and directed the mine surveying practice. He was the first to introduce a system for the independent laboratory teaching of students. He also established the water management system at Selmecbánya. His fame reached foreign countries during his lifetime, as he was a member of the Prussian Scientific Academy at the age of 35.

The nineteenth century

Professor József Szabó (1822–1895) was one of the first teachers of hydrogeology and Professor Károly Than (1864–1908) introduced the teaching of chemistry.

Institutum Geometrico-Hydrotechnicum, which later became Budapest University of Technology and Economics (BME), was founded in Buda in 1782, and was the first institute in Europe to train engineers at university level. It became one of Central Europe's largest institutions for technical education, where water specialists were educated. The institution received the rank of university in 1871 and was named as Hungarian Royal József Nádor Technical University. It acquired the right to issue diplomas subject to a comprehensive examination in engineering. An independent water engineering department was established in 1884.

A new university was founded in Kolozsvár, in the centre of Transylvania, in 1872. It had a modern structure and it was the first university in the country to have a faculty of natural sciences. It was named as Ferenc József University of Arts and Sciences.

The twentieth century

Oil and natural gas research related to thermal and medicinal waters coupled with geophysical research, further enhanced the hydrogeological knowledge of the country. It also generated interest in the hydrogeology of sedimentary basins. The water engineering department of the Technical University of Budapest (BME) was divided into two parts, creating the water engineering and water management departments. Diplomas in water engineering have been awarded since 1950.

The Academia Monstancia founded at Selmecbánya (now Banska Stiavnica) in 1762 is the legal predecessor of the University of Miskolc (ME). Hydrogeological engineering training was introduced at the university in 1958 and offers training for hydrogeologists and engineer-geologists.

Both geology and hydrogeology are taught at the Eötvös Loránd University of Arts and Sciences (ELTE), earlier Pázmány Péter University of Arts and Sciences, one of Hungary's oldest universities.

REFERENCES

Deák, J., Deseő, É., Böhlke, J.K. & Révész, K. (1996) *Isotope Hydrology Studies in the Szigetköz region, Northwest Hungary Isotopes in Water Resources Management.* Vienna, IAEA. pp. 419–432.

Emszt, K. (1913) Bericht über die Tätigkeit des Chemischen Laboratoriums der kgl. ungar. Geologischen Reichsanstalt im Jahre 1911. *Jahresbericht der königlich ungarischen Geologischen Reichsanstalt für 1911.* pp. 236–243. (225–245.)

Erdélyi, M. (1979) Hydrodynamics of the Hungarian Basin. *Proceedings 18.* Budapest, VITUKI (Research Institute of Water Management).

Halaváts, Gy. (1896): A magyarországi artézi kutak története, terület szerinti eloszlása, mélységök, vizök boségének és hofokának ismertetése (in Hungarian, The history of the atresian wells in Hungary, the spatial distribution of them, review of their depth, yield and tempertaure). Az 1896. évi ezredéves kiállítás alkalmából (World Exhibition). Budapest 1896.

Horusitzky, H. (1924) Részlet Budapest székesfováros Duna balparti terülte földtani, talajtani és vízi viszonyainak ismeretéhez (in Hungarian, Details for the geological, pedological, and groundwater settings of the left-hend side bank of Danube, Capital Budapest). A Szent István Akadémia Mennyiségtani és Természettudományi Osztályának felolvasásai (Public Reading at the Academy of Szent István, Division of Mathematics and Natural Sciences), Budapest. 14 p.

Kitaibel, P. (1829) *Hydrographica Hungariae praemissa auctoris vita.* Tom. 1. Ed. Joannes Schuster. Pestini, Vol. 1, 2.

Liebe, P. (2002.) Felszín alatti vizeink (in Hungarian, Our groundwaters). Tájékoztató (Handout). Környezetvédelmi és Vízügyi Minisztérium (Ministry for Environments and Water). Kármentesítési Program (Remediation Programme). VITUKI Hidrológiai Intézet (Water Resource Center, Institute of Hydrology). p. 56.

Lóczy, L. (1912) A kissármási gázkitörés (in Hungarian, Explosion of gas at Kissármás). *Földtani Közlöny (Geological Bulletin)*, 1. füz, 1–11.

Major, P. (1972) A kisalföld medence felszín alatti vizei természetes izotóp tartalmának vizsgálata (Environmental isotope study of groundwater in the Little Hungarian Plain). Budapest, VITUKI (Water Resource Center, Istitute of Hydrology). Annual report for 1970.

Martel, D.J., Deák, J., Dövényi, P., Horváth, F., O'Nions, R.K., Oxburgh, E.R., Stegena, L. & Stute, M. (1989) Leakage of helium from the Pannonian Basin. *Nature*, 342, 908–912.

Marton, L. & Mikó, L. (1990) Izotóphidrogeológiai kutatások az Alföldön (in Hungarian, Isotophydrological studies at the Great Plain). Magyar Állami Földtani Intézet (Hungarian Geological Survey), Évi Jelentés (Annual Report of the year) 1988. évrol. pp. 136–152.

Mazalán, P. (1939) A mélyfúrású kutak helyes létesítési módja és az ezzel kapcsolatos teendok (The proper development of deep wells and activities in connection with it). *Vízügyi Közlemények (Hydrological Bulletin)*, 3–4.

Pattantyús-Ábrahám, G. (1935) Gázos kutak üzemi jellemzoi I-II. (The operational characteristics of wells with solved gas). *Bányászati és Kohászati Lapok (Journal of Mining and Mettalurgy)*, 83, 209–253.

Rónai, A. (1961) Az Alföld talajvíztérképe (The groundwater map of the Great Hungarian Plain). *Magyar Állami Földtani Intézet (Hungarian Geological Survey)*, Budapest.

Rónai, A. (1985) Az Alföld negyedidoszaki földtana. (The Quaternary of the Great Hungarian Plain). *Geologica Hungarica. Series Geologica*, Tom. 21. Budapest. p. 445.

Schmidt, E.R (1939) A Kincstár csonkamagyarországi szénhidrogénkutató mélyfúrásai (in Hungarian, The hydrocarbon exploratory deep boreholes of the Treasury of Little Hungary). Magyar Állami Földtani Intézet Évkönyve (Annual Report of the Hungarian Geological Survey), 34. pp. 205–267.

Schmidt, E.R. (chief ed.) (1961) Magyarország Vízföldtani Atlasza (Hydrogeological map of Hungary). Magyar Állami Földtani Intézet (Hungarian Geological Survey), Budapest.

Schmidt, E.R. (chief ed.) (1962) Magyarázó Magyarország Vízföldtani Atlaszához (Explanatory text to the Hydrogeological map of Hungary). Magyar Állami Földtani Intézet (Hungarian Geological Survey), Budapest. p. 73.

Stocker, L. (1721) *Termographia Budense*, Buda, Hungary.

Stute, M. & Deák, J. (1989): Environmental isotope study (^{14}C, ^{13}C, ^{18}O, D, noble gases) on deep groundwater circulation systems in Hungary with reference to paleoclimate. *Radiocarbon*, 31 (3), 902–918.

Stute, M., Sonntag, C., Deák, J. & Schlosser, P. (1992) Helium in deep circulating groundwater in the great hungarian plain: flow dynamics and crustal and mantle he fluxes. *Geochemica et Cosmochemica Acta*, 56, 2051–2067.

Szádeczky-Kardoss, E. (1941) A Dunántúli-Középhegység karsztvizének néhány problémájáról (in Hungarian, About some problems of the Transdanubian Mountain). *Hidrológiai Közlöny (Hydrological Bulletin)*, 21, 67–92.

Sümeghy, J. (1929) Geothermal gradients in the Great Hungarian Plain. (In German). A Magyar Királyi Földtani Intézet Beszámolója 1928. Budapest.

Tóth, J. (1963) A theoretical analysis of groundwater flow in small drainage basins *Journal of Geophysical Research*, 68.

Tóth, J. & Almási, I. (2001): Interpretation of observed fluid potential patterns in a deep sedimentary basin under tectonic compression: Hungarian Great Plain, Pannonian Basin. *Geofluids*, 1 (1), 11–36.

Vígh, F. (1944) Az esztergomi szénmedence hidrológiai viszonyai és a víz elleni védekezés módozatai (in Hungarian, The hydrological settings of the coal basin at Esztergom, and methods of dewatering). *Bányászati és Kohászati Lapok (Journal of Mining and Mettalurgy)*, LXXVII (14), 215–222. és 15, 227–239.

Weszelszky, Gy (1922) A geotermikus grádiensrol (About the geothermal gradient). *Hidrológiai Közlöny (Hydrological Bulletin)*, 2. évfolyam 1. sz., 5–13.

A brief history of Indian hydrogeology

Dr. Shrikant D. Limaye
Director, Ground Water Institute (NGO), Pune, India

ABSTRACT

Groundwater has been recognised in India as an important resource for getting domestic and irrigational water supply for thousands of years. In Vedic scriptures (around 8000 BC) it has been praised as 'springs bestowing health, happiness and peace on the community'. The history of groundwater in India could be divided into three periods; 1. Pre-historic and Historic; 2. British period (1818 to 1947 AD) and 3. Post-Independence or Post-1947 period. In all these periods using groundwater in various rock-types was associated with religious and cultural aspects of the society. India has about 15% of the world's population, but has only 6% of the world's water resources and 2.5% of the world's land. Both land and water resources must, therefore, be carefully managed in a sustainable manner. Although hard rock areas, occupying about 66% of the country, are poor aquifers, groundwater has assumed an important role in the history and culture of the society even in these areas. Dug wells in hard rocks are typically 10 to 12 m deep but in historical times some of them have gone up to 30 m depth, desperately looking for water.

INTRODUCTION

India, being the cradle of one of the oldest civilisations, has a long history of exploration and use of groundwater resources. Initially, the civilisation was established close to perennial rivers like the Indus and Ganga (Ganges). Later on, due to the spread of civilisation away from major perennial rivers into the valleys of smaller seasonal rivers and streams, groundwater became an important source for drinking water supply and for irrigation. Even in those days it was known that groundwater was of better quality than surface water and hence better for drinking purposes. By the side of a flowing river, women used to dig small pits in the sandy banks, a few meters away from the river and take filtered water in their pots for domestic use. This practice has continued till today.

Groundwater development in Pre-Historic and Historic Periods up to the British occupation

The Holy Vedic Scriptures dating back to 8000 BC have references to dug wells or open wells fitted with a wooden pulley and rope attached to an earthen pot. Dug wells have also been mentioned in the epics of Ramayana and Mahabharata. The technique of siting suitable locations for wells date back up to Manu (1000 BC), while in the 4th century AD, a sage named Varah Mihir from Magadh (presently the State of Bihar in India) prepared the first text-book, named 'Drikgargal Shastra', on locating wells. This book is included in 'Bhrugu Samhita' and its printed editions are available

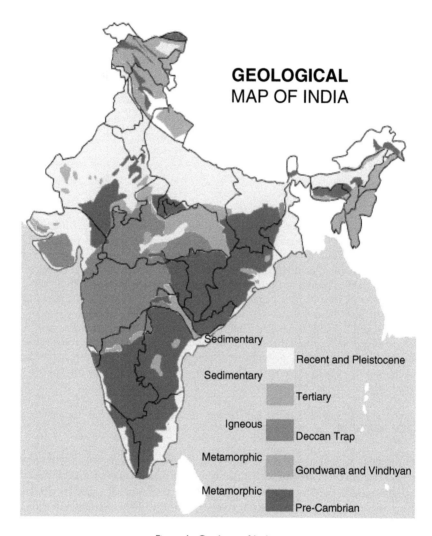

Figure 1   Geology of India.

even today. The methods of well siting in this book depend on indicators of shallow groundwater, like certain types of trees, anthills, or the nature of weathered rocks. The art of water divining which started in pre-historic times continues till today and almost every village or group of villages has a 'water-man' or 'diviner' for advising farmers on the location for well digging/drilling. Although they claim to have 'divine' power, many water diviners work on the basis of their keen observations of successful or failed wells in a region. Some however, strongly believe in their superhuman power.

During the historic period, water was brought to major cities by concealed, earthenware pipes from surface water reservoirs or from dug wells located at higher elevation from the cities. Part of Pune (Poona) city, in the basaltic terrain of western India, used to get its supply from an earthen pipeline connected to a dug well. This dug well of about 15 m depth was located at a distance of about 6 km from the city and at an elevation 35 m higher than the city. It was covered with a dome to protect the water. In flat terrain, large diameter dug wells (known as baoli or step wells) had decorative underground buildings surrounding them in which steps were provided for people to go down to the water level. Separate pathways were also constructed for the animals used for bringing water in leather bags from the well. One such famous well is at Adalaj near Ahmedabad, the capital of Gujarat State in western India.

In dry parts of Rajasthan state in western India, dug wells used to go down to about 50 to 60 m depth. Water used to be lifted out using bucket, rope and wooden pulley. Most of the forts used to depend upon water stored in tanks during the rainy season and on seasonal springs. In coastal areas with shallow water tables, animal powered Persian wheels mounted with a string of earthen pots were commonly used.

Groundwater development during the British period (1818–1947)

With the establishment of Geological Survey of India in 1851 during the British period, scientific studies in groundwater started, initially in the highly productive alluvial terrain and later in the lower-yielding hard rock aquifers. In the early days various terms were in vogue like Underground water, Subsurface water, Geo-hydrologist, Ground-water Geologist, etc., until the terms groundwater and hydrogeologist became popular. Early papers on groundwater occurrence and exploration in the basaltic, hard rock terrain (Deccan trap), which occupies about 500 000 km^2 in western India, were written by pioneers like Limaye (1940), Rao (1947) and Deshpande (1949). One of the earliest problem-oriented reports in alluvial terrain, dealing with the groundwater balance and sustainability of groundwater development, was published by Limaye in 1958. This report discussed the sustainable availability of tube well water supply to Chandigarh, the new capital city, which was being established for the State of Punjab in northern India.

Groundwater development since Independence

After gaining independence in 1947, the planners in India initially concentrated on surface water projects. But during the severe droughts of 1951, 1952, 1962, 1965–1967 and of 1972, the limitations of surface water development became evident and more attention was focused on groundwater development.

Surface water development has always depended on Government funds and the canal water has been given to farmers for irrigation at highly subsidized rates. In surface water schemes it has not thus been possible to recover even the operation and maintenance costs from the beneficiaries, let alone the capital and interest. This was not so in the pre-British period, when the local chiefs or kings used to give some money and encourage the public to build tanks and manage and maintain the irrigation facilities. In contrast, groundwater development has always been the result of the individual motivation of farmers and has been carried out using private funds. The farmers have always tried to maximize the agricultural produce from the limited quantity of groundwater available for irrigation from their dug well or bore. Productivity per cubic meter of groundwater used for irrigation is, therefore, more than that of surface water. The

Government has worked not as an investor or manager but as a catalyst or facilitator in groundwater development in India. Some of the factors influencing groundwater development from 1970 to 1980 were:

- Central Government and State Government departments and institutions employed more hydrogeologists and engineers for scientific groundwater studies, taking the watershed as a basic unit.
- Farmers had access to Institutional finance for well digging/drilling, at low rates of interests, in schemes financed by Commercial Banks under refinance from the Agricultural Refinance Corporation (ARC). Nationalisation of major commercial banks accelerated the dispersion of bank branches in the rural sector. International finance from the World Bank – IDA was available for groundwater development in many States. Bilateral aid projects for groundwater development for irrigational and domestic use were also taken up by Government Departments, NGOs and Charitable Trusts.
- Initially, pumping of groundwater was done with diesel engine pump-sets. Then many States took up the Rural Electrification Program and even the remote villages got connected to the national or regional electricity grid. Farmers were given electric connections at their wells, with a concession in the electric tariff.
- Initially, bore well drilling in hard rocks was carried out with calyx rotary rigs using 5 H.P. diesel engines. In basaltic terrain, drilling up to 30 m depth with such a rig often took up to a month. In granitic areas, more time was required. However, since 1972, down-the-hole hammer rigs became popular. With these high power rigs using compressed air, drilling a bore of 150 mm diameter in hard rocks up to 60 m depth, was possible within one day. With the advancements in pumping technology, farmers were able to pump water from such deep bore wells. This enabled the farmers to use groundwater from deeper aquifers for irrigation. Drilling up to 100 m depth was taken as the limit for tapping useful supplies of groundwater in the hard rock areas. In Government's programs for rural drinking water supply, a 150 mm diameter bore well up to 60 m depth, installed with a hand-pump, became a standard.

In the 1950s experts from the USA started visiting India and working with the Geological Survey of India (GSI) on groundwater studies in selected regions of the country. A Central groundwater Organization (CGWO) was first established. Later on, under a Technical Aid Mission from the USA, several drilling rigs and drilling staff were sent to India to train local engineers and hydrogeologists deputed from GSI in exploratory water well drilling. This was the beginning of the Exploratory Tube well Organization (ETO). After the first two phases of exploratory drilling, the ETO started recruiting its own technical staff. From the late 1960s to 1970, ETO was assisting various states in the installation of production wells. The "Engineering Geology and groundwater" series of publications from the GSI had included about 30 papers dealing with groundwater by 1970. The ETO also helped in executing specialised groundwater studies, sponsored by international aid agencies, in different states. Thus the UNDP projects in Rajasthan State; Canadian aid projects in Andhra Pradesh and Karnataka; the Swedish aid project in Ponnani basin; the Netherlands aid project in Bihar; and the British aid project in Betwa basin were carried out. The ETO helped in the development of a methodology for the rapid assessment of groundwater potential of an area (Agashe, 2002). Later on in 1971–72, the ETO was transformed into Central groundwater Board (CGWB) and the groundwater wing of GSI was merged into it. CGWB has now about 18 regional offices in various parts of the country, plus about 16 divisional offices and 9 unit offices. It employs about 630 technical people.

The establishment of the Central groundwater Board (CGWB) in 1970 accelerated the pace of groundwater related studies in various parts of the country. In the mid-1970s, groundwater development for irrigation in different states received a big boost through the World Bank (International Development Agency – IDA) project of providing institutional loans from commercial Banks to farmers for well digging/drilling for irrigation. As a prerequisite to these projects, state level groundwater Boards or Development Agencies were established and scientific studies on the availability of groundwater were conducted in various States. The emphasis thus shifted from availability of groundwater in a particular farm or a village to groundwater availability in a region, in a sub-basin or in a river basin. Topics such as aquifer input-output, modelling, sustainability of groundwater supply, estimation of recharge using various techniques, recharge augmentation, watershed development, soil and water conservation, groundwater quality, pollution etc. started gaining importance in hydrogeologicalal field work and in discussions at the scientific meetings and seminars. Many universities in the country started courses related to groundwater and graduates in geology and engineering started getting jobs in groundwater development Agencies and Boards. CGWB opened its zonal or regional offices in various zones in India and also state units in some states to conduct problem-oriented work.

Development of geophysical techniques for groundwater exploration in India

The use of resistivity and seismic refraction methods started appreciably in the late 1950s with imported instruments. The earlier resistivity meters were DC potentiometers in which the stray potential in the ground had to be balanced first, before introducing the DC current into the ground through aluminum or stainless steel current electrodes. $Cu\text{-}CuSO_4$ non-polarizing electrodes were used for potential measurements. Both Wenner and Schlumberger electrode configurations were used, the former preferred for profiling and the latter for sounding. In rural areas the DC instruments worked satisfactorily but in urban areas variations in stray potentials created great difficulty in measurements. Twelve channel refraction seismometers were used in hard rock areas for finding the depth of overburden or the thickness of the shallow, phreatic aquifer. These were also used in engineering geology for foundation exploration. The Central Water and Power Research Station (CWPRS) at Khadakwasla, near Poona (Pune) in Maharashtra State, was amongst the pioneers in the use of geophysical instruments. In the private sector, the first resistivity meter was purchased by the groundwater Institute in Pune in 1958 and used for groundwater surveys in Dharwar metamorphic rocks (shales, phyllites and schists) in south India (Limaye & Limaye, 1959).

In the mid-sixties, the National Geophysical Research Institute (NGRI) was established at Hyderabad, the capital of Andhra Pradesh State. Some divisions in NGRI were devoted to groundwater related work including exploration, assessment, use of tracers for recharge estimation and aquifer modelling. The Instrumentation Division at NGRI developed AC Resistivity Meters and single-channel refraction seismometers at affordable prices for Indian hydrogeologists. The private sector also joined in manufacturing geophysical instruments, especially the resistivity meters. With the advent of integrated circuits, the weight, size and prices of low frequency AC resistivity meters dropped and the reliability of measurements increased. A typical resistivity meter was easily available in the market for US $ 500 to 800.

This, however, had one undesirable side-effect. Many enthusiastic people, without qualifications or training in geophysics, purchased resistivity meters and started advising farmers on locations for well digging, by taking resistance measurements using their own configurations. Their failures were responsible for creating skepticism in rural communities about the resistivity method of groundwater exploration.

In addition to the curve matching technique used in developed countries for the resistivity data interpretation from Wenner soundings, a new method, called 'inverse slope method', was propounded and successfully used by two geophysicists at NGRI. In this method, the inverse of resistance (R) is plotted against electrode separation (a) and the slopes of straight-line segments of the graph are measured. This method has been found useful even in highly heterogeneous strata like glacial till.

A major hydrogeological meeting at the national level took place in 1958, when the Central Board of Geophysics organized a Symposium on groundwater, in New Delhi. A few years before this Symposium, electrical resistivity surveys had already become a popular tool for exploration of groundwater. Deshpande and Sengupta (1956), Limaye and Limaye (1959) were some of the early papers on the application of resistivity surveying in hard rock areas. In 1972, the national seminar on 'Hydrogeology of hard rock areas in relation to regional development of groundwater' attracted a large number of participants. However, the International Symposium on groundwater Development at Chennai (Madras), held in 1973, was probably the first major international event in India in the field of hydrogeology. Four volumes of the Symposium proceedings contain a large number of papers by Indian authors, and give a clear picture of the state of the art in Indian hydrogeology. The sessions of this symposium included topics such as groundwater exploration (geological & geophysical methods), hydrogeology of aquifers, hydrogeological processes, groundwater movement and well hydraulics, sea water intrusion, artificial recharge, groundwater development case-histories, groundwater quality, management and optimization.

PRESENT DAY

In this historical background, it is worthwhile to briefly take a look at the current situation. The main problem for India is that it has 15% of the world's population, but that this has to be sustained with only 6% of the world's water resources and 2.5% of the world's land. Both land and water resources must, therefore, be carefully managed in a sustainable manner. The national water policy has given first priority to domestic water supply, followed by irrigation, industry and environment. Rather than attending to only supply-side management, technical people are now getting oriented towards demand management including the economic use of water, conservation, reuse and re-circulation. Water related topics are finding prominent places in science text-books for schools, universities and adult education programs. NGOs are playing an increasingly important role in promoting people's participation in water resources management.

About two-thirds of India is occupied by hard rocks and the remaining portion by unconsolidated and semi-consolidated sediments. The average annual precipitation is 4000 billion cubic metres (BCM). Average surface water resources are estimated at 1869 BCM per year, of which 690 BCM can be stored. In addition, the groundwater potential is estimated at 432 BCM. Against this water availability of 2301 BCM, the population of the country is likely to be between 1.5 to 1.8 billion by 2050. The per capita water availability will then be around 1400 cubic meters per year (assuming no changes in water availability due to climate change), making India a water-stressed country.

The groundwater potential of about 432 BCM is dispersed across the country in the wide spectrum of soft and hard strata having different characteristics grain size, consolidation, texture, structure, weathering and fracturing. The groundwater situation is also complicated by the varied physiographic and hydro-meteorological conditions in landscapes ranging from snow clad mountains in the north to the arid region in the north-west, to the coastline of over 7000 km length in the west, south and east.

There is continuous stress on groundwater resources due to increasing water demands. The number of groundwater abstraction structures has increased from merely 4 million in 1951 to 17 million in 1997. Consequently, the irrigation potential created by groundwater abstraction has increased from 6.0 million hectares (Mha) in 1951 to 36.0 Mha in 1997. Stress on groundwater resources has caused problems related to over-exploitation, such as declining groundwater levels, seawater intrusion, quality deterioration, etc. in some watersheds covering an area of about 0.2 million km^2. Augmentation of groundwater resources in such watersheds through activities for recharge promotion and demand management has become necessary.

REFERENCES

Agashe, R.M. (2002) Central groundwater Board, 261 Shaniwar Peth, Satara 415 002, Private Communication.

Deshpande, B.G. (1949) A note on rural water supply in Deccan trap area in Mewashi estate, West Khandesh District. Geological Survey of India (Unpublished Report).

Deshpande, B.G. & Sengupta, S.N. (1956) Geology and groundwater in Deccan traps and application of geophysical methods. *Bulletin of Geological Survey of India*, Sr. B.8:24.

Limaye, D.G. (1940) Laboratory tests on water bearing capacity of Deccan trap and associated rocks. *Journal of Geology, Mining and Metallurgical Society of India*, 12 (2), 31–50.

Limaye, D.G. (1958) Tube well water supply for Chandigarh capital. *Proceedings of the Symposium on groundwater*. Calcutta, Central Board of Geophysics. Publ. no. 4. pp. 79–94.

Limaye, D.G. & Limaye, S.D. (1959) Electrical resistivity testing for groundwater in Dharwar (metamorphic) rocks of Sandur area. *Proceedings of the Symposium on Geophysical Exploration*. Hyderabad, National Geophysical Research Institute. Publ. No. 5. p. 218.

Rao, D.V. (1947) Manual on groundwater and well sinking. Publication by Hyderabad State Government. p. 291.

History of hydrogeology in Japan

Yasuo Sakura[1], Katsumoto Momikura[1], Ryuma Yoshioka[2],
Makoto Nishigaki[3], Kazuki Mori[4], Noriharu Miyake[5] &
Mamoru Saito[1]

[1]Department of Earth Sciences, Chiba University, Yayoi-Cho, Inage-ku, Chiba, Japan
[2]Toyama Prefectural University, Kurokawa, Kosugi-cho, Imizu, Toyama, Japan
[3]Okayama Univ, Okayama, Japan
[4]Nihon Univ, Dept Geosyst Sci, Setagaya Ku, Tokyo, Japan
[5]Shimizu Corporation, Koto Ku, Tokyo, Japan

INTRODUCTION

The history of research and study on groundwater resource refers to that focused on its development and conservation. The Japanese population was 30 million around 100 years ago; however it has increased up to 120 million at present. During this period, progress in groundwater study and research has reflected the variation in the demand for groundwater.

People have taken good care of groundwater from a long time ago, because it provides stable water resources. The use of the springs as outcrops of groundwater dates back to the Jomon Era (10 000 B.C–500 B.C.). Digging techniques for Manbo (canat), which have a horizontal well to make an artificial outcrop of groundwater, and the construction of shallow wells have the long history dating back to the Yayoi Era (500 B.C.–300 A.D). Thereafter, facilities for water abstraction have been improved and deep and collecting wells have been widely used following the progress of the development of groundwater exploration and deep well construction. Finally, the development of pumping techniques has resulted in the increase of groundwater demand, overexploitation and occurrence of groundwater problems.

The history of groundwater development and conservation in Japan during the last 100 years shows the history of its development for the first 60 years and that of seeking sustainable development for the last 40 years. This paper describes the developing stages for the first half as follows; 1) Dawn of groundwater development, 2) Modernization of groundwater intake facilities and beginning of well hydraulics, 3) Japanese dark ages and fledgling for next era, and 4) Huge groundwater development after World War II. And then, we describe the groundwater problems for the last half as follows; 1) Groundwater problem due to its overexploitation, 2) Two laws on groundwater and its regulation of local government, 3) Development of subsurface dam and 4) Groundwater pollution.

GROUNDWATER DEVELOPMENTIN JAPAN

Dawn of groundwater development

When the modern western culture arrived in Japan at the end of the Edo Era (1867 A.D.), the facilities for groundwater abstraction were springs in the cliffs and the toes

Figure 1   Example of the Kazusa system of well boring in Ichihara, Chiba.

of alluvial fans; shallow dug wells in the lowland; maimaizu wells, whose shape is similar to a snail shell or cone shaped hollow on the diluvial upland; manbo (canat), which has horizontal wells and collecting ditches in the alluvial fan; and artesian wells drilled by the Kazusa System (Shibasaki *et al.*, 1995).

The Kazusa System is a device with an iron tube and a chisel attached to the edge of bamboo splinters utilizing bamboo's elasticity. It was innovative in the sense that it used muddy water and clay to prevent the well from collapsing. This technique was used not only for water wells, but also for agriculture, petroleum, natural gas, coal and hot springs, before the import of the modern machine drilling system from the western countries. Hishida (1955) states that this technique was used to reach a depth of 500 m in 1883 and the method was published as a *"Note and Report on the Kazusa System of Deep Boring for Water (2nd edition)"* written by F. J. Norman in Calcutta and Simla Thachenk co., 1902. After its completion, it spread across Japan and transferred to north China and Manchuria. The principle of this technique developed in Japan is the same as that of the power-driven percussion method (Figure 1).

A major Challenge at that time (late 19th century) were the increased groundwater resource exploitation due to the increasing population, automatic pumping and development of techniques for drilling to deeper depths or of wider diameter to increase the inflow. Under these circumstances, Naumann (1883) produced the first hydrogeological report in Japan on *"Planning for improvement of well water in Sakai City, Osaka prefecture"* in the first Annual Report of the Geological Survey of Japan, in which he suggested a method to prevent sea water intrusion. Suzuki (1889) produced the first Japanese hydrogeology paper on the *"Hydrogeology in Tokyo City"*, a paper which was recently introduced by Nagase (2007).

Since 1910, there were many reports published by the Geological Survey of Japan on the hydrogeology of various regions in Japan; however, most of those were descriptions of the facts and far from responding to social needs.

In 1913, a well drilling company imported the first rotary boring machine from the USA and drilled a well for the ice making industry in Shinjyuku-ward, Tokyo, which was 12.5 in. ($\sim$317 mm) in diameter, 158 m in depth, had a flowing rate of 54.5 m^3/day and a pumping rate of 5400 m^3/day (Nissaku, 1981). As these new systems withdrew large amounts of deep artesian groundwater which had low circulation rates compared with shallow unconfined groundwater, subsidence would be subsequently caused.

MODERNISATION OF GROUNDWATER INTAKE FACILITIES AND BEGINNING OF WELL HYDRAULICS

After the First World War, Japan underwent a period of modernisation and industrialisation and population increase. As a result, water demand increased. Because these new industries had no surface-water rights, they required access to groundwater resources, so demand for drilling wells increased remarkably. During the latter half of this period, political measures associated with the economic growth of the war-oriented industrial sectors before World War II accelerated this trend.

Research and training for groundwater in the agricultural area in the Ministry of Agriculture and Commerce, produced good results if researcher and trainee determined the boring site from geological survey and information of existing wells. A report on "Kouchi (Farming Land)" produced by the Ministry of Agriculture and Commerce at the time had articles on: estimation of geological structure from regional reconnaissance survey and land survey; determinations of well location and depth; test boring; well maintenance and warnings against excess well drilling. Improvement in well drilling techniques made further progress, in which methods to produce shallow large diameter wells with pumping rates of 5000–10 000 m^3/day were developed.

The success of the first modern rotary boring in Shinjyuku-ward (Tokyo) acted as a trigger, with deep wells constructed in many industrial areas around Metropolitan Tokyo, Yokohama, Nagoya and Osaka. Increases in the number of deep wells subsequently caused the lowering of hydraulic heads and subsidence.

With the increase in the number of groundwater abstractions and records on groundwater surveys and withdrawals, an increasing number of papers and books were published, including many treatises on groundwater. From 1931 to 1933, Iwanami, a famous book publisher in Japan, published the "Iwanami Courses of Lectures", in which Suzuki (1931), Fukutomi (1933) and Abe (1933) contributed the "Groundwater Compendium", "Groundwater", "Hydrology" respectively. These were the first books on water in Japan. Subsequently, Kimijima (1934), Sakai (1941), and Yoshimura (1942) published treatises on groundwater research.

As for hydrogeological study, Kanbara (1929) produced a book titled "Geology and Hydrology in Mt. Fuji" in which the special case of groundwater in the volcanic region was described, and the abundant water in the base of Mount Fuji identified.

Since the Association of Japanese Geographers was born from the Division of Geography within the Geological Society of Japan, groundwater studies and research started to be carried out systematically in the Geography departments of many universities. This meant that the study of groundwater, developed in geography departments, gained independence from that of geology. Topographical detailed studies by Imamura (1938) and Yoshimura (1938) contributed to the progress of groundwater studies in that time.

Within commercial well drilling, the custom by which the drilling worker had the responsibility for the failure of a new well was unchanged. Therefore, to provide the understanding of the well dimensions and pump design needed to of get the required abstraction amounts, many papers on well hydraulics were published in the fields of geophysics, civil and agricultural engineering (e.g. Yoshida, 1928 and 1931; Tamachi, 1931 and 1938; Monobe, 1935; Homma, 1935). At this time, Nomitsu and Yamashita (1943) derived their hydraulical theory independently to that of the non-equilibrium formula presented by Theis (1935).

Increases in the numbers of deep wells and the severe withdrawal of deep groundwater caused huge drawdowns in groundwater heads and consequent subsidence in the lowlands of Tokyo and Osaka which had overlying thick clay layers. In the Tokyo lowland, the subsidence was recorded as a noticeable phenomena during the Taisyo Era (1912–1926).

Figure 2 shows changes in groundwater head and land subsidence in the subsidence areas of Tokyo and Osaka. The Tokyo data are based on changes in groundwater head recorded in the 380 m deep observation well of the Earthquake Research Institute, University of Tokyo. From this record, it is evident that groundwater heads declined by around 10 m from the beginning of the observation period in 1933 to 1943 (during World War II), then rose by up to 5 m, before resuming a declining trend from 1951. The cumulative land subsidence in the lowland, Koto-ward, Tokyo is also included in Figure 2. From the beginning of the observation period until 1973, land subsidence occurred continuously. Its rate was greatest from the 1920s to 1930s and from the 1950s to 1960s when the construction of deep wells was increasing, and it reduced during the periods of decreasing pumping.

Recently, underground railroad stations have been constructed deeper than 50 m below the surface in the metropolitan city of Tokyo. On the other hand, regulation of groundwater pumping has been enforced, following the land subsidence, since the 1970s. The recovery of deep groundwater heads is remarkable (Figure 2). As a result, it has been necessary to prevent the rebound of groundwater heads from damaging the underground constructions due to the rebound effect (Sakura et al., 2003).

In contrast, in Osaka, land subsidence was identified as a trigger of the Tango Earthquake in 1927 and a contributory factor to the devastating coastal flooding caused by the Muroto typhoon in 1934 (Shibazaki et al., 1995). The cumulative land subsidence in the lowland of Osaka is also shown in Figure 2. Shibasaki et al. (1995) describe the history of land subsidence in Osaka as follows. Hirono and Wadachi (1939) of the Research Institute for Disaster Prevention constructed a land subsidence observation well in cooperation with the Public Works Department of Osaka City. This observation well, equipped with the double tube system, could precisely measure the subsidence and groundwater head simultaneously. In the following year, an observation well equipped with a hydraulic head self-recorder was installed in Osaka. Wadachi and Hirono (1940) carried out research from 1939 to 1942 based on the records from such observation wells. They found an obvious correlation between the changes in hydraulic

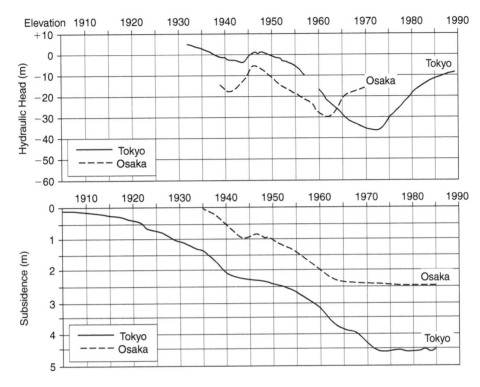

Figure 2 Changes in hydraulic head and land subsidence in Tokyo and Osaka (modified from Momikura and Ono, 1995).

heads and the rate of land subsidence. They concluded that land subsidence is caused by compaction of soft clay layers following the decline of hydraulic heads, and that groundwater pumping was the most important cause of land subsidence. This theory was broadly verified from the fact that land subsidence ceased when the use of groundwater decreased after the destruction of factories during World War II. However, these understandings did not inform the peaceful era immediately afterwards, as land subsidence again became a serious problem when industry began to recover (Hirono, 1953).

The Japanese dark age and the foundatios for the next era

During the period in which Japan had headed toward militarism, it had urged its young hydrogeologists to go to other Asian countries to assist in the development of their water resources. The scientific and technological understandings and experiences which had been accumulated until that time were, therefore, excised in the Japanese colony and occupied areas at that time. As a result, there were no significant groundwater studies within the country at this time. In these bad circumstances, local geographical studies on groundwater in the Kanto plain (Yoshimura, 1938, 1942) and an agricultural engineering study on groundwater (Kachi, 1943) marked the beginning of the next period of hydrogeology in Japan. Furthermore, it was notable that Kachi's idea marked the start of the development of subsurface dams which are constructed rapidly and energetically now in Japan.

Huge groundwater development after World War II

The end of World War II and the confusion due to its defeat had taken a lot out of the Japanese people. M. L. Brashears, a special advisor on groundwater in the Bureau of Natural Resources within the General Headquarters (GHQ) of Allied Forces, pointed out that the Japanese groundwater community needed the establishment of new groundwater theories and the propagation of research methods. The advice by GHQ described as follows; Japanese students in science and technology courses found few opportunities and incentives to take systematic educations and experiences. Most Japanese Universities taught geology, geophysics, hydraulics, hydrology, meteorology, civil engineering and so on, but they did not provide the education in which these disciplines could connect to hydrogeology directly. Therefore, there were less than 10 scientists and researchers in Japan who had confidence in hydrogeology. Nobody knew the progress of groundwater hydraulics during the previous 15 years (Brashears, 1951; Yamamoto, 1953).

Japan at that time was urged to increase its food self-sufficiency and to provide opportunities for demobilised soldiers from overseas. As part of these actions, urgent irrigation projects were started from 1946. In the Ministry of Agriculture, the groundwater research group was organised with the aim of selecting appropriate reclaim lands and securing water resources for drinking water. This group, consisting of hydrogeologists, geophysical prospectors and groundwater engineers, carried out groundwater surveys throughout the nation and took a position of leadership in the Japanese community of groundwater research.

This group was developing the research agency in association with not only exploration of water resources, but also the construction designs of dams and tunnels related to disaster preventions. Those studies and research results were published annually from the local bureaus of nine regions in Japan and the Institute of Agricultural Engineering. This group has advanced the cultivation of human resources contributing to the further development of hydrogeology in Japan. In that group, S. Yamamoto promoted the strengthening of its organisation in the Ministry of Agriculture, went abroad to study with the US Geological survey and provided a breath of fresh air to the theory of groundwater analysis at that time. N. Kurata and T. Murashita moved to the Geological Survey of Japan, within the Ministry of Commerce and Industry, from the Ministry of Agriculture in 1951 and they organized the research group on industrial water to study groundwater resources as an industrial water supply throughout the nation. These results were published as reports of regional research on industrial water supply and as hydrogeological maps.

In the coastal industrial regions in the 1950s, economic activities were increased by rebuilding following the wartime damage and also by special procurement associated with the Korean War. The pumping of industrial water was continuous and its amount reached several thousands m^3 per well using the borehole-pump or turbine-pump. Such excess pumping resulted in sea water intrusion and land subsidence. N. Kurata and colleagues, who watched the miserable state of groundwater problems, intended to plan and establish the law of industrial water supply through their groundwater research.

In the 1960s, the Japanese economy entered the period of rapid growth, including the expansion of urban areas and the urbanisation of rural areas due to the introduction of industry. As a result, water demand increased all over the country and most of new users, who were without water rights, required groundwater as their water resources. As withdrawal of groundwater increased with accelerating speed, groundwater problems such as sea water intrusion and land subsidence occurred frequently.

In addition to the increase in water demand, no less important was the fact that submersible motor-pumps imported from West Germany became widely used. In the 1950s, borehole pumps which dominated the market could not be used with water levels below a depth of 25 m, but submersible motor-pumps could be used even when the water level was below a depth of 150 m. It meant that we could withdraw huge amount of groundwater over a much larger area.

The appropriate assignment on the points of weakness within the Japanese hydrogeological community by M. L. Brashears gave an impulse to many hydrogeologists, and acted as a blasting fuse to advance hydrogeology widely by the bureaucracy-led system.

The development of water resources in the dry areas, such as the volcanic mountain foot, and the improvement of wells for irrigation in connection with land improvement projects were carried out in the Ministry of Agriculture. In contrast, exploration for natural gas reservoirs and groundwater research in the coastal industrial regions were also carried out in the Ministry of Commerce and Industry. These activities produced many papers and reports. In this era, the Japanese Association of Groundwater Hydrology and the Japan Society of Engineering geology were established in 1959 and 1960, respectively, and activities for groundwater research and study were accelerated.

The "*Groundwater Survey*" book was written by Yamamoto (1953) and "*Hydrogeology*" was by Kurata (1955). "*Agricultural Hydrology*" (Kaneko, 1957), "*Key Theory of Groundwater*" (Murashita, 1962, "*Electric Logging Technology in Water Wells*" (Yamaguchi, 1963), "*Groundwater Survey Methods Using Radio-isotopes*" (Ochiai, 1965) and "*Groundwater Study*" (Sakai, 1965) were published one after another. It was a good time for Japanese Hydrogeology. Above all, "*Groundwater Survey*" (Yamamoto, 1953) which was cleaved through empirical facts and rationality provided hydrogeological students with a textbook and manual.

THE TRIAL AND ERROR APPROACH TO CONSERVATION OF GROUNDWATER RESOURCES

Groundwater problems due to its overexploitation

A large amount of groundwater withdrawal had caused decreases in groundwater hydraulic heads, cessation of flowing wells, sea water intrusion, groundwater pollution and land subsidence. These problems advanced with the social problems in the 1960s. The theory by Wadachi and Hirono in which land subsidence was caused by excessive groundwater pumping was verified by the empirical evidence of WW2; however, the experience of over pumping, incidence of land subsidence, control of pumping and recovery of groundwater heads were repeated in various regions, because the academic society and industrial community could not understand the theory sufficiently.

These circumstances can be seen in the changes since 1955 (Figure 2). Hydraulic heads which had been decreasing annually by 1 m shifted to increasing rapidly since 1962 in Osaka, and increasing gradually since 1965 and rapidly since 1973 in Tokyo, respectively. After the hydraulic heads had started increasing, land subsidence soon ceased, but the compaction of the soft clay layer was irreversible and the amount of compaction had reached 4.5 m in Tokyo, 3.4 m in Osaka, 1.5 m around Nagoya and 1 m in Shiroishi, Saga Prefecture respectively.

Two laws on groundwater and its regulation of local government

The change to rising hydraulic heads and reducing subsidence in the coastal and industrial regions in Tokyo and Osaka had appeared after two laws for the water use by industry in 1956 and building in 1962, and groundwater regulations of local government were established successively. For the law for industrial water use, government intended to subsidise the introduction of industrial surface water supplies so that industries would not need to use deep groundwater any more. As for the law of building water use, the government also intended to control borehole diameter and the depth by introducing a notification or licencing system and regulating the pumping amount. As the smaller diameter and the deeper boreholes were outside of the control subjects of this law, it was less than perfect.

In Osaka, the mechanism of land subsidence had been better understood since the 1930s and industrial water supplies of surface water were introduced in the 1950s. After the flood damage of the 2nd Muroto Typhoon had occurred in lowlands with an elevation of 0 m (sea level) in 1961, Osaka City legislated to stop all pumping wells shallower than 600 m below sea level, which resulted in the successful recovery of hydraulic heads and the cessation of land subsidence. After the cessation of land subsidence in the coastal region of Osaka, the city expanded toward its suburbs due to the urbanisation and industrialisation on the fringe of Osaka.

In Tokyo, the damage caused by subsidence, such as the lifting phenomena of wells and buildings, uneven settlements of bridges and subsurface constructions, and flooding by high tides were almost similar to those in Osaka. Because of the continuing land subsidence, the Metropolis of Tokyo had no choice but to change its plans of countermeasure projects against high tides. Unfortunately, the provision of surface water industrial water supplies was delayed by more than 10 years compared to Osaka. After the regulation of groundwater withdrawal, following the introduction of the Law of water use for industry in 1963 and that for building in 1963, all water use for buildings had changed from groundwater to surface water. Since 1965, the hydraulic heads in lowland of Tokyo started to increase; however, we had to wait until 1972 when the rising hydraulic heads and reducing rates of subsidence had appeared clearly following the expansion of the area in which groundwater withdrawal was regulated by local government.

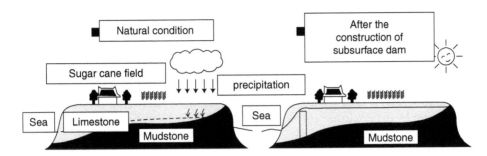

Figure 3 Concept of the subsurface dam.

Development of subsurface dams

Kachi (1943) had a plan for the construction of a subsurface dam (Figure 3) of a moderate size for irrigation at the alluvial fan in Nasunogahara, Tochigi Prefecture. His plan was not realised, but was considered to be attractive. Kawasaki *et al.* (1993), Nagata *et al.* (1993) and Sakura *et al.* (2003) reported the geotechnical development of the subsurface dam project in Japan.

The Ministry of Agriculture, Forestry and Fisheries (MAFF) of the Government of Japan has conducted the subsurface dam development programme on the islands to develop groundwater resources from 1974 to today. Subsurface dam can dam the groundwater flow in an aquifer and retain groundwater that would have flowed into the sea. An irrigation project with subsurface dams of a considerable size is now under implementation, based on the know-how obtained through the programme. The agriculture on the Ryukyu Islands in the south-west of Japan is deeply dependent on unstable rainfall. Though the mean rainfall on the Ryukyu Islands is 2000 mm/yr, it is very unstable year by year, varying from 1000 to 3000 mm/yr. The annual amount of rainfall on the islands depends on the passage of typhoons. When no typhoons pass through the islands, a severe drought may hit the area. The geology of the majority of the islands consists of an elevated coral reef limestone, known as the Ryukyu Limestone, which is highly pervious. The majority of the rainfall infiltrates the ground due to the high permeability of the limestone. The fluvial system on the islands is, therefore, less developed and characterised by small catchments and small rivers. The groundwater discharges into the sea without being used through the permeable Ryukyu Limestone and seawater intrudes into the coastal aquifer of Ryukyu Limestone due to excessive pumping up of the groundwater.

The subsurface dams under construction on the Ryukyu Islands have two main purposes. The first is to dam up and store groundwater, which would quickly discharge into the sea, by constructing cut-off walls, and to use it effectively for agricultural purposes. The second is to prevent saltwater intrusion into fresh reserved water near the seacoast and to separate saltwater and inland groundwater by cut-off walls. As an example, a schematic diagram of a saltwater cut-off type subsurface dam is shown in Figure 3.

Groundwater pollution

There are many kinds of groundwater pollutants such as heavy metals, chlorinated solvents, nitrate, radioactive substances and pesticides. Water Quality Bureau (2000) describes groundwater pollution in Japan. At present, about 30% of water for the urban activity comes from groundwater. Recently, groundwater pollution by trichloroethylene, tetrachloroethylene and other pollutants was revealed in accordance with the progress of urbanisation and industrialisation. According to the surveys of groundwater by the Environment Agency in 1993, groundwater pollution was detected in 1151 areas. As a result of the monitoring by prefecture governments, we know that groundwater pollution has been increasing every year.

General groundwater surveys were conducted in 1498 municipalities in the fiscal year of 1994. These surveys found groundwater contamination by trichloroethylene (11 out of 3996 wells in excess of assessment standards) and other pollutants.

The surveys conducted in the 1980s found widespread groundwater contamination in Japan. Based on this survey, the Environment Agency amended the Water Pollution Control Law in June 1988 stipulating the prohibition on infiltrating discharges containing toxic substances and the monitoring of groundwater by prefecture governments, which is subsidised by the Environment Agency. After this amendment, the remediation technology for groundwater reached a practical use level and although groundwater contamination was still discovered in many areas, the importance of the remediation of polluted groundwater based on a legal system had been pointed out.

In February 1996, the Central Environment Council submitted a report concerning purification measures in order to prevent pollution groundwater. Based on this report, the Environment Agency amended the Water Pollution Control Law in May 1996. The amended law, which is enforced since the fiscal year of 1997, stipulates that the prefecture governor can order that the polluter to remediate contaminated groundwater.

The Environment Agency established the Environmental Quality Standards (EQS) for groundwater in March 1997, aiming at further promotion of the comprehensive conservation of the groundwater environment. These Environmental Quality Standards are applied to all groundwater, with the same EQS values for 23 substances established for protecting human health and public water resources. It was established to "make every effort to be attained and maintained immediately" because it is related to human health. From now on, conservative or precautionary administration of groundwater is conducted with the aim of attaining and maintaining this environmental standard.

Groundwater pollution was and is caused by the lack of attention for the surrounding environment or a lack of awareness. Therefore, the steady efforts by wide disciplinary specialists on groundwater will be a good way to solve such problem and many hydrogeologists are also required to participate.

CONCLUDING REMARKS

Groundwater is the first underground resource developed by man. It was carefully protected and fostered since early times. It has the following characteristics; 1) Cool in summer and warm in winter, 2) Good quality, 3) Steady quantity, 4) Free use without water rights, and 5) Cheap water resources collected by simple equipment.

New industries in Japan which had no water rights could not develop without the use of groundwater. Groundwater had contributed not only to engineering industries but also the modernisation of all industries.

Asking for efficiency, convenience and amenity in association with industrialisation and modernisation resulted in a lack of attention being paid for the natural environment. As for the use of groundwater, these activities led to excessive pumping, abnormal decreases in hydraulic head and land subsidence. Land subsidence in Japan has a long history. Terada, a famous scientist and essayist, focused on it in 1915. Miyabe carried out measurements of compaction of surface layers in the 1930s. Wadachi and Hirono carried out research from 1939 to 1942 based on the records from such observation wells (Shibasaki *et al.*, 1995). They found an obvious correlation between the change in hydraulic heads and the rate of land subsidence. From this evidence, they concluded that land subsidence was caused by the compaction of the soft clay layer following the decline of hydraulic heads, and that the pumping of groundwater was the most important cause of land subsidence. Regardless of such verification, we had the same experience in various regions due to the original sin driven by modernisation and industrialisation.

If the safe critical water level or safe yield were determined, paying attention to the environment and socio-economical investigations, and if we could withdraw groundwater within the safe yield, we might receive the full benefit of groundwater sustainably. Most of the regions within Japan chose the easy way in which water resource was converted to surface water so that land subsidence occurred due to the withdrawal of groundwater. It led to a lack of respect for groundwater and fostered groundwater pollution. In order to break out of this cycle, the continuation of groundwater use is assumed to be the first step for conservation of groundwater and protection of groundwater pollution.

Finally, we would like to review the use and conservation of groundwater to aim for the sustainable co-existence of harmonising regions with the environmental ecosystems.

This short history of hydrogeology in Japan is based on Momikura and Ono (1993), Kurata (1955), Shibasaki *et al.* (1995) and Yamamoto (1995). We would like to express our gratitude to the achievements of our predecessors on hydrogeology.

REFERENCES

Abe, S. (1933) *Hydrology*. Iwanami Course of Lecture on Geology, 70p. (in Japanese).

Brashears, M.L. (1951) *Ground-water Situation in Japan*. Natural Resources Section Prelim. Report. Rep. No. 51.

Fukutomi, T. (1933) *Groundwater*. Iwanami Course of Lecture on Geology, 88p. (in Japanese).

Hirono, T. & Wadachi, K. (1939) On the land subsidence in western Osaka area 1. Report of Disaster Science Institute, No. 2, pp. 344–407 (in Japanese).

Hirono, T. (1953) Land subsidence due to pumping of ground water. *Journal of Geography*, 62, 143–159 (in Japanese).

Hishida, T. (1960) Idea of Kazusa system. *Information for Cultural Resources in Chiba Prefecture*, pp. 2–16 (in Japanese).

Homma, M. (1935) New method on the theory of groundwater flow.

Imamura, G. (1938) Measurements on groundwater and its application, *Kagaku (Science)*, 8 (2), 50–51 (in Japanese).

Kachi, K. (1943) *Enhancement of Groundwater, Irrigation and Drainage*. Chijin Shokan, 224p. (in Japanese).

Kambara, S. (1929) *Geology and Hydrology of Mt. Fuji*. Hakushin Kan, 482p. (in Japanese).

Kaneko, R. (1955) *Agricultural Hydrology*. Doboku Zasshisya, 295p. (in Japanese).

Kawasaki, S., Sugawara, T., Miyakita, J., Komatsu, M., Nagata, J., Nagata, S., Enami, N., Nishijima, T. & Azuma, K. (1993) Geotechnical development of subsurface dam project in Japan. In: Sakura, Y. (ed.) *Selected Papers in Environmental Hydrogeology*. Verlang Heinz Heise. pp. 215–228.

Kimijima, H. (1934) *Groundwater*. Maruzen. 434p. (in Japanese).

Kurata, N. (1955) *Hydrogeology*. Asakura Shoten, 308p. (in Japanese).

Momikura, K. & Ono, K. (1993) Underground water. In: Editorial Committee, Geological Society of Japan (ed.) *Hundred Years of Geology in Japan*, 419–430. (in Japanese).

Monobe, N. (1935) *Hydraulics*. Iwanami Syoten. 351p.

Murashita, T. (1962) *Key Theory to Groundwater*. Syoukoudou, 155p. (in Japanese).

Nagase, K. (2007) The first paper on hydrogeology in Japan, Toshi Suzuki "Hydrogeology in Tokyo City" Journal of Geography Vol. 1, No. 3, 1889. *Journal of Geography*, 116, 721–724 (in Japanese).

Nagata, S., Enami, N., Nagata, J. & Katoh, T. (1993) Design and construction of cutoff walls for subsurface dam on Amami and Ryukyu Islands in the most southwestern part of Japan. In: Sakura, Y. (ed.) *Selected Papers in Environmental Hydrogeology*. Verlang Heinz Heise. pp. 229–245.

Nauman, E. (1883) *Planning for improvement of well water in Sakai, Osaka*. Annual Report of Geological Survey, Japan. 1, Geology Part. pp. 3–22 (in Japanese).

Nissaku Cooperation, Editing Committee of Company History, 1981. *Company History for 70 Years*. Nissaku Cooperation. 200p. (in Japanese).

Nomitsu, T. & Yamashita, K. (1943) An advance in the theory of wells. (The 2nd report.) Changes in hydraulic heads and coefficient of elasticity following the beginning and the cessation of pumping in a horizontal well. *Geophysics*, 7, 21–40 (in Japanese).

Ochiai, T. (1965) *Groundwater Survey Methods Using Radio-isotopes*. Shokodo. 234p. (in Japanese).

Sakai, G. (1941) *Survey Method on Groundwater*. Kokon Shoin. 241p. (in Japanese).

Sakai, G. (1965) *Groundwater Study*. Asakura Shoten, 418p. (in Japanese).

Sakura, Y., Tang, C., Yoshioka, R. & Ihibashi, H. (2003) Intensive use of groundwater in some areas of China and Japan. In: Liamas, R. & Custodio, E. (eds.) *Intensive Use of Groundwater*. A.A. Balkema Publishers. pp. 337–353.

Shibasaki, T. & Research Group for Water Balance (eds.) (1995) *Environmental Management of Groundwater Basins*. Tokai University Press. 202p.

Suzuki, T. (1889) Hydrogeology in Tokyo City. *Journal of Geography*, 1, 77–83 (in Japanese).

Suzuki, M. (1931) *Compendium on Groundwater*. Iwanami Course of Lecture on Geography, 78p. (in Japanese).

Wadachi, K. (1940) *On the land subsidence in western Osaka area 2*. Report of Disaster Science Institute, No. 3. pp. 454–488 (in Japanese).

Yamaguchi, H. (1952) *Electric Logging Technology in Water Wells*. Shokodo, 254p. (in Japanese).

Yamamoto, S. (1953) *Survey Method of Groundwater*. Kokon Shoin, 240p. (in Japanese).

Yamamoto, S. (1955) 50 years after World War II and groundwater. *Chikasui Kijyutsu (Groundwater Technology)*, 37, 4–11 (in Japanese).

Yoshimura, S. (1942) *Groundwater*. Kawade Science Pocket Edition, 258p. (in Japanese).

Yoshimura, S. (1938) Abnormal rise of water table due to heavy rainfall in Musashino Plateau. *Kagaku (Science)*, 8 (10), 399–400 (in Japanese).

Water Quality Bureau. (2000) *Water Environment Management in Japan*. Japan, Water Quality Bureau, Environmental Agency, 41p.

Yoshida, Y. (1928) New study on well radius of influence.

Yoshida, Y. (1931) Use of groundwater as a resource, No. 1 Groundwater flow in the unconsolidated rock.

Tamachi, M. (1931) Relation between soil and water, in particular, on capillary phenomena and infiltration.

Tamachi, M. (1938) Theory and experiment of discharge from the drilling well.

The development of groundwater hydrology in The Netherlands between the mid-19th century and the late-20th century

Jacobus J. de Vries

Emeritus Professor of Hydrology and Hydrogeology, Vrije Universiteit Amsterdam, The Netherlands

INTRODUCTION

High water tables and brackish and polluted groundwater are among the specific problems that people normally face in coastal lowlands. Not surprisingly, the scientific development of groundwater hydrology in the Netherlands has always been closely associated with the needs to cope with these specific problems.

The Netherlands is situated at the mouth of the rivers Rhine and Meuse. It has a moderate maritime climate which is characterised by an average annual temperature of 9.5°C, a mean annual precipitation of about 800 mm and an average evapotranspiration in the order of 500 mm. The subsurface is built up by a large Pleistocene fluvial fan that dips in a coastward direction beneath Holocene deltaic and marine tidal deposits, including peat layers. The Pleistocene sediments consist predominantly of medium grained sand with a thickness, ranging from zero near the eastern border to more than 200 m near the coast. This large aquifer is covered by Holocene confining layers of clay and peat, which reaches an average thickness of 20 m near the coast. Most of the area is flat and characterised by shallow groundwater tables, which range from less than 1 m during the winter period to less than 2 m in summer. An exception are a few ice-pushed ridges in the central part of the country, and the hilly area in the extreme south-east.

A major part of the Netherlands can be considered as man-made in the sense that a substantial component of the country has been made inhabitable by reclamation of marshland, lakes, lagoons and estuarine areas. As a consequence, water management has always been an integral part of life in this coastal lowland. Drainage of marshland, protection against floods, disposal of surplus water and the struggle against encroaching seawater have been a continuous cause of care and attention for more than 1000 years. In fact, the problems have aggravated in the course of time due to land subsidence by drainage and the creation of low-level areas by reclamation of lakes and sea embayments. More than 25% of the country now lies below sea level and 65% of the land surface would be flooded if not protected by coastal dunes and dikes (Figure 1).

Concern for water and the study of its behaviour were for a long time focused on surface water; subterranean water with its remarkable appearance in springs and

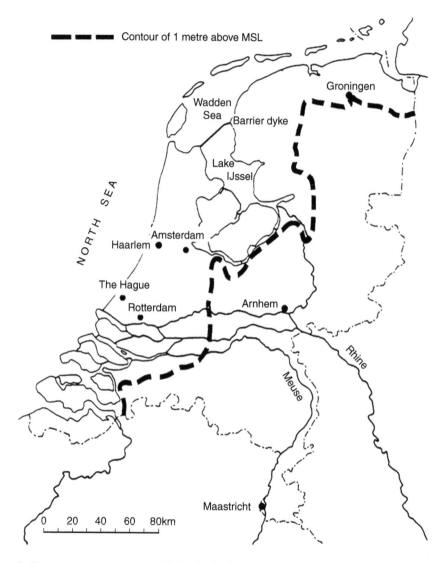

Figure 1   Physiographic outline of the Netherlands; the 1 m elevation contour forms the approximate
boundary between the sandy Pleistocene and the Holocene clayey and peat deposits of the
coastal plain. Several polders in the latter area are situated more than 6 m below mean sea
level. Lake IJssel is the reclaimed former Zuiderzee, an embayment of the North Sea.

free flowing artesian wells remained a mysterious phenomenon of obscure origin and
destiny and the domain for water diviners. Scientific interest in groundwater emerged
when problems with the supply of clean drinking water became manifest in the early
19th century with the growth of the cities and the outbreak of waterborne epidemic
diseases, notably cholera and typhoid. Water in the canals and shallow wells used to
be contaminated because of a lack of adequate sewage disposal, whereas deep wells
often produced brackish water.

Successful drillings for artesian water in France and England in the first half of the 19th century – featuring the famous 548 m deep artesian well at Grenelle (Paris) in 1841 – drew the attention of scientists as well as the government. One wondered if fresh artesian water could also be a solution for public and industrial water supply in the Netherlands. Several attempts were subsequently made in the period between 1830 and 1850 to reach this intriguing resource – without success. Illustrative is the drilling at the Nieuwmarkt in Amsterdam: This work started in 1836 and one finally decided to give up in 1843 after reaching a depth of 172 m, without obtaining any indication of artesian pressure. Although not successful for water supply, these well-sampled boreholes provided important information on the subsurface – notably through the able lithological and paleontological description of the samples by Pieter Harting (*cf.* Harting, 1852).

Theoretical understanding of groundwater movement remained very poor until the mid-19th century and it would subsequently take more than 50 years to develop a clear idea of the origin of groundwater, and to formulate and apply the basic laws of groundwater flow to a proper understanding of actual field conditions. Notably the exploration for clean drinking water in the coastal dune belt, with its limited resources, stimulated the development of conceptual ideas on the origin and dynamics of groundwater. These conceptions eventually merged with the main stream of physics into a comprehensive theory at the beginning of the 20th century. Subsequently groundwater hydrology in the Netherlands developed in close connection with the solution of groundwater flow problems pertaining to groundwater extraction for drinking water supply, large reclamation and infrastructure works, and water management for agriculture. This included leakage of brackish water into deep polders, seepage through dikes, well flow in leaky aquifers, the dynamics of the fresh-salt water interface and groundwater table control by land drainage.

A more elaborated and complete overview of the evolution of scientific groundwater hydrology in the Netherlands is given by De Vries, 1982 (in Dutch). For the technical, socio-economic and political context of the history of water management and land reclamation in the Netherlands, the reader is referred to Van de Ven (ed.), 2003. Dufour (2000) presented an overview of the groundwater situation in the Netherlands.

EARLY CONCEPTS AND DEVELOPMENTS

Drainage and the origin of groundwater

Although land drainage by ditches had been practised for hundreds of years, a clear perception of the actual flow processes involved did not emerge before the end of the 19th century. In 1857 a committee of the Royal Agriculture Society published a discussion on suggestions for the proper application of tile drainage based on the "rules of science" (*cf.* De Zeeuw, 1966). Evaluation by the chairman, the noted soil scientist and geologist W.C.H. Staring, gives a good insight to the prevailing "rules of science" at that time. A proposal for a drain depth of between 130 and 150 cm was rejected with the argument that in such a case the drain would lie permanently below the groundwater table and, therefore, would hardly participate in the drainage process. Evidently, the general perception was that infiltrating rainwater would move in vertical direction through the unsaturated zone until it reached the area of influence of the drain. Substantial participation in the drainage process by the saturated zone below the groundwater table was obviously not considered. The Committee therefore – not surprisingly – categorically refutes the recommendation by some experts (notably from Great Britain) to increase drain spacing with an increase in drain depth.

The doctrine that the volume below the groundwater table would only play a minor role in groundwater circulation was also manifest in discussions about recharge of the saturated zone. The idea prevailed that at least part of the groundwater would originate from condensation of water vapour. This hypothesis was undoubtedly inspired by ideas of philosophers of antiquity, who based their opinion on the existence of wells and springs in arid areas without evidence of infiltration and recharge by rainfall. The belief in a substantial contribution to the water balance by condensation was supported in the second half of the 19th century by observations on evaporation by way of inadequate evaporation pans and lysimeters. Over-estimation of the evapotranspiration component on several experimental sites suggested an excess of annual evapotranspiration over annual rainfall. The noted meteorologist C.H.D. Buys Ballot (1879) therefore assumed that a substantial part of the water consumption by vegetation must originate from condensation.

The German geologist Otto Volger (1877), who categorically denied any relation between rainfall and groundwater, took an extreme view in this respect. He even argued that concern for contamination of groundwater was a fear for phantoms, which would lead to unnecessary costs with the expensive construction of extraction wells for public water supply outside urban areas. An opposing idea was proposed in the Netherlands by the geologist-paleontologist T.C. Winkler, who distrusted the self-purifying capacity of the soil. He calculated the accumulated numbers of buried dead bodies since earth's creation and concluded that groundwater could be nothing more than diluted dissection poison. He therefore propagated the use of water from the coastal dunes, which he rightly considered as originating from local rain-water that had infiltrated in pure and uncontaminated sand (Winkler, 1867).

Percolation experiments and Darcy's Law

The first systematic investigations that were carried out in the Netherlands to gain insight into the behaviour of groundwater flow, were performed by the physician, geologist and professor of natural history, P. Harting. Harting conducted percolation experiments on soil samples from Amsterdam's subsurface, in the framework of his geological studies of this area, and deduced (Harting, 1852; Figure 2):

> "It can be concluded without a large failure, that the quantity of water percolating through a clay layer in a time unit increases or decreases at the same rate with increase or decrease of the length of the water column resting upon this layer".

Harting was evidently aware of the basic principles that would lead to Darcy's Law (1856), but since he was working with an outflow under atmospheric pressure, he wrongly neglected an elevation head term. Subsequent experiments with sieve fractions led Harting to the relation between pore diameter and permeability and he recognised the analogy between groundwater percolation through a porous medium and the flow through a capillary tube, thus satisfying the Hagen-Poisseuille Law.

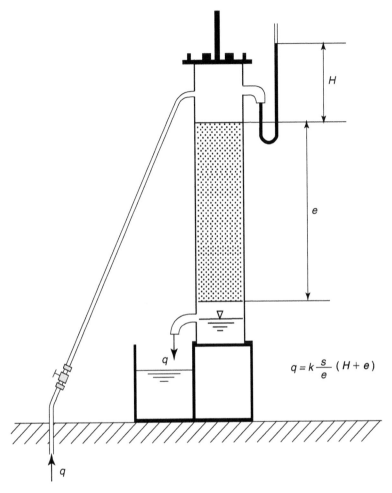

$$q = k\frac{s}{e}(H + e)$$

Figure 2 Principle of the percolation experiments by Harting and Darcy. Harting (1852) wrongly deducted from his experiments the relation:

$$q = ks\frac{H}{e}$$

Where q is flow rate; k is the permeability factor; s is the surface area of the column. The correct formula was given by the French engineer Henry Darcy (1856):

$$q = ks\frac{H+e}{e}.$$

Harting (1877, 1878) extended his percolation experiments in relation to proposals for large reclamation works in the Zuiderzee – a north-south inland extension of the North Sea that reached as far as Amsterdam. This work, that eventually started after World War I, included enclosure of this inland sea by a dike of 30 km length and the reclamation of part of the newly formed fresh water lake (Lake IJssel) behind the barrier dam (Figure 1). Soil samples were obtained by drilling in the bottom of the Zuiderzee as early as 1875, and their permeability was subsequently investigated to assess the rate of upward seepage that could be expected in the projected deep polders. The fear of project failure due to excessive leakage was real and based on bad experience with previous reclamation of the Naardermeer and Béthune polders. These polders, which were situated near the margins of the higher Pleistocene grounds, had to be turned into lakes again because seepage rates in the order of 20 mm per day could not be coped with. Harting was aware of the important role of the Holocene confining layer against upward seepage and advised to leave this bed as undisturbed as possible. He further applied sand-box models to simulate the reduction in pressure and seepage from the fringe towards the centre of the polders. The results of his experiments were reassuring.

State of knowledge in the late-19th century

It can be concluded that Darcy's Law was properly understood and applied in laboratory experiments at the end of the 19th century. Extrapolation of this knowledge to field conditions, however, was hindered by lack of a clear perception of the actual flow pattern and the concept of hydraulic head in the sense of pressure head and elevation head. This can be illustrated by the view of a commission of scientists and engineers, as expressed in their report on problems associated with exploitation of the coastal dunes for Amsterdam's public water supply (Rutgers van Rozenburg *et al.*, 1891). They rightly argued that recharge of the dune area originates from rainfall minus evapotranspiration, and they arrived at a replenishment of 40% of the precipitation. This, to our present knowledge good assessment, was based on a study of rainfall, evaporation and groundwater level fluctuations.

However, their perception of the actual flow process was less lucid. Groundwater was extracted at that time from the dunes by a series of drainage canals and the Commission assumed horizontal flow through the section between the groundwater table and a horizon through the bottom of the canal (Figure 3). They were rather vague

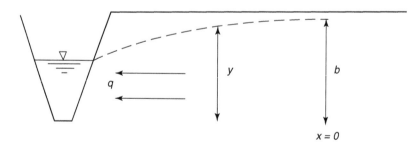

Figure 3 Horizontal flow to a drainage canal. The Commission of 1891 that investigated the groundwater situation in the Amsterdam dune water catchment presented for this flow case, the following formula:

$$\frac{qx^2}{kcb^2} + \frac{y^2}{b^2} = 1$$

where k is the quotient of hydraulic conductivity and porosity c; q is discharge per time unit; y is thickness of that part of the subsurface that participates in the discharge process at distance x from the canal; $y = b$ if $x = 0$. This formula is similar to the well-known Dupuit equation, which applies to a situation where the flow to the canal originates from another, more elevated, canal at $x = 0$ (Dupuit, 1863).

about the area below the canal bottom and believed that, in general, groundwater flow would not extend more than a few meters in depth. Their perception was that flow would normally be restricted by a less pervious layer, or in the absence of such horizon it would abate due to cohesion of the water particles and their adhesion to the soil. Flow at greater depth would therefore change into a "trembling" of the water particles. This concept was derived from the authoritative book by the German professor Otto Lueger (1883, 1895). Lueger's ideas were based on the wrong doctrine that groundwater under free water table conditions could never flow in an upward direction (*cf*. section 3.1).

Similar obscure ideas existed concerning the extension of fresh-water reserves under the dune area. The Commission knew that the subsurface consisted predominantly of sand to at least a depth of 100 m, and they assumed that fresh water in the upper layers had removed the original salt water to greater depth by hydrostatic pressure during formation of the dunes. A gradual change from fresh to brackish water was supposed and they pleaded for a systematic exploration to determine the depth of this interface. No reference was made to Badon Ghijben's principle, published in 1889 (*cf*. section 3.1).

Following Lueger's ideas of horizontal flow, the Commission produced an early drainage formula (Figure 3). Discrepancy between calculated groundwater flux to the canals and the measured much higher discharge by the canals, subsequently inspired the engineer with the Amsterdam Dune Water Supply Company, J.M.K. Pennink, to postulate a much larger part of the subsurface participating in the groundwater drainage process (*cf*. section 3.1).

DAWN OF SCIENTIFIC HYDROGEOLOGY

Dune water research

In the middle of the 19th century, after endless problems with polluted and brackish water and many plans to improve the situation, Amsterdam went to the coastal dunes for its principal drinking water supply; 1853 marked the first groundwater extraction from dunes near Haarlem, 30 km west of Amsterdam. Lack of insight into the vertical extent of fresh water occurrence in the dunes and the fear of salinisation initially led to a groundwater production system with drainage canals. The dune water was hence piped to Amsterdam, which led to a further concern that in case of war an enemy would cut off the town from its water supply. Consequently, the officer of the Army Corps of Engineers, Captain W. Badon Ghijben (Figure 4), was commissioned to explore the Amsterdam subsurface for an emergency source of fresh water. Badon Ghijben indeed discovered a few suitable fresh water pockets surrounded by salty water and advised drilling near the northwestern fringe of the town, near the village of Sloten. Amsterdam was at that time situated along the Zuiderzee and its surface was more or less level with this inland-sea, which was then still open to the North Sea (*cf*. Figure 1). In relation to possible encroachment of Zuiderzee water, Badon Ghijben stated (Drabbe and Badon Ghijben (1889):

"The pressure head on the groundwater column inside the coast is lower than outside at any place where the water level inland is below sea level, and thus sea

Figure 4 Willem Badon Ghijben (1845–1907). (Photograph Royal Military Academy) Badon Ghijben (Badon is part of the family name) joined the army in 1862 and became a member of the Army Corps of Engineers. He worked on several defense constructions and was a specialist in the typical Dutch water-fortress system in which forced inundation formed the most important part of the defense (the "flooded earth policy"). He wrote an authoritative book on the extensive water and fortress defense line that protected Holland until World War I and, among others, was a lecturer at the Royal Military Academy. His early retirement because of ill health, with the rank of colonel followed in 1902. Badon Ghijben was not aware of the earlier formulation of the fresh-salt water interface equilibrium by Joseph DuCommun in the USA in 1818; he was even not aware of having presented a new idea at all, as he later commented.

water must intrude. Equilibrium is only reached if the lighter inland water is a little above sea level. The situation is quite different, however, along the North Sea, where the groundwater in the coastal dunes always shows a higher level than the sea. Assuming a difference $= a$, and considering a specific density of North Sea water of 1.0238, there will be an equilibrium between saline water in the outer area and fresh water in the inner area at a depth of $a/0.0238 = 42\ a$".

He thus claimed that the interface between fresh and salt water under the dune area would be found at a depth of not less than 42 times the elevation of the groundwater table above sea level (Figure 5). Badon Ghijben's ideas did not receive proper recognition until the German engineer A. Herzberg (1901) independently reached the same conclusions, after discovering fresh water extending to a depth of 60 m under

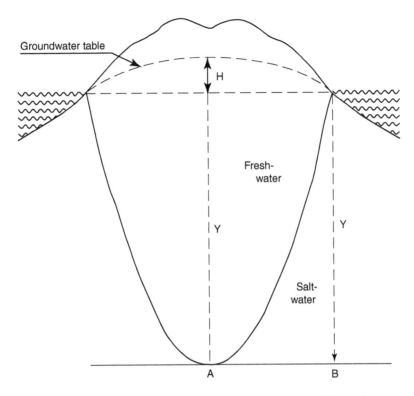

Figure 5 The position of the fresh-salt water interface according to the Ghijben-Herzberg principle. The weight of the column of fresh water at A: $(y+H)\rho_f$, equals the weight of the column of salt water at B: y. ρ_s, where ρ_f and ρ_s are the specific weight of fresh water (1 g/cm^3) and salt water (1.0238 g/cm^3) respectively, so that $y = 42\,H$. The average maximum height of the water table H in the Netherlands coastal dune area is in the order of 5 m.

the North Sea island Norderney. Subsequent drilling in the Amsterdam dune catchment revealed a fresh water lens with a thickness of more than 150 m and in 1903 the extraction of groundwater by deep wells began (*cf.* Pennink, 1904a).

The Ghijben-Herzberg principle is based on the assumption of hydrostatic conditions that actually cannot exist. Deviations from the theoretical model therefore increase with complexity of the dune structure and the associated groundwater flow pattern. Engineer C.P.E. Ribbius, director of the Delft Water Supply Company, was the first to suggest the dynamic character of a fresh groundwater lens (Ribbius, 1903/1904). He showed a clear conceptual insight into the groundwater flow pattern, including the occurrence of upward bending flow lines under free water table conditions (Figure 6). Ribbius realised that this was not in accordance with Lueger's theory as laid down in his authoritative book on public water supply (Lueger, 1895; *cf.* section 2.3). He therefore cautiously stated that he did not want to challenge Lueger's theory in general, but that the very special conditions of the dune area did allow for this deviation (!)

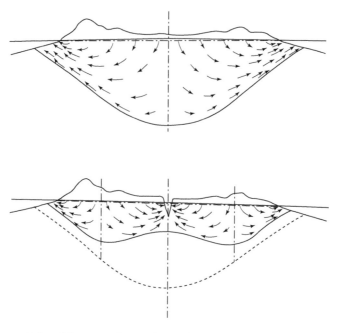

Figure 6 Conceptual idea of the regional groundwater flow pattern in the dune area and the influence of a drainage canal, according to Ribbius (1903/1904).

More rigorous in his opinion of Lueger's doctrine was Director J.M.K. Pennink of the Amsterdam Municipal Water Works (which in 1896 evolved from the private Dune Water Company). As mentioned in section 2.3, Pennink (Figure 7) noticed that measured flow through the drainage canals was a factor three higher than the groundwater flux calculated with Darcy's flow formula on the basis of horizontal flow (cf. Figure 3). This brought him to the idea that the thickness of the subsurface participating in the groundwater discharge was possibly underestimated in the then current theory. He therefore postulated radial and upward bending flow lines in a vertical section perpendicular to the channels. Pennink splendidly demonstrated the flow pattern below partially penetrating channels from both field evidence and experiments with parallel-plate models for viscous flow (Pennink, 1904b, 1905). His models consisted of parallel glass plates, initially filled with liquids and in later experiments filled with sand. His field experiments were carried out using rows of piezometers to measure water pressure at different distances from the canals and at various depths. From these observations he constructed the spatial pattern of hydraulic head and the connected flow lines (Figure 8), then proving his principle of radial flow. Pennink thus understood that the driving force in a point of the flow field is formed by the gradient of the sum of the elevation head and the pressure head.

Pennink's experiment also clearly explained the increase in hydraulic head with depth under areas of discharge. This phenomenon, as well as strongly up-welling of water in excavations at the foot of dunes, had often been misinterpreted as an indication for the occurrence of artesian water veins beneath the dunes. The artesian water

Figure 7   Johan M.K. Pennink (1851–1936). (Photograph Amsterdam Waterworks) Pennink was an engineer with the Amsterdam Dune Water Company and its successor the Amsterdam Municipal
Water Works, from 1890–1900; he subsequently became its director. He not only combined
management with scientific and experimental work, but also developed technical schemes for
safeguarding of the increasing water demand for the town. His most important design was a
system for artificial recharge of the dunes with water from the River Rhine, only executed half
a century later after serious salinisation problems. Pennink experienced long-lasting conflict
with the Amsterdam municipality because he stubbornly refused to extract more water from
the dunes than a percentage of the quantity he rightly assumed to be the rainfall replenishment. Council members blamed him for being more interested in scientific research than in
water supply and he was finally dismissed as director in 1916, but stayed on as an advisor.
Pennink received the highest award by the Royal Institute of Engineers for his scientific and
technical work.

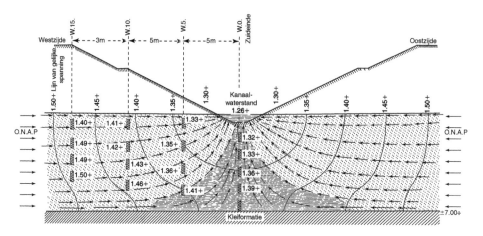

Figure 8 Flow net around a drainage canal based on hydraulic head observations in piezometers; redrawn from Pennink, 1905. (Kanaal waterstand = canal water level relative to NAP = Dutch ordnance datum, approximately mean sea level; lijn van gelijke spanning = line of equal hydraulic head relative to msl.).

hypothesis formed an alternative explanation for the existence of the thick fresh water pocket. Not all scientists and engineers at that time accepted the Ghijben-Herzberg principle and the autochthonous origin of the fresh water lens by infiltrating rainwater. The origin of the postulated artesian water was sought in higher ground of the eastern ice pushed ridges or even further to the east in Germany and Belgium. Pennink fiercely rejected an alien provenance of the dune water by artesian veins and simulated the formation of fresh water lenses beneath coastal dunes with his parallel-plate analogue with fluids of different density, to verify the Ghijben-Herzberg principle (Figure 9).

He further demonstrated with this parallel-plate model the risk of over-production and the up-welling of salt water, and stubbornly refused to extract more water from the dunes than 70% of the replenishment by rainfall. He rightly assessed this exploitable amount to be in the order of 250 mm per year. This policy, however, caused a water shortage in Amsterdam and a long-lasting conflict with the municipality. Pennink therefore proposed – as early as 1901- artificial recharge of the dune catchment with water from the River Rhine, for which he designed a detailed scheme. The municipality, however, was confused by the ongoing scientific debate, and considered his plans too expensive and perhaps unnecessary, should the inexhaustible artesian water source indeed exist. It was to be another 15 years before Pennink's view became generally accepted, but artificial recharge of the Amsterdam dune catchment with river water started only in the 1950s in the aftermath of severe salinisation problems.

One of Pennink's opponents was his colleague, the director of the Water Supply Company of The Hague, engineer Th. Stang. The Hague extracted much more dune water from a smaller catchment than Amsterdam, sanctioned by Stang's belief in a considerable extra replenishment by condensation and a postulated clay layer that would separate and definitely protect the fresh water from salt water intrusion. Director Stang, who also refuted the Ghijben-Herzberg principle, argued that when abstraction

Figure 9 Experiment on the Ghijben-Herzberg principle with a viscous parallel-plate model with fluids of different density; according to Pennink, 1905. (Drijvend Duinwater = floating dune water.).

was kept lower than the replenishment – which he estimated at 750 mm – salinisation would be "a physical impossibility and in the next centuries not to be expected" (Stang, 1903). The Amsterdam municipality was impressed by Stang's performance and invited him to act as reporter on one of the many commissions that were installed to propose a solution for Amsterdam's drinking water problems. Pennink reacted fiercely to Stang's unsound analysis in a counter report in which he moreover ventilated his personal opinion of him in an annihilating – not to say insulting – way. He characterised his opponent, in fact, as a self-conceited charlatan and finally advised the municipality "not to take the responsibility to base any decision whatever, on such rotten grounds" (Pennink, 1907).

Another antagonist of Pennink's ideas was the geologist Reinier D. Verbeek[1], who – for more than 25 years – propagated and defended the idea of an inexhaustible artesian groundwater resource below the dunes. His misinterpretation was mainly based on the occurrence of groundwater with an upward flow component in the vicinity of the drainage canals and the up-welling groundwater at the foot of the dune belt. Verbeek exerted considerable influence on politicians and the general public with brochures

---

[1] Not to be mistaken for his geologist-cousin, R.D.M. Verbeek, who was, among others, renown for his analysis and description of the Krakatau volcanic explosion in 1883.

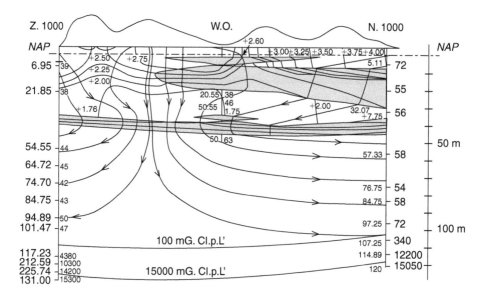

Figure 10 Geo-hydrological section through the dune area near Castricum. This figure exemplifies one of the many geo-hydrological sections through the coastal dune area, according to a report by the National Bureau for Drinking Water Supply, published in 1916 by its director J. van Oldenborgh. It shows the chloride content, hydraulic head distribution (corrected for density) and flow lines. These profiles and a regional "finite-difference" numerical model to assess the influence of groundwater extraction depth on the lowering of groundwater head and the associated threat of salinisation, were most probably established by the engineer with the Bureau, J. Kooper. Van Oldenborgh and Kooper were both educated as officers in the Army Corps of Engineers.

and letters to newspaper editors (e.g. Verbeek, 1905). In fact, as late as 1921, the Minister of Works asked the Royal Netherlands Academy of Arts and Sciences for advice with respect to the origin of fresh groundwater in the dune area. The background of this request was an action by landowners, who argued that expansion of the dune catchment and expropriation of their estates for The Hague's water supply would be unnecessary if subterranean recharge by artesian water proved to be a reality. At that time however, overwhelming evidence for an autochthonous origin of dune water from extensive exploration of the whole coastal dune area had already become available (Figure 10), and the Academy, through their members, the professors of geology Eug. Dubois and G.A.F Molengraaff, refuted the artesian water hypothesis (Molengraaff and Dubois, 1921). Parliament subsequently criticised the Minister for becoming involved in a scientific discussion, probably because the landowners had used the delay to raise their selling price.

Water balance and recharge mechanisms

Section 2 notes that inadequate evaporation measurements led to confusion about the water balance. A. Elink Sterk, chief-engineer with the Haarlemmermeer polder, was the first to indicate the conceptual problems and data inaccuracies of evaporation pans and lysimeters. From comprehensive water balance studies of the Haarlemmermeer polder, he derived an annual evapotranspiration of 481 mm, with a rainfall of 784 mm and a subsurface inflow from the surrounding areas of 150 mm (Elink Sterk, 1897/1898). This, to our present knowledge rather good approximation, was in contrast with the results of systematic lysimeter experiments carried out at the Oudewetering Observatory of the Rijnland Water Authority, since 1876. Unlikely results were obtained from these instruments, particularly when water had to be supplied in dry seasons to keep the vegetation alive. Evapotranspiration, for example, exceeded precipitation with 496 mm during the dry year of 1893. These data caused Elink Sterk to turn fiercely against these Oudewetering experiments and to have them terminated in 1901. He showed his aversion with the comment that he "hoped the instruments would soon disappear from man's memory".

Another engineer with the water authorities, H.E. de Bruyn (1903), carried out lysimeter experiments and analyses of groundwater level fluctuations in the dune area for a period of eight years. He arrived at an acceptable annual figure of 350 mm for annual average groundwater recharge. Meanwhile, the hypothesis of groundwater recharge by water vapour condensation remained a matter of controversy. Apart from the confusing results from evaporation studies, the fact that even during and after heavy rainfall no groundwater was observed to flow into an open pit above the groundwater table, led to denial by some investigators of the reality of infiltrating rainwater as a means of recharge (*cf.* section 2.1). A clear explanation for this observation and refutation of the condensation concept as theory for water in the unsaturated zone, were given by the geologist and mining engineer J. Versluys (Figure 11) in his 1916 PhD thesis at the Delft Technical University, on capillary phenomena. He made clear that water in the unsaturated zone exerts a negative pressure relative to atmospheric pressure, and thus cannot flow out at the surface in an excavation above the groundwater table.

Versluys' work forms one of the first comprehensive treatises on water in the unsaturated zone. Although his approach was of a qualitative nature, it was based on sound physical principles and extensive observational work produced by the US Dept. of Agriculture (Versluys, 1916a). In the same year, Versluys also correctly explained the origin of the often detected and discussed sodium-bicarbonate ($NaHCO_3$) in coastal groundwater areas, through his hypothesis on cation exchange in the zone where saline groundwater had been flushed by fresh water (Versluys, 1916b; Figure 11). The sodium-bicarbonate type of groundwater subsequently became an important tracer in reconstructing the evolution of fresh-saltwater interaction in coastal areas (*cf.* Section 4.4).

The emergence of a basic physical theory

The physicist A.H. Borgesius (1912) generalised Pennink's empirical work on the flow pattern by referring to the mathematical analogy between a groundwater flow field and an electro-magnetic field. He thus simulated a number of analogous flow patterns based on such methods as superposition, imaging and refraction. However, Borgesius showed poor perception of the reality of groundwater hydrology by arguing that the flow-line pattern around drainage galleries – as applied in The Hague's dune water

Figure 11 Jan Versluys (1880–1935). (Photograph Amsterdam University) Versluys was educated as a mining engineer and was the first "all-round" hydrogeologist in the Netherlands. He was familiar with the work of well-known contemporary scientists such as King, Slichter, Boussinesque and Forchheimer. His 1916 PhD thesis on capillary phenomena in the soil forms one of the first systematic treatises on this subject. In the same year he presented the correct explanation for the often-observed occurrence of sodium-bicarbonate in ground-water in the coastal area, according to the process of cation exchange by flushing of saline water by fresh water:

Na clay $+ Ca(HCO_3)$ water $---->$ Ca clay $+ NaHCO_3$ water

After several positions in drinking water supply and mining in the Netherlands, the Dutch East Indies and Suriname, Versluys joined the Bataafsche Oil Company in 1927 and brought oil and gas exploration and exploitation to a scientific level. In 1918 he became the first lecturer in hydrology in the Netherlands, at the Delft Technical University; in 1932 followed his appointment as professor of economic geology at Amsterdam University.

extraction – would make salinisation almost impossible[2]. The noted physicist and Nobel laureate, Prof. H.A. Lorenz showed – in a companion paper – that a combination of Darcy's flow formula with the continuity equation results in the general differential equation that governs groundwater flow (Lorenz, 1913). This equation is

---

[2]This misconception and his playing down on Pennink by pretending surprise that Pennink had seriously reasoned against the odd and unmatched concepts of Lueger, obviously reflects Borgesius' earlier involvement in a study that was prepared in support of the practise of The Hagues Water Supply and the ideas of its director, Th. Stang (*cf.* De Vries, 1982).

an expression of the well-known Laplace equation that also holds for electro-magnetic fields and for heat conduction. Lorenz further presented a treatise on the lowering of groundwater level around drains and up-coning of the fresh-salt water interface, and adds the comment that over-exploitation of the catchment would – in any case – finally lead to salinisation. In the next year Pennink published his well known Salinisation Report, in which he convincingly demonstrated that the boundary between fresh and salt water was slowly moving up and that several deep wells had already been affected by salinisation (Pennink, 1914).

The French Joseph Boussinesq (1877) and the Austrian Philipp Forchheimer (1886) were probably the first to combine the continuity condition with a flow equation for the general differential equation for steady groundwater flow. The transient variety appeared probably for the first time in the work of Boussinesq in 1904. Forchheimer derived several solutions of flow cases, generalising the formula for horizontal flow, derived by Jules Dupuit (1863; *cf.* Figure 3) and the pumping well formulae developed by Adolph Thiem (1870). The American Charles S. Slichter (1899) arrived at the same results and extended Forchheimer's approach by including a vertical flow component. Remarkably, he did not mention Forchheimer, although he showed to be familiar with the German authors by referring to Thiem, Volger and Lueger. F.H. King (1899) – in a companion paper – gave a general description of groundwater flow under the influence of topography and presented a sketch for the flow pattern near a stream similar to Pennink's case (Figure 8). King correctly argued that the flow of water beneath a channel or depression is forced upward under hydrostatic pressure, developed by the inflow and accumulation of water percolating downward through the surrounding higher land. In contrast to Pennink however, King did not verify his ideas by field experiments.

Slichter (1902) applied King's sketch to discuss the flow pattern in a framework of general theory. He emphasised that it would be misleading to compare groundwater flow with pipe flow and stream flow, because the frictional resistance in ground-water is not transmitted by the fluid layers. In this connection he considered the influence of an undulating impervious base for the flow pattern in an aquifer and stated:

"The contention of some German hydrographers (he then refers to Lueger; deV.) that there can be no motion in a region like ASB (referring to a figure with a concavity ASB in the impervious base of the aquifer; deV.) must be entirely abandoned. Water must circulate in all parts of the enlargements in the porous medium, for the same reason that heat would be conducted over similar enlargements in a conductor". He further specified: "If it were not for the ever present controlling influence of gravity the motion would be entirely analogous to the flow of heat or electricity in a conductive medium".

Pennink was obviously not familiar with this theoretical approach, or at least did not know how to connect the results of his experiments with this theoretical knowledge. In 1914 however, Forchheimer's comprehensive volume "Hydraulik" appeared and this textbook, which contains a chapter on groundwater hydraulics, became henceforth widely used in the Netherlands.

APPLIED RESEARCH ON THE BASIS OF THEORETICAL DEVELOPMENTS

Introduction

It can thus be stated that an era of speculation and scientific controversies had come to an end in the period around World War I, through development of a sound theoretical basis for groundwater hydrology and an adequate dissemination of this knowledge. Groundwater hydrology subsequently evolved by way of interaction between field observations and mathematical analysis, by solving the general differential equation of groundwater flow for appropriate schematisations and boundary conditions. This approach produced several original Dutch contributions to groundwater hydrology, particularly in connection with the typical Netherlands conditions pertaining to groundwater flow in leaky aquifers and control of shallow groundwater tables.

The problem of leakage and water table control became notably manifest after World War I in connection to large-scale land reclamation works in the Zuiderzee area (*cf.* section 2.2). This included groundwater control in excavations for large hydraulic structures, upward leakage of brackish groundwater in the newly created deep polders, seepage through and below dikes and an adequate regulation of the groundwater table by drainage. Concern for the groundwater table also arose in connection with the higher Pleistocene sandy soils in the eastern part of the country, where vegetation and crops are partly dependent on capillary transport from the shallow water table to the root zone. Improved drainage and acceleration of runoff from wetlands in that area during the first half of the 20th century, had led to unexpected desiccation during the summer season when evapotranspiration normally exceeds rainfall.

Well flow

The first original Dutch contribution to the field of groundwater hydraulics was a theoretical study of steady flow around a well or circular polder in a leaky aquifer, by J. Kooper in 1914. Kooper arrived already at a mathematical solution with Bessel functions, which proved to be very useful in groundwater exploration and management. The solution for this flow case was subsequently elaborated by G.J. de Glee (Figure 12) in his 1930 PhD thesis at the Delft Technical University, and is generally referred to as the "De Glee formula" (Figure 13). Kooper and De Glee were both engineers with the National Institute for Drinking Water Supply (RID). This organisation – which was founded in 1913 to stimulate and support central drinking water supply, notably in the rural areas – remained the central agency for groundwater exploration and research for more than 50 years.

The engineer with the Amsterdam Water Works, L. Huisman, and mathematician J. Kemperman, subsequently elaborated the Kooper-De Glee formula for an aquifer between two semi-pervious layers (Huisman and Kemperman, 1951). G.A. Bruggeman of the RID finally developed the general solution for flow through multiple aquifer systems (Bruggeman, 1969). Further mention should be made of the remarkable phenomenon of a rise in water pressure during the first minutes of pumping tests in some areas in the Netherlands – the so-called "Noordbergum effect", named after the village where this phenomena was first observed. An explanatory model of this effect, on the basis of three-dimensional subsurface deformation, was developed by A. Verruijt

Figure 12 Gerrit J. de Glee (1897–1975). (Photograph De Glee family) De Glee began his career in 1920 as engineer with the water supply company of the town of Tilburg. He joined the National Bureau for Drinking Water Supply in 1925 and published his well-known formula for steady flow to a well in a leaky aquifer in 1930 as part of his PhD thesis. In that same year he became the first director of the Water Works of the Province of Groningen. His equation for flow to a well in leaky aquifer (see Figure 13), which in fact was based on earlier work by J. Kooper (1914).

(1969) in his work on elastic storativity. The considerable Dutch expertise on flow in leaky aquifers under a variety of conditions was compiled in textbooks by Verruijt (1970, 1982) and Huisman (1972).

Leakage and dikes

Expected upward leakage of fresh and brackish groundwater in the projected deep polders of the Zuiderzee reclamation (total area of 170 000 ha) was a matter of concern for the Department of the Zuiderzee Works. An early theoretical treatment of the flow below a dike was presented by J.M. Burgers, professor of aero- and hydrodynamics at the Leyden University. Burgers produced steady and transient solutions for radial flow from partially penetrating canals into a low-lying confined aquifer (Burgers, 1926; Figure 14). This approach, however, did not take into account the resistance against upward leakage through the semi-confining covering layer. J.P. Mazure,

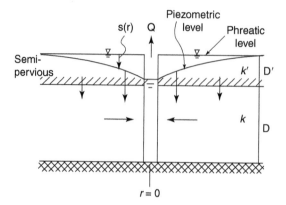

Figure 13 Steady-state radial symmetrical flow to a well in a leaky aquifer; according to Kooper, 1914 and De Glee, 1930.

$$s = \frac{Q}{2\pi T} K_0 \left(\frac{r}{\lambda} \right)$$

$$\lambda = \sqrt{Tc}$$

Where s is the lowering of hydraulic head at distance r from the well through a groundwater extraction Q; T and 8 are transmissivity and leakage factor of the aquifer respectively; c is the vertical flow resistance, which equals D'/k'; K_0 is a modified Bessel function of zero order. This equation, and its variety for flow around a large diameter excavation or polder, have been widely used for pumping test analysis, evaluation of the impact of groundwater extraction, and to assess groundwater inflow into deep polders and excavations.

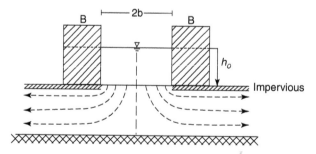

Figure 14 Groundwater flow rate q below a dike in a confined aquifer, through a head difference h_0; according to Burgers, 1926:

$$q = \frac{kh_0}{\pi} \ln \frac{8b}{B}$$

Burgers also derived a solution for the transient flow with storativity S, caused by an instantaneous change in canal level h_0; T is transmissivity:

$$q(t) = h_0 \sqrt{\frac{TS}{\pi t}}.$$

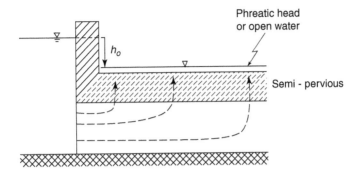

Figure 15 Flow below a dike in a leaky aquifer with horizontal flow in the aquifer and vertical flow in the semi-pervious confining layer, according to Mazure (1932). The diffuse upward leakage $v(x)$ at distance x from the dike is (for symbols see Figure 13):

$$v(x) = \frac{h_0}{c} e^{-x/\lambda}$$
$$\lambda = \sqrt{Tc}.$$

engineer with the Zuiderzee Works, subsequently proposed a simplified flow pattern but more realistic hydrogeological schematisation by including a leakage factor. He assumed horizontal flow in the aquifer and vertical flow through the confining layer, which resulted in a simple exponential function for the steady-state hydraulic head distribution (Mazure, 1932; Figure 15). This formula proved to simulate the actual observed situation rather well in the first Zuiderzee reclamation in the 1930s, the Wieringermeer polder (Wieringermeerrapport, 1936), and has subsequently widely used in the Netherlands.

Another outstanding piece of work was produced by C.G.J. Vreedenburgh, who was the first to derive the exact mathematical solution for the flow through a dam under free water table conditions (In: De Vos, 1929). He moreover solved the problem of flow through an anisotropic medium and simulated two-dimensional groundwater flow in an electrolyte-tank on the basis of the analogy between Darcy's Law and Ohm's Law (Vreedenburgh, 1935; Vreedenburgh and Stevens, 1933, 1936). Vreedenburgh – at that time – was professor at the Technical University of Bandung (former Dutch East Indies), and his studies basically originated from the problem of seepage through dams and levees in irrigation fields. Both Burgers and Vreedenburgh applied complex function theory to solve the two-dimensional Laplace equation.

Further mention should be made of J.H. Steggewentz' 1933 PhD thesis on the propagation of ocean tides in coastal aquifers, which included an early solution for the delayed reaction of the vertical flow near the water table with respect to water pressure oscillations. This flow resistance by the capillary fringe was supposed to explain the relatively fast propagation of the water pressure in phreatic aquifers that Steggewentz observed in the coastal dune area. It was only in the 1950s that this concept appeared in the international literature as the "delayed yield" principle, in connection with the early phase of transient well flow (Boulton, 1954). Several Dutch contributions to the analysis of ocean tides in leaky and compressible aquifers were subsequently produced.

A systematic elaboration of these flow cases and their application to parameter determination was carried out by the Dutch-Canadian G. van der Kamp in his 1973 PhD
thesis at the Amsterdam Vrije Universiteit.

Brackish groundwater and paleohydrologic evolution

Research in the framework of the Zuiderzee Works included the prediction of
diffuse upward leakage of saline groundwater into the new polders. Extensive mapping of the fresh-salt groundwater distribution was therefore carried out as part of
this investigation; initially by drilling and from the 1950s by geo-electric surveys
(Wieringermeerrapport, 1936). A. Volker of the Department of the Zuiderzee Works,
observed a gradual decrease in salt concentration from the surface of the Holocene
clayey deposits to the underlying Pleistocene (Figure 16). He made plausible that this
profile originated from salt water diffusion from the Zuiderzee since the Middle Ages,
when this embayment was formed by encroachment of a large tidal inlet into this former peat area. This means that in general no vertical infiltration of salt water by density
currents (free convection) had intruded in this area, except for deeply incised tidal inlets
and other eroded surfaces. Similarly Volker explained the gradual increase in salt content with depth in the Pleistocene aquifer as the result of salt diffusion from the Early
Pleistocene marine deposits into the overlying fluvial deposits. This could have taken
place during the last ice-age (the continental Weichselian) – that lasted about 100,000
years – in a flat area with hardly any horizontal flow (Figure 17; Volker, 1944, further
elaborated in Volker and Van der Molen, 1991).

　　These studies proved that a marine transgression not necessarily leads to salinisation of the underlying aquifers if protected by a low-conductivity horizon. The first
transgressions during the Early Holocene were preceded by a rise of the groundwater table and extensive peat growth on the sandy Pleistocene surface. Obviously this
peat layer, that now forms a consolidated and impervious horizon below the younger

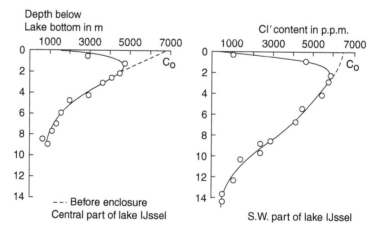

Figure 16 Diffusion profile in the Holocene clayey subsurface of the former Zuiderzee, according to
Volker, 1944.

clayey deposits, protected large areas from seawater intrusion. Hydrochemical studies (e.g., Geirnaert,1973; De Vries, 1981) made plausible that salinisation of the Pleistocene aquifer during the Holocene transgressions, mainly took place through deeply incised tidal inlets and by plumes of saline/brackish water, that forced their way up from greater depths, below deep polders. Fresh water infiltration used to be topography driven and proceeded from the high dune belt, which was formed on top of the coastal barriers about 1200 years ago, and from other relatively high elevated areas into adjacent low lying polders (Figure 18).

Many studies of groundwater flow involving fresh and salt water followed in the course of time by analytical as well as numerical methods – initially supported by electric analogons and subsequently by computer simulation models. An example of the first category was the 5000 grid points Electric Network Analogon for Groundwater Flow (ELNAG). This model was developed in the 1960s for simulating steady as well as transient fresh and salt groundwater flow in connection with a large scale

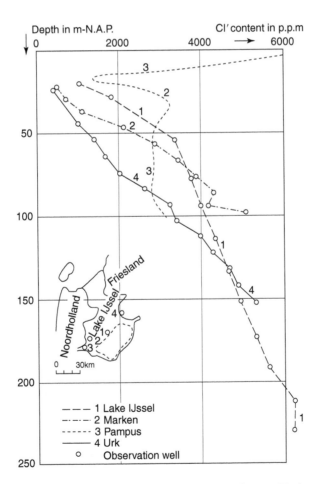

Figure 17 Chloride profiles in the Pleistocene aquifer of the former Zuiderzee, according to Volker, 1944.

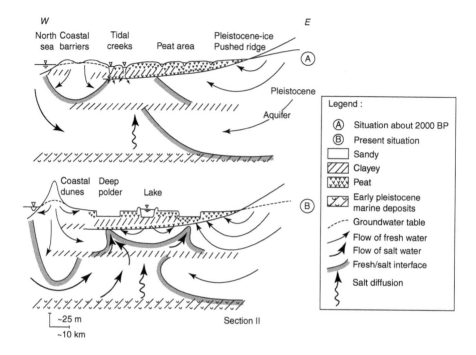

Figure 18 Schematised reconstruction of the evolution of the distribution of fresh and salt water under the influence of a change in topographic conditions through land drainage and reclamation, according to De Vries, 1981.

water management and flood protection project (the so called Delta Works) in the south-western estuarine area of the Netherlands. This water management and flood protection scheme, carried out from the 1950s to the 1980s, was a response to the devastating flood that struck the Netherlands in 1953. An overview of the many Dutch contributions to the problem of moving fresh and salt water in the 1960s and 1970s is given in a publication by the Committee on Hydrological Research – TNO (1980).

Dunes and groundwater extraction

Salinisation of the coastal dune area by over-exploitation of groundwater for drinking water supply proceeded until artificial recharge with river water began to reverse this process after World War II. Different exploitation strategies had been exercised in the preceding period to prevent salinisation. These strategies were mainly based on a choice between deep extraction from a semi-confined part of the subsurface or shallow extraction from the upper phreatic aquifer. Mazure (1943) produced an analytical solution for predicting the evolution of the fresh water pocket for different scenario's. His approach was further elaborated by Todd and Huisman (1959) for the Amsterdam dune water catchment, making use of the first main-frame electronic computer at the Amsterdam Mathematical Centre for the extensive numerical calculations. More advanced analyses, including the flow of salt water below the fresh water pocket towards the deep polders behind the dunes and the influence of artificial recharge, were subsequently carried out when the computing power increased (Venhuizen, 1968).

Further theoretical analyses of transient flow were initiated to predict the ground-water movement near the proposed infiltration channels. This work was carried out by the engineer with the Amsterdam Water Works, J.H. Edelman, and published in his 1947 PhD thesis at the Delft Technical University. His work included analysis of (i) the propagation of water level change in the channel into the adjacent aquifer and the associated infiltration regime, (ii) transient well flow, and (iii) a hexagonal finite difference model for regional groundwater flow, including density driven flow. His basic solution of transient flow in the vicinity of channels is similar to Burgers' (1926) result (*cf.* Edelman, 1972). Huisman and Olsthoorn elaborated the experience with artificial recharge and groundwater recovery in the dune area in a well-known textbook (Huisman and Olsthoorn, 1983).

Another aspect of the dune area concerned the replenishment by rainfall, and focused on the classical question of the influence of afforestation of the dunes. There-fore large lysimeter experiments have been carried out since 1936 in the dune area near Castricum. These experiments, with 4 lysimeters of a size of 25×25 m and a depth of 2.5 m each, were instigated by J. van Oldenbourgh. Van Oldenbourgh, who began his career as the first director of the RID (*cf.* Figure 10), became subsequently the first director of the dune water supply works of the province of Noord-Holland with, among others, a well field near Castricum. The lysimeter plots were respectively sowed with deciduous trees, pine trees, natural dune shrub and one was left barren. For the period 1947–1981 when the trees had reached maturity – the following amounts of drainage water, as percentage of rainwater, had been collected (average annual precipitation of 820 mm):

Bare sand and mosses 76%
Dune shrubs 42%
Deciduous trees 37%
Pine trees 17%

These figures were derived from the comprehensive study of the hydrodynam-ical and hydrochemical evolution of groundwater in the dune area that formed P.J. Stuyfzand's 1993 PhD thesis at the Amsterdam Vrije Universiteit.

Land drainage, groundwater table control and regional studies

The long tradition in the Netherlands on land drainage had not led to much insight in the drainage process itself, nor were the criteria clear on which the drainage practice was based. Still in the 1930s, this important aspect of water management was mainly driven by a combination of traditional experience and trial and error. Scientific research on land drainage and soil improvement began as part of the vision to develop the new Zuiderzee polders in a rational manner. Moreover, the occurrence of desiccation problems that accompanied the land drainage and stream canalisation projects in the Pleistocene sandy region, made clear that a systematic and scientific approach was essential. This research program began in the 1930s at the Experimental Station and Soil Science Institute in Groningen, by S.B. Hooghoudt (Figure 19).

Figure 19 Symen B. Hooghoudt (1901–1953). (Photograph Soil Science Institute Groningen).

Hooghoudt, who was a chemist by training, performed a formidable task by developing methods to determine the hydraulic properties of the soil, both in situ and in the laboratory, and by deriving drainage formulas for various cases pertaining to the Netherlands conditions. The first category included a kind of slug test in a shallow auger hole, the second comprised analysis of groundwater flow towards drainage channels in layered conditions, including the influence of a radial flow component near a partially penetrating channel (Hooghoudt, 1937, 1940; Figure 20). The basic theory is mainly based on a rather simple superposition and imaging of radial symmetrical well flow, but the practical generalisations were the product of extensive observations and experiments on plot scale as well as on farm land parcel scale. By the 1940s the perception of land drainage and the tools for implementation had arrived a scientific level. Hooghoudt moreover initiated, and partly performed, a program for systematic mapping of the hydrogeological properties of the subsurface to obtain a sound basis for groundwater management and land improvement projects.

After World War II, Hooghoudt ascertained himself of the help of physicists and geologists (notably L.F. Ernst and N.A. de Ridder, respectively) to improve the theoretical and practical foundation of his work. Ernst concentrated on the radial flow component and analysed the non-linear character of the drainage process in the hierarchical drainage network on the Pleistocene sandy area (Ernst, 1956, 1978; Figure 21).

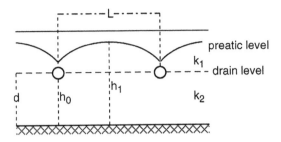

Figure 20 Section through a parallel system of drainage channels for a two-layer case, for which Hooghoudt's well-known 1940 formula reads:

$$NL^2 = 8k_2 d(h_1 - h_2) + 4k_1(h_1^2 - h_0^2)$$

Where N is the flux through the phreatic surface; d is *equivalent thickness* of the aquifer in which the influence of the upward bending and contracting flow lines near the channels is included (d can be derived from nomograms).

Ernst's basic equation for the radial flow component is similar to Burgers' formula for steady state flow (*cf.* Figure 14). De Ridder carried out the first systematic regional hydrogeological surveys within projects to evaluate and improve the agricultural water management conditions. Both Ernst and De Ridder joined the Institute for Land and Water Management Research in Wageningen (ICW), where all water management and hydrological studies related to agriculture were concentrated from 1955 onwards. Hydrological research at the ICW focused on the unsaturated zone and associated evapotranspiration and crop water requirement. Notably the influence of water table depth on capillary transport and crop yield was an important issue in relation to the question of an optimum water table depth.

A related problem was the rate of damage- notably reduction in crop yield – caused by groundwater level lowering due to groundwater extraction for public water supply. The need to evaluate and predict the influence of groundwater exploitation on the groundwater table had stimulated the National Institute for Drinking Water Supply (RID) in the 1950s to carry out rather unique experiments with horizontal viscous parallel-plate models for transient flow in phreatic aquifers (Santing, 1957; Figure 22). Subsequently, Ernst (1971) developed a well flow formula to determine the drawdown in areas where the groundwater extraction was compensated for by recharge from a non-linear stream system. De Ridder's association with international development co-operation through the International Institute for Land Reclamation and Improvement (ILRI), prompted the compilation, together with G.P. Kruseman, of their well known text book on the analysis and evaluation of pumping tests (Kruseman and De Ridder, 1970, 1990).

Catchment-scale studies

Another aspect of land drainage that previous to the 1950s had not received much scientific attention, concerned the required discharge capacity to reach desired water table depth, with a given frequency. Observations revealed a strong relation between

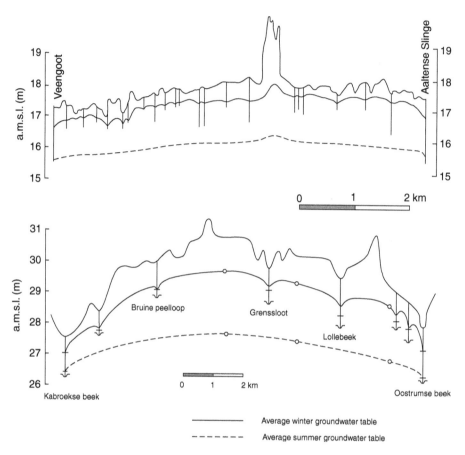

Figure 21   The contracting and expanding hierarchical drainage network system in the Pleistocene area through seasonal groundwater table fluctuation, according to Ernst; cf. De Vries, 1994.

groundwater table depth (storage capacity) and the required drainage capacity. This resulted in design-criteria which range from less than 3 mm/day for the higher areas with an average groundwater depth of more than 140 cm, to more than 12 mm/day for polders and low brook valley plains with water table depths of less than 40 cm. Subsequent observations of the runoff regime and groundwater depletion during dry periods were carried out in a number of small stream catchments. Analysis with the transient linear reservoir formulas for groundwater drainage by J.W. de Zeeuw and F. Hellinga (1958) and D.A. Krayenhoff van de Leur (1958) at the Wageningen Agriculture University, proved to be rather adequate for these areas with predominantly groundwater discharge (De Zeeuw, 1966; De Jager, 1966).

The close connection between the seasonal fluctuation of the groundwater table and the number and order of stream branches that participate in the drainage process (cf. Figure 21) is reflected by a shift in the observed values of the recession coefficient. J.J. de Vries, in his studies at the ICW and the Amsterdam Vrije Universiteit, recognised the drainage system on the sandy area as a continuum of groundwater and surface water

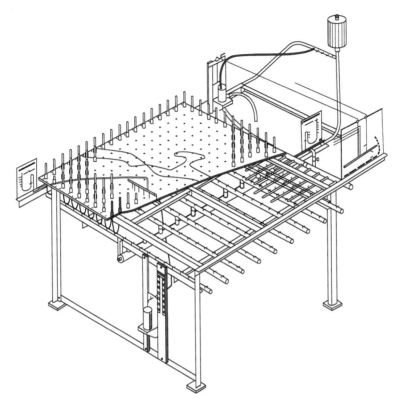

Figure 22 Principle of a viscous horizontal parallel plate model for transient flow under free water table conditions, according to Santing, 1957.

in which stream systems were initiated by groundwater sapping-erosion at the intersect of groundwater table and topography. On the basis of this concept, he developed an analytical model to explain the evolution and morphology of the natural stream network as dependent on the geological, geomorphologic and climatic conditions (De Vries, 1974, further elaborated in 1994).

The regional studies of the ICW focused in the 1960s on the province of Gelderland. For the first time all aspects of hydrology were integrated, involving almost all organisations associated with water, subsurface and agriculture. Detailed studies were carried out in the polder Tielerwaard-West in the reclaimed flood plain of the River Rhine, and in the catchment of the small Leerinkbeek River in the sandy plain of East Gelderland. A regional hydrogeological study of the whole of East Gelderland by the ICW led to the first regional computer-simulation model for the saturated zone in the Netherlands (Ernst *et al.*, 1970). This model enabled to include the close interaction between the stream network and the associated groundwater flow systems as well as their mutual connection with the topography. A key factor in the hydrologic-morphologic interaction is that small topographical differences causes relatively large variations in the shallow groundwater table, and thus in storage capacity. This in turn results in large differences in stream density and drainage regime (De Vries, 1974).

Advancement in the knowledge of transport in the unsaturated zone and the associated evapotranspiration, as well as progress with computer-simulation models for saturated as well as unsaturated flow, led subsequently to an attempt to analyse and model the whole hydrological system of Gelderland. More than 80 specialists of 30 organisations were involved – initially co-ordinated by J.H. Colenbrander and subsequently by E. Romijn. (Both well-known as Secretary-General of the International Association of Hydrological Sciences IAHS and President of the International Association of Hydrogeologists IAH, respectively.) One of the achievements was the development of a computer model that coupled groundwater extraction, water table decline and reduced evapotranspiration and crop yield (De Laat et al., 1975). The Gelderland project, which produced 109 reports and publications, marked the beginning of an integrated approach to water management in the Netherlands. The general objective of integrated water management is (a) harmonisation of the natural characteristics of the water system and the requirements of the various interested parties, and (b) balancing the partly complementary and partly conflicting objectives of the different parties (Committee on Hydrological Research-TNO, 1981).

LEGISLATION

Water management and water use with its various aspects of safety, health and agriculture have always been the concern of different ministries, provincial authorities and traditional regional water boards, and have accordingly been regulated by a variety of water legislation. In 1968 the Ministry of Transport, Public Works and Water Management published the first overall policy document on a coherent water management in the Netherlands. This first document mainly covers the issues related to fresh water distribution and protection against salt water. Subsequently interests shifted and broadened to quality management and nature conservation, whereas in the post 1985 period a comprehensive vision evolved, including all sectors of society.

No general regulations with respect to groundwater had been formulated until the mid-20th century. Before that time every landowner could sink a well and extract water as long as no damage was done to somebody else's property, which meant that the limitations were set by private law. This situation posed a problem to water supply companies who could never be sure of continuity. Therefore the Groundwater Act Water Supply Companies was enacted in 1954 which meant that drinking water companies needed a license for any abstraction, including conditions and damage compensation; other groundwater abstractions were explicitly excluded. In 1981 an overall Groundwater Act was initiated which includes all abstractions and activities related to infiltration and groundwater recharge. The provinces are, according to this law, the responsible authorities for permission, registration and reporting. Quality aspects are mainly related to protection of recharge areas. Other groundwater quality issues are dealt with the Soil Protection Act of 1987, which includes regulations to prevent subsurface pollution and remediation of contaminated soils. Within this legislation, the provinces have a duty to set up a groundwater monitoring and management plan (see also Dufour, 2000).

EVALUATION AND CONCLUDING REMARKS

Scientific groundwater hydrology in the Netherlands emerged around the turn of the 19th century in close association with the growing demand of groundwater for public water supply. Suitable fresh groundwater resources in the western and most densely populated part of the country were mainly restricted to the belt of coastal dunes, and were accordingly prone to depletion and salinisation. A proper and sustainable exploitation thus required a sound knowledge of the origin and replenishment of the fresh water reserves as well as understanding of the dynamics of the fresh/salt water interface and the flow towards the extraction means. It took a period of more than half a century of speculations, observations and theoretical considerations before a proper understanding on the basis of an adequate physical theory emerged. The scientific debates often led to strong personal controversies, sometimes with political implications because of the societal relevance of the subject. Several Dutch pioneers performed trail-blazing and outstanding work during the inception period before World War I.

The basic theory, that became operational in the period around World War I, enabled to consider a groundwater flow case as a boundary-value problem that could be solved as a mathematical equation. The Dutch hydrogeological conditions, which are characterised by relatively homogeneous aquifers and simple and artificially maintained boundary conditions, proved very suitable for this theoretical approach. A number of problems related to groundwater extraction, groundwater level control, land reclamation and civil engineering works were subsequently solved. This has led to several original Dutch contributions to the theory of groundwater flow in situations were groundwater and surface water are closely related under phreatic as well as leaky aquifer conditions. This development took mainly place in the period following World War I.

Until World War II most groundwater research was carried out by engineers rather than by geologists. This can be explained by the fact that the geology of the Netherlands is relatively simple and engineers used to have a better background for the above-mentioned quantitative approach. Moreover, most groundwater problems were related to soil and water engineering projects. The National Institute for Drinking Water Supply (RID) has played a major role in this development. It remained the central organisation for groundwater exploration and research for both drinking water supply and groundwater management connected with infrastructure works, for more than half a century.

The situation changed after World War II when government strongly stimulated reactivation and improvement of agriculture. The focus of groundwater hydrology then shifted to the unsaturated zone, crop water requirement and to water table control. Therefore, from the 1950s, scientific research became concentrated in the Institute for Land and Water Management Research (ICW), involving agriculture engineers and soil physicists.

Awareness of pollution and environmental degradation arose in the 1970s, and water became increasingly considered as an environmental component, requiring an integrated approach of quantity as well as quality of surface water and groundwater, and their interaction with the geological environment. This resulted in broadening of the spheres in which hydrologists became employed, including an increasing involvement of earth and environmental scientists in hydrogeologic research and exploration.

This was, among others, facilitated by the initiation of an education program in hydrogeology and geographical hydrology at the Amsterdam Vrije Universiteit in the late-1960s. Another concurrent development was the establishment of the Institute for Groundwater Survey (DGV-TNO) in 1968, with as main task to carry out a regional inventory and monitoring of groundwater data and to produce a groundwater map 1:50,000 with explanatory reports.

The shift of interest from agriculture to environment and the associated need to adapt the water management-infrastructure accordingly, became notably manifest in the 1980s. In line with this evolution, the RID merged with the National Institute of Public Health and Environment and the ICW became part of a larger organisation for agriculture and environment (initially the Staring Centre and subsequently the Alterra Research Centre). Concurrently, the DGV-TNO was transformed into the Netherlands Institute for Applied Geosciences (TNO-NITG) with a broader mission. This institute became the central organisation for groundwater investigations, monitoring and data management from the mid-1980s. Their position strengthened considerably when the National Geological Survey joined this organisation in the 1990s. This meant that earth scientists, in general, became increasingly involved in hydrogeology and associated subsurface engineering problems. Evidently, the focus of groundwater hydrology in the Netherlands shifted from *geohydrology* in the first half of the 20th century to *hydrogeology* in the second half of this century.

REFERENCES

Borgesius, A.H. (1912) Grondwaterbeweging in de omgeving van bronnen. *De Ingenieur*, 27 (49), 995–1008.

Boulton, N.S. (1954) The drawdown of the water table under non-steady state conditions near a pumped well in an unconfined formation. *Proceedings of the Institute of Civil Engineers*, 3, 564–579.

Boussinesq, J. (1877) Essai sur la théorie des eaux courrantes. *Mém. Présentés par Divers Savants à l'Acad Sci*, t. 23 (1), 1–680.

Boussinesq, J. (1904) Récherches théoriques sur l'écoulement des nappes d'eau infiltrées dans le sol. *Journal de Mathématiques Pures et Appliquées* 10, 5–78.

Bruggeman, G.A. (1969) The reciprocity principle in flow through heterogeneous porous media. *Int Assoc Hydraulic Research, Proc Symposium Haifa*. pp. 136–149.

Burgers, J.M. (1926) Grondwaterstrooming in de omgeving van een net met kanalen. *De Ingenieur*, 41 (32), 657–665.

Buys Ballot CHD. (1879) Hoe zal men de verdampingshoeveelheid bepalen voor polders. *Versl Meded Kon Ned Akad v Wetensch, afd Natuurk*, 2nd reeks, deel 14, 27–51.

Committee on Hydrological Research – TNO. (1980) Research on possible changes in the distribution of saline seepage in the Netherlands. *Proc and Information 26, The Hague*. 216 p.

Committee on Hydrological Research – TNO. (1981) Water resources management on a regional scale. *Proc and Information 27, The Hague*. 173 p.

Darcy, H.P.G. (1856) *Les fontaines publiques de la ville de Dijon*. Paris, V Dalmont. 647 p.

De Bruyn, H.E. (1903) Lysimeter-waarnemingen en de hoeveelheid drinkwater die de duinen dienovereenkomstig kunnen geven. Hand. 9th Ned Natuur- en Geneesk Congres. pp. 148–154.

De Glee, G.J. (1930) Over grondwaterstroomingen bij wateronttrekking door middel van putten. PhD dissert, Delft University of Technology; Waltman, Delft. 175 p.

De Vos, H.C.P. (1929) Eenige beschouwingen omtrent de verweekingslijn in aarden dammen. De Waterstaats-Ingenieur 17 (11):335–354

De Jager, A.W. (1965) Hoge afvoeren van enige Nederlandse stroomgebieden. PhD dissert Wageningen University of Agriculture, Pudoc, Wageningen. 167 p.

De Vries, J.J. (1974) Groundwater flow systems and stream nets in The Netherlands. PhD dissert Vrije Universiteit Amsterdam; Editions Ropodi, Amsterdam. 226 p.

De Vries, J.J. (1981) Fresh and salt groundwater in the Dutch coastal area in relation to geomorphological evolution. Geologie en Mijnbouw, 60, 363–368.

De Vries, J.J. (1982) Anderhalve eeuw hydrologisch onderzoek in Nederland (1830–1980). Amsterdam, Editions Rodopi. 195 p.

De Vries, J.J. (1994) Dynamics of the interface between streams and groundwater systems in lowland areas, with reference to stream net evolution. Journal of Hydrology, 155, 39–56.

De Zeeuw, J.W. (1960) Hoe draineerde men 100 jaar geleden? Tijdschr. Kon. Ned. Heide Mij, 71 (11–14), 50–56, 59.

De Zeeuw, J.W. (1966) Analyse van het afvoerverloop van gebieden met hoofdzakelijk grondwater afvoer. PhD dissert Wageningen University of Agriculture; Meded Landbouwhogeschool, 166–5. 139 p.

De Zeeuw, J.W. & Hellinga, F. (1958) Neerslag en afvoer. Landbouwk Tijdschr, 70, 406–422.

Drabbe, J. & Badon Ghijben, W. (1889) Nota in verband met de voorgenomen putboring nabij Amsterdam. Tijdschr Kon Inst v Ingenieurs, Verhandelingen 1888/1889, 8–22.

DuCommun, J. (1818) On the causes of fresh water springs, fountains, etc. American Journal of Science, 1st Ser, 14, 174–176.

Dufour, F.C. (2000) Groundwater in the Netherlands: Facts and figures. Netherlands Institute of Applied Geosciences – TNO, Delft, 96 p.

Dupuit, A.J.E.J. (1863) Études théoriques et pratiques sur le movement des eaux dans les canaux découvertes et à travers les terraines perméables, 2nd ed. Paris, Dunod. 364 p.

Edelman, J.H. (1947) Over de berekening van grondwaterstroomingen. PhD dissert Delft University of Technology, 77 p.

Edelman, J.H. (1972) Groundwater hydraulics of extensive aquifers. ILRI Bull 13, Wageningen. 216 p.

Elink Sterk, A. (1898) Over regen, verdamping en kwel in den Haarlemmermeerpolder. Tijdschr Kon Inst v Ingenieurs, Verh 1897/1898, 63–72.

Ernst, L.F. (1956) Calculations of the steady flow of groundwater in vertical cross sections. Netherlands Journal of Agricultural Sciences, 4, 102–131.

Ernst, L.F. (1971) Analysis of groundwater flow to deep wells in areas with a non-linear function of the subsurface drainage. Journal of Hydrology, 14, 58–180.

Ernst, L.F. (1978) Drainage of undulating sandy soils with high groundwater tables. Journal of Hydrology, 39, 1–50.

Ernst, L.F., de Ridder, N.A. & de Vries, J.J. (1970) A geohydrologic study of East Gelderland (Netherlands). Geologie en Mijnbouw, 49, 457–488.

Forchheimer, Ph. (1886) Uber die Ergiebigkeit von Brunnen-Anlagen und Sickerschlitzen. Zeitschrift der Architekten- und Ingenieur-Verein, 32, 539–564.

Forchheimer, Ph. (1914) Hydraulik. Leipzig, 566 p.

Geirnaert, W. (1973) The hydrology and hydrochemistry of the lower Rhine fluvial plain. PhD dissert Leiden University; Leidse Geol Meded, 49, 59–84.

Harting, P. (1852) De bodem onder Amsterdam, onderzocht en beschreven.Verhandelingen 1st Kl Kon Ned Inst, 3rd reeks 5, 73–232.

Harting, P. (1877) De geologische en physische gesteldheid van den Zuiderzee-bodem, in verband met de voorgenomen droogmaking. Versl Meded Kon Ned Akad v Wetensch, afd Natuurk, 2nd reeks 11, 302–325.

Harting, P. (1878) Nieuwe proeven over de doordringbaarheid van zand en van klei door water, en beschrijving van een zandschifter. *Versl Meded Kon Ned Akad v Wetensch, afd Natuurk,* 2nd reeks 13, 228–264.

Herzberg, A. (1901) Die Wasserversorgung einiger Nordseebäder. *Journal fņr Gas- beleuchtung und Wasserversorgung,* 44, 815–819; 842–844.

Hooghoudt, S.B. (1937) Bijdrage tot de kennis van eenige natuurkundige grootheden van den grond, deel 6. *Versl Landbouwk Onderz,* 43 (13) B, 461–676, The Hague.

Hooghoudt, S.B. (1940) Bijdrage tot de kennis van eenige natuurkundige grootheden van den grond, deel 7. *Versl Landbouwk Onderz,* 46 (14) B, 515–707, The Hague.

Huisman, L. (1972) *Groundwater Recovery.* London, Macmillan. 336 p.

Huisman, L. & Kemperman, J. (1951) Bemaling van spanningsgrondwater. *De Ingenieur,* 62 (13) B, 29–35.

Huisman, L. & Olsthoorn, T.N. (1983) *Artificial Recharge.* Pitman. 320 p.

King, F.H. (1899) Principles and conditions of the movements of groundwaters. In: *19th Annual Report of the US Geological Survey,* part 2. pp. 59–295.

Kooper, J. (1914) Beweging van het water in den bodem bij onttrekking door bronnen. *De Ingenieur,* 29 (38), 697–706; 29 (39), 710–716.

Kraijenhoff van de Leur, D.A. (1958) A study of non-steady groundwater flow with special reference to a reservoir coefficient. *De Ingenieur* 70 (19), 87–94.

Kruseman, G.P., de Ridder, N.A. (1970) Analysis and evaluation of pumping test data. ILRI, Bull. 11, Wageningen. 200 p (revised edn 1990, 377 p).

Lorentz, H.A. (1913) Grondwaterbeweging in de nabijheid van bronnen. *De Ingenieur,* 28 (2), 24–26.

Lueger, O. (1883) Theorie der Bewegung des Grundwassers in den Alluvionen der Flussgebiete. Stuttgart. 66 p.

Lueger, O. (1895) Die Wasserversorgung der Städte; Band I. Darmstadt. 834 p.

Mazure, J.P. (1932) Invloed van een weinig doorlaatbare afdekkende bovenlaag op de kwel onder een dijk. *De Ingenieur* 47 (13) B, 41–43.

Mazure, J.P. (1943) Enkele vergelijkende berekeningen betreffende de gevolgen van boven- en diepwateronttrekking in het duingebied. *Water* 27 (13), 117–124.

Molengraaff, G.A.F. & Dubois, Eug (1921) Prae-advies over de vraag van den Minister van Arbeid waaraan de aanwezigheid van artesisch grondwater in de duingronden te danken is. *Versl Meded Kon Akad v Wetensch* 30 (4/5), 208–215.

Pennink, J.M.K. (1904a) Investigations for ground-water supplies. *American Society of Civil Engineers, Transactions,* V 54 (D), 169–181.

Pennink, J.M.K. (1904b) De prise d'eau der Amsterdamsche duinwaterleiding. *De Ingenieur* 19 (13), 213–223.

Pennink, J.M.K. (1905) Over de beweging van grondwater. *De Ingenieur* 20 (30), 482–492 + 42 diagrams.

Pennink, J.M.K. (1907) Advies van den directeur der Gemeente Waterleidingen aan de Heren Burgemeester en Wethouders van Amsterdam op het rapport aan de Commissie van Bijstand, enz . . . , uitgebracht door den heer Th. Stang. Stadsdrukkerij, Amsterdam. 54 p.

Pennink, J.M.K. (1914) Verzoutingsrapport. Rapport omtrent het stijgen van het zoute-water, het toenemen van het chloorgehalte van het duinwater en het verminderen van de zoetwater-voorraad in de duinwaterwinplaats. Stadsdrukkerij, Amsterdam. 66 p. + 26 appendices.

Ribbius, C.P.E. (1903–1904) De duinwatertheorie in verband met de verdeeling van het zoete en zoute water in den ondergrond onzer zeeduinen. *De Ingenieur* 18 (15), 244–248; 19 (4), 71–77.

Rutgers van Rozenburg, J.W.H., Forster, J., van Lennep, H.S., de Vries, H., Henket, N.H. & Mees, A.W. (1891). Rapport der commissie van onderzoek inzake de duinwaterleiding van Amsterdam. Stadsdrukkerij, Amsterdam. 121 p. + 21 appendices.

Santing, G. (1957) A horizontal scale model based on the viscous flow analogy of studying groundwater flow in an aquifer having storage. IASH Publ 44. pp. 105–114.

Slichter, C.S. (1899) Theoretical investigations of the motion of groundwater. In: *19th Annual Report of the US Geological Survey*, part 2. pp. 295–384.

Slichter, C.S. (1902) The motions of underground waters. US Geological Survey, Water Supply Paper 67. pp. 11–106.

Stang, Th. (1903) Het verzouten van de Haagsche duinwaterleiding is in de eerste eeuwen ondenkbaar. Hand 9th Ned Natuur- en Geneesk Congres. pp. 427–443.

Steggewentz, J.H. (1933) De invloed van de getijbewegingen van zeeën en getijrivieren op de stijghoogte van het grondwater. PhD dissert, Delft University of Technology; Meinema, Delft. 138 p.

Stuyfzand, P.J. (1993) Hydrochemistry and hydrology of the coastal dune area of the western Netherlands. PhD dissert Vrije Universiteit Amsterdam; KIWA, Nieuwegein, 366 p.

Thiem, A. (1870) Die Ergiebigkeit artesischer Bohrlocher, Schachtbrunnen, und Filtergallerien. *Journal für Gasbeleuchtung und Wasserversorgung*, 14, 450–467.

Todd, D.K. & Huisman, L. (1959) Groundwater flow in The Netherlands coastal dunes. Journal of Hydraulics Division, Proceedings ASCE, 85, Hy 7, 63–82.

Van der Kamp, G.S.J.P. (1973) Periodic flow of groundwater. PhD dissert Vrije Universiteit Amsterdam, Editions Ropodi, Amsterdam. 121 p.

Van de Ven, G.P. (ed.) (2003) Man-made lowlands: History of water management and land reclamation in The Netherlands. Matrijs, Utrecht, 293 p.

Van Oldenborgh, J. (1916) Mededeelingen omtrent de uitkomsten van door het Rijksbureau voor Drinkwatervoorziening ingestelde geo-hydrologische onderzoekingen in verschillende duingebieden. *De Ingenieur* 31 (17), 458–467; (26), 474–498.

Venhuizen, K.D. (1968) De vorm van het zoetwaterlichaam onder de Amsterdamse Waterleidingduinen. Meeting Hydrologisch Colloquium, 6 maart 1968.

Verbeek, R.D. (1905) Artesisch drinkwater voor Amsterdam en 's-Gravenhage. Erven Bohn, Haarlem, 104 p.

Verruijt, A. (1969) Elastic storage of aquifers. In: de Wiest, R.J.M. (ed.) *Flow through Porous Media*. New York, Academic Press, 530 p.

Verruijt, A. (1970, 1982) *Theory of Groundwater Flow*. London, Macmillan. 190 p.

Versluys, J. (1916a) De capillaire werkingen in den bodem. PhD dissert Delft University of Technology, Versluys, Amsterdam. 136 p.

Versluys J. (1916b) Chemische werking in den ondergrond der duinen. Versl Meded Kon Ned Akad v Wetensch, afd Wis- en Natuurk, 24, 1671–1676.

Volger, O. (1877) Die wissenschaftliche Lösung der Wasser-, im besondere der Quellenfrage mit Rücksicht auf die Versorgung der Städte. *Zeitschr der Ver Deutscher Ingenieure*, t 21 (11), 482–502.

Volker, A. (1944) De hydrologie van het diepere grondwater onder het IJsselmeer. *Hydrologisch Colloquium*, HC 38, 17 p.

Volker, A. & van der Molen, W.H. (1991) The influence of groundwater currents on diffusion processes in a lake bottom: An old report review. *Journal of Hydrology*, 126, 159–169.

Vreedenburgh, C.G.J. (1935) Over de stationaire waterbeweging door grond met homogeen-anisotropische doorlaatbaarheid. *De Ingenieur in Ned Indië* 2 (6), 140–143.

Vreedenburgh, C.G.J. & Stevens, O. (1933) Electrodynamisch onderzoek van potentiaal-stroomingen in vloeistoffen, in het bijzonder toegepast op vlakke grondwaterstroomen. *De Ingenieur* 48 (32), 187–196.

Vreedenburgh, C.G.J. & Stevens, O. (1936) Electric investigation of underground water flow nets. *Proc Int Conf Soil Mech and Found Eng, Cambridge (Mass)*. pp. 219–222.

Wieringermeerrapport. (1936) Geo-hydrologische gesteldheid van de Wieringermeer. Rapp Meded Zuiderzeewerken 5, The Hague. 131 p.

Winkler, T.C. (1867) *Ons drinkwater en onze duinen*. Haarlem, Erven Bohn. 66 p.

The history of hydrogeology in Norway

Kim Rudolph-Lund
Editor, NGI, Ullevål Stadion, Sognsveien, Oslo, Norway
With contributions from:
Bioforsk, Holymoor Consultancy, NGU, NGI, NIVA, NVE, SWECO, UMB

ABSTRACT

The early development of hydrogeology was one of necessity. The art of drilling wells to supply groundwater to farms and communities gradually included an understanding of the science of hydrogeology. This activity was expanded and institutionalised by national authorities. Since the 1950s, the Geological Survey of Norway (NGU) was the dominant driving force behind the expansion of hydrogeology, often working with the Norwegian Water Resources and Energy Directorate (NVE) and the Agricultural University of Norway (NLH). Major groundwater mapping programmes were conducted during the 1970s and 1980s, supplying groundwater to large municipalities around the country. Results from these studies indicated that 25 to 30% of Norway's population could be supplied with groundwater, in contrast to the 13% using groundwater at the time.

Centers of hydrogeological study, research and consulting were established around the country during the 1970s and 198's, when higher institutions of learning began offering advanced degrees in hydrogeology. Activities in the large public consulting companies, notably Noteby and NGI, represented the expanding recognition of hydrogeology as an important earth science.

THE HISTORICAL BACKGROUND FOR HYDROGEOLOGY IN NORWAY

Abrief introduction

To understand the history of hydrogeology in Norway, a brief introduction to the country's geology, hydrology and demography is useful. Norway is situated on the western margin of the Precambrian Fennoscandian shield. The Caledonian orogenic belt runs as a backbone through the country, from the southwest to the northeast. In Permian and Carboniferous times, the region around Oslo was affected by extensional rifting and volcanism. Therefore, practically all of the country is underlain by igneous, metasedimentary or metamorphic bedrock.

Only very small residues of Mesozoic and Tertiary rocks are preserved above sea level in Norway today. However, some geologists believe that deep, subtropical weathering processes during the Mesozoic may have been responsible for the characteristic mineralogy (e.g. swelling clays) that is observed in the largest fracture zones in the crystalline bedrock of some areas of Norway. It is the mineralogy which in turn controls the hydrogeological behavior of those zones (Olesen & Rønning, 2008).

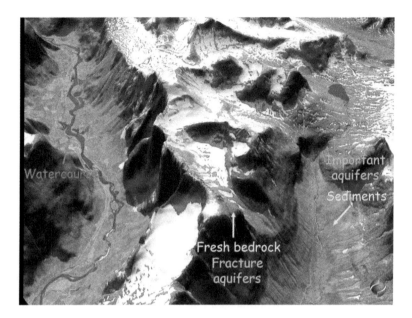

Figure 1   Groundwater sources in Norway.

During the most recent Ice Age, which ended about 10 000 years ago, the mountains were heavily eroded and the bedrock was deeply scoured by glaciers, resulting in the deep valleys and fjords which are visible today as typical features in the Norwegian landscape. Many of Norway's large, deep lakes are also a result of the glaciers, overlying local and regional structural features in the bedrock.

The glaciers also left behind drift deposits of variable types and thicknesses: upland areas are mainly overlain by thin deposits of till and peat, and lower-lying areas and valley bottoms are dominated by glaciofluvial gravel, sand and alluvial sediments. Due to post-glacial uplift, which was the result of isostatic rebound, previously submerged areas along the coast and the fjords are today covered by marine silts and clays.

There has always been relatively easy access to surface water resources in Norway. Annual precipitation ranges from 250 mm in rain-shadow area behind the mountains, to almost 4000 mm locally along the Atlantic coast. Norwegian surface waters have therefore been the main focus of water resources management, resulting in comparatively less interest in obtaining new knowledge and developing new technology related to groundwater resources. However, in the occasional cases when surface water quality is an issue, the inherent advantages of using groundwater come to light.

When people first settled in Norway after the Ice Age, they lived by hunting and fishing, before gradually starting to keep livestock and cultivate the land. The first farms came into existence from about 500 BC. Arable land was sparse in rural areas and farms were widely spread in the countryside. Traditionally, each farm had their private water supply from a brook, spring or shallow dug well. Towns and cities were usually fed by water from a lake or a river.

Peoples, myths and legends

Historically, Norway has been inhabited by several groups of peoples. The Sami (or Lapps) are usually considered to be the oldest indigenous people of Scandinavia and northern Russia, speaking a language of the Finno-Ugric group. The Northmen (also called Norsemen or Vikings), on the other hand, represent a northern Germanic branch of the Indo-European linguistic/cultural group that probably ultimately had its origins in the Near East or Central Asia. For many families, daily life was a struggle. It was a constant challenge combating the harsh northern climate while trying to provide enough food, water and shelter.

To make sense of their natural surroundings, the ancestors of the Northmen turned to their mythology. Lightning was explained as originating from the mighty god Thor's thunder-hammer called Mjølner. In Norse mythology groundwater springs were essential for life. The roots of the Norse "tree of life" "Yggdrasil", it was said, reached down to three subterranean wells – the Hvergelmer spring (the source of all rivers), Mimur's spring (a source of wisdom and creativity) and Urd's spring (the southernmost well, with warm waters). As in many other parts of the world, springs were given religious significance and their waters were believed to have healing powers. In Norse mythology, many springs were dedicated to Thor, also recognised as the god of rain and the protector of peasants, farmland and fertility.

When the Viking King Olav ("the Saint") forced Christianity upon his subjects in the early part of the 11th Century, as a smart trick, the springs of Thor were converted to Saint Olav's springs, while maintaining their alleged healing powers. Today, these springs can still be found in several places, including at Nidaros Cathedral in Trondheim where King Olav is buried.

The link between hygiene and health

The latter part of the 19th Century is recognized throughout Europe as the time when the link began to be made between hygienic living conditions, drinking water quality, sanitation and health. In around 1854, John Snow (1813–1858) connected occurrences of cholera to drinking water supplies in London. By 1882, this awareness had permeated into Norwegian literature, with Henrik Ibsen's play "En Folkefiende" (An Enemy of the People). In the play, the hero (or anti-hero, depending on your viewpoint), Dr. Stockmann, discovers that the springs feeding the town's tourist baths have become contaminated with industrial waste from a local tannery "All this filth up in Mølledalen – all of it, stinking to high heaven – it's contaminating the pipes leading to the well-house, and same poisonous muck is seeping out on the beach, too ...!" Dr. Stockman became an environmental activist, antagonizing local businessmen, politicians and the press, and finally reaped the community's retribution.

The modern Nordic influence

The modern years of hydrogeology in Norway were dominated by the first water well drillers. Well drilling in Europe was common in the middle of the 19th century. By 1890, when Norway was still part of a political union with Sweden, the Swedish Diamond Bedrock Drilling Company AB drilled most of the wells in Norway, Sweden and Finland. Much of the early work in Norway was inspired by the work of Swedish geologists who drilled exploration wells in bedrock along the coasts of Sweden and Finland in the 1890s as a part of ore exploration near Arkø, Sweden.

In early exploration drilling, it was natural to encounter groundwater at various depths, all shallow by today's standards. One well drilled in granite/gneiss to the depth of 35 meters gave 450 liters per hour of clear, potable water. It was reported that none of the other wells were drilled deeper than 30 m, but all gave good results. Based on these data, it was concluded that it was not necessary to drill deeper than 30 meters because it was at this depth the water amassed which had infiltrated through the fractured bedrock. Everywhere in Sweden and Finland where drilling took place at the time, water was encountered at a constant depth, slightly over 30 meters, with a production capacity of between 200 and 2000 liters per hour. In retrospect, the fractures encountered at these depths were probably relatively open horizontal "stress release" fractures near the surface.

Influenced by the Swedish geologists, the Norwegian Geological Survey of Norway (NGU) began drilling water wells for a number of fishing communities in northern Norway. All of the wells were drilled to between 30 and 45 meters and gave good results, with the exception of one well outside Henningsvær that produced saltwater. Around this time it was recognized that water-bearing fractures could be found deeper than the 30 m assumed by the Swedes to be the lower limit for productive wells. These deeper fractures were observed in mountainous terrain and folded bedrock (Helland, 1898).

Helland's background as a geologist led him to begin performing measurements at the turn of the twentieth century to find out how much water could be produced from different rock types. His results showed that in Norway, granite yielded 4–9 liters per m^3, Kolsås sandstone yielded 30 liters per m^3, Grip/Vardø sandstone yielded 10–21 liters per m^3, Dønna gneiss yielded 15 liters per m^3, Tonsåsen syenite yielded 23 liter per m^3, and Stamsund syenite yielded 4.4 liter per m^3. Helland also looked to Sweden where he found that the Bokedal gneiss produced 1.4 liters per m^3. These determinations should probably be interpreted as estimates of specific yield of between 0.1 to 3%. This was the first study in Norway which correlated rock types and groundwater resources.

The Swedish Diamond Bedrock Drilling Company AB, represented by Backe and Bonnevie in the Norwegian capital Kristiania (renamed Oslo in 1925), had a monopoly on the drilling market until 1920 and remained active in Norway until the late 1950s. In their 1916 brochure, written after twenty years of well drilling in Sweden and Norway, the company stated, "The wells we drill in bedrock have a 90% chance of giving more than 100 liters/hour. Only 2% of all bedrock wells have been considered dry wells!" (This is effectively the "non-parametric statistical" approach favored by recent Scandinavian hydrogeologists working in hard-rock terrain, and the yield associated with the 90% success rate is remarkably consistent with today's estimates (Banks et al., 2005)). The company advised to drill to 60 meters. "If the well is dry, the drilling rig should be moved 30 m to the side and you will most certainly find water", it was stated. At the time a new saying began to emerge in the water well drilling business: "The amount of water produced from bedrock wells can be roughly estimated, but not precisely calculated."

Concerning the occurrence of groundwater in sediments, one geologist wrote in the early 1920s that he doubted there were any places in Norway with gravel and sand accumulations similar to Finland and Sweden that would produce enough water for whole cities (Rekstad, 1922). The conclusion was based upon observations from the first large groundwater abstraction from sand and gravel in Åbo, Finland in 1916. This operation yielded only 32 l/s. This observation must be seen in the light of that period's assumption that groundwater was recharged from precipitation: an awareness of the possibilities that induced infiltration from surface water could offer did not dawn until the 1950s.

By 1940, attitudes to finding groundwater in sediments had changed. It was reported by NGU that in the Norwegian countryside where the bedrock was covered by sediment, "it was possible to find water for small enterprises – a small farm or a little settlement – even though Norwegian geology offers little groundwater". The most common types of wells were those dug in saturated sand/gravel or so-called "flåbrønner", designed to store water within less permeable material. These wells collected mostly surface water and precipitation which fell directly in the well; groundwater comprised only a small portion of the well water due to the surrounding less permeable material. The construction of deep wells with the help of cement rings or stones was an advanced engineering feat not without danger. In Kvam, however, a 37 m deep stone well with a pumping chamber was constructed to supply water to a soapstone mill. This piece of well drilling history has regrettably been destroyed.

Norwegian expertise takes over

The first Norwegian firm which started with well drilling was the Norwegian Diamond Drilling Company (NDDC) around 1912. This firm had earlier drilled for minerals for the Norwegian Government until the Government bought their own diamond-drilling machine and began their own drilling programme. NDDC used a German produced Urbanik drilling rig which required two men for taking 4 inch continuous cores. (Interestingly enough, it was an American who trained Norwegians in the use of the Urbanik.) This method drilled from 1.5 to 3 m per day, removing 100 mm cores ranging from 1 to 1.3 m.

After the German invasion on April 9, 1940, Norway was under military occupation and civil rule of a German commissioner in collaboration with a pro-German puppet government. In these war years, NDDC was forced to begin drilling wells for the German invaders. The forced enlistment of the NDDC drilling rig by the Germans was possible because of a complicated personal situation in the family of the rig's owner. The owner's daughter was married to an Englishman who was able to maintain his freedom only as long as the rig drilled wells for the Germans! The owner, however, probably did not lose too much sleep over unsuccessful water wells. This difficult situation continued until the occupation ended May 8, 1945 after the capitulation of German forces in Europe. Memory of the German occupation lingers long in Norwegian consciousness, but it is likely that German scientists and engineers introduced a number of highly innovative techniques during the wartime period. Indeed, one of the earliest applications of Leo Casagrande's large-scale electro-osmotic soil dewatering and stabilisation was carried out during the construction of the massive U-boat base (the *Dora* Complex in Nyhavna) in Trondheim, using 40 V and 20–30 A per well (Rasmussen & Haigler, 1953).

Figure 2 Cable-tool percussion drilling.

After World War II, there was disagreement in NDDC from 1945 to 1946 about whether they should continue with the Urbanik or switch to cable-tool percussion. As a result, an employee formed the new company O. Jansen Machine and Well Drilling. By 1950 a number of new drilling firms had also established themselves in Norway. Diamond drilling equipment, which until that time was the dominant drilling technique, was replaced by cable-tool percussion methods. This new method could drill up to 6 m per day. These machines dominated the Norwegian market for water well drilling during the 1950s and 1960s, drilling faster than the older machines and requiring only one man to do the drilling.

Drilling with cable-tool percussion was performed using a chisel and drilling rod. The rod weighed 300–400 kg and was lifted up and released 30 to 60 times per minute, depending upon the adjustable lifting distance. The crushed material was removed from the borehole with a dart valve bailer. There was no difference in the techniques used in drilling in loose sediments and bedrock. Drilled wells at this time were test pumped for 48 hours.

Drilling with cable-tool percussion was, however, very difficult and physically demanding work. Especially when drilling in bedrock, the chisel often needed to be repaired. With its 10 m high drilling tower and heavy rod, dangerous situations sometimes occurred. Occasionally, the drilling rig even tipped over; in some rare cases, workers were injured or killed.

The hydraulic top hammer was used by a number of well drillers, before it was replaced by drilling rigs using pneumatic down-the-hole (DTH) hammers in the early 1970s. The hydraulic top hammer wells were notoriously crooked and seldom drilled deeper than 60 m. The DTH drilling rigs could drill to greater depths than previously possible. The drilling capacity with these new rigs resulted in a ten-fold increase in drilling and started a new revolution within the well drilling industry.

By the middle of the 1970s the ODEX drilling machine was introduced into the Norwegian market. ODEX equipment enabled the drilling and casing of deep holes simultaneously in all types of formation, even those with large boulders. The method is based on a pilot bit and eccentric reamer, which together drill a hole slightly larger than the external diameter of the casing tube. This enables the casing tube to follow the drill bit down the hole.

The first compressed air machines were foreign made, but in 1976 the first Norwegian product NEMEK (Nestetogmekaniske) from Vinje in Telemark was introduced. This machine has since dominated the Norwegian market. The NEMEK machines had their own diesel engines, but much of the equipment still needed to be connected to an air compressor.

New developments improved the capabilities of the well drillers. Directional drilling gave the driller the ability to rotate from vertical to horizontal wells. To increase the water production in a well, controlled explosion, an old and accepted method, was also used. This method was replaced over the years by hydraulic fracturing which involved injecting water, at extremely high pressure, through a packer. Eventually, the water pressure becomes so great that it exceeds in-situ gravitational and tectonic stresses, and the tensile strength of the rock, and it literally "cracks" the rock, producing a new fracture (or, at the very least, opening up an existing fracture). Hopefully, the new fracture will connect into a wider, water-bearing, fracture network, enhancing the yield of the well.

With the increasing demands on well drilling, there was also a movement to establish a certification of well drilling during the 1980s. This led to voluntary certification of well drillers, with the intention of maintaining a certain standard for rendered services. The movement was coordinated by a Well Drilling Committee (WDC), a subcommittee of the Water Resources Committee in Norway at the time. The WDC had numerous representatives: one (the leader) came from the Norwegian Institute of Public Health, one from the Norwegian Municipal Alliance, one from Ramnes Municipality, and one from the Ministry of the Environment. Apart from the raising of standards for well drilling, WDC's work also helped to establish many well drillers in Norway.

THE GEOLOGICAL SURVEY OF NORWAY

The new science of hydrogeology

In 1951, the Geological Survey of Norway (NGU), established in Oslo in 1858, began a new program led by Per Holmsen examining the results from well drilling by NDDC in Norway. In 1952 a well drilling register was established after the Danish pattern. By 1958, the number of registered drillings in bedrock had reached 3000. The average

Figure 3 S. Skjeseth (right) performing a pumping test in the field. Notice his attire which was considered appropriate at the time.

depth of the drillings was 40 m, but some reached as deep as 226 m; 90% of the drillings were in southern Norway.

Steinar Skjeseth, a paleontologist by trade, took over the new Hydrogeology Section in 1953 (Figure 3), after NGU Director Føyn expressed his dissatisfaction with mounting drilling costs. It was hoped that young Skjeseth was the right man to reign in the increasing costs of the burgeoning drilling program. However, Skjeseth soon proved to be more interested in developing the new science of hydrogeology rather than limiting the associated costs of drilling. The end result, nevertheless, was that the hydrogeology section at NGU, with its new leader, increased the focus on and raised the standards for the new science of hydrogeology in Norway.

By 1958 over 600 new wells were drilled annually in Norway. The knowledge garnered during the well drilling operations was deemed so important that a new initiative was taken to start the publication series "Notes from the water drilling archives". In all 14 separate monologues were published in this series, the last one being published as late as 1966 (Meddelelser fra vannboringsarkivet, Nr. 1–14).

NGU moves – hydrogeologists stay behind

The decision to move NGU from Oslo to Trondheim was taken by the Norwegian Government on February 28, 1957. There was some indecision as to whether NGU should move to the Norwegian Institute of Technology (NTH) or to a new building that would be constructed at Østmarka in Trondheim. The Ministry of Industry finally decided on May 14, 1957 that NGU should be moved to the new building. In fall 1961 the move was completed, but not all of the employees moved to Trondheim. The Hydrogeology Section, with its five professionals, remained in Oslo. It was reasoned that since most of wells were drilled in southern Norway, it would be more practical to keep the section in Oslo, close to the action.

The Hydrogeology Section was headed by Skjeseth and had five employees. Skjeseth was known as a driven professional, a gifted lecturer, intense, and outgoing. He continuously tried to transfer his own knowledge not only to his fellow hydrogeologists, but also to other professional groups he came in contact with during his work. Two things Skjeseth emphasized were: make observations and solutions together, and immerse yourself in the facts in order to understand the problem. By the end of the 1950s, Skjeseth and two other employees were working full time with bedrock wells.

In 1958 there were only 15 bedrock wells using casing. None of these 100 mm diameter wells used filters. In this year two new wells were drilled in unconsolidated sediments: one for Rena Packaging Factory, and one at Sagtjernet for Elverum Municipality. Norsk Dypbrønnsboring successfully installed the wells using available material and equipment. The wells were drilled using an arm attached to a tractor which drove back and fourth. The work was extremely difficult in the rocky soil. But with the successful completion of these two wells, a new era began for Norwegian water supply that, especially for the eastern Norway region, lasted until 1985.

Drilling for water in bedrock became a common strategy for water supply by the end of the 1950s. At the farming exhibition at Ekeberg in 1959, the Hydrogeology Section participated by drilling a well to 70 m in order to supply water to the farmers at the exhibition.

In addition to constantly updating and analyzing the data in their drilled well register, from 1960 to 1970 NGU focused increasingly on finding the best locations for water wells. This resulted in groundwater being quickly accepted as a viable alternative to surface water. This new acceptance, for using groundwater produced from sand and gravel aquifers, applied not only to small country villages, but also to towns and cities.

In the Hydrogeology Section, the number of employees increased with the work load. A new move saw the section sharing offices with the hydrogeological section of the Norwegian Water Resources and Energy Directorate in the summer of 1977. The number of employees at this new "Oslo office" grew to 18 by 1984, 12 of whom worked for Hydrogeology Section. By this time, water supply studies had topped one hundred, with over half of these being conducted for the local municipalities.

Groundwater mapping

International Hydrological Decade

During the International Hydrological Decade (IHD), a program running from 1965 to 1974 conducted under the auspices of the United Nations, the Norwegian national IHD committee started groundwater level measurements in three "representative basins": Filefjell (high mountain terrain with a thin cover of till), Sagelva (a thick cover of till) and Romerike (lowland terrain with glacifluvial deposits). Particular emphasis was given to the Romerike area where groundwater was plentiful and of economic importance. NGU coordinated activities and was active in data collection and analysis. At the end of the IHD, the Romerike drainage basin had all together three streamflow recording gauges, about 180 tubes and about 20 wells used for recording groundwater levels.

Hydrogeological mapping

In addition to NGU's consulting for the public, the production of groundwater maps gradually became an important activity at the survey. Hydrogeological maps, with descriptions, of Bergen (1115I) and Drøbak (1814II) at a scale of 1:50 000 were published in 1978 and 1979, respectively. These were color maps with bedrock geology where detailed fault analysis and registered data from drilled water wells gave the basis for the interpretation of expected water yield from bedrock locations. NGU performed test drilling in selected areas with unconsolidated sediments, and yields greater than that from bedrock were indicated on the maps. Because of the high cost of producing the maps, due to extensive field and interpretation work, their production was ultimately discontinued by the survey.

Groundwater resource mapping

At about the same time, the survey began producing another series of groundwater resource maps called "Ground water in unconsolidated sediments" at 1:50 000 scale, the so called "blue series". Despite its title, this was a series of black and white maps, with descriptions, that synthesized the results from earlier groundwater surveys in unconsolidated sediments within map quadrangles. The earlier results were supplemented with new field assessments and exploratory drillings to characterise the potential groundwater quantity and quality in sand and gravel deposits.

The map text in the blue series comprised a general description of groundwater and a special section about the possibilities for extracted groundwater from unconsolidated sediments, along with short descriptions of the groundwater potential in the bedrock in that area. The area profiles also contained groundwater quality analysis, drilling profiles with grain size distribution curves for sampled sediments, and groundwater quantities extractable using 5/4″ sand filter well points. A total of 41 groundwater resource maps were produced from 1976 to 1987, and is considered the best groundwater series that has ever been produced in Norway.

Trial mapping in Oppland and Finnmark

By the 1980s, NGU had received requests from numerous municipalities and property owners around Norway to focus their mapping efforts more on the needs of the end users, for example using administrative boundaries rather than the limits of the map quadrangles. The end users also wanted to give more input concerning the content of the maps.

In an answer to these requests, NGU decided to make a sample map of the municipalities of Oppland and Finnmark. Each municipality was contacted prior to the mapping in order to prioritise areas with a need for improved water supply, then field work was carried out in each of the prioritised areas and test drillings were made on unconsolidated sediments with a potential for groundwater extraction. The bedrock in the prioritised areas was also surveyed and the potential water quantity was indicated based on bedrocks maps and data from NGU's well drilling archive.

Reports were made for all of the municipalities (26 in Oppland and 19 in Finnmark) with maps that indicated, for the prioritised areas, the location and description of unconsolidated sediments and bedrock with the potential for ground-water extraction. The disadvantage with the maps was the exclusion of proved and possible groundwater occurrences outside the prioritised areas.

Groundwater in Norway (GiN) Project

Another project which had considerable impact on hydrogeology in Norway was the "Groundwater in Norway" or GiN project (Ellingsen & Banks, 1993). This project was initiated in 1989 by the Ministry of the Environment and then coordinated by NGU. The program included the development of new methodologies, index mapping, registration and evaluation of groundwater occurrence and pollution threats, in addition to new initiatives to inform municipalities and county administrations. Fifteen of Norway's counties participated in the project. The non-participating counties were: Oppland and Finnmark (which had already been mapped during their own project, as mentioned earlier), Møre and Romsdal (which had already been surveyed by the Sogn and Fjordane University College), and Oslo (which already had a good water supply).

Mapping in the individual counties was performed by a formal working group composed of:

- one hydrogeologist
- one specialist in Quaternary geology
- one specialist in structural geology.

In addition to NGU personnel, hydrogeologists and geologists from Sogn and Fjordane University College, Telemark University College, the University of Bergen and several consulting companies were involved in the mapping.

The different municipalities in the counties were also divided into A and B municipalities depending upon the urgency of the need for local surveys. Field work was carried out in A-municipalities, while in B-municipalities the mapping was based upon existing geological information at NGU. Mapping was performed in areas prioritised by the municipalities with the exception of the counties Østfold and Vestfold where a general evaluation was performed for possible aquifers.

Based upon the mapping in the individual municipalities, the possibility for groundwater supply from bedrock or available unconsolidated sediments was characterized as "Good", "Possible" or "Poor" according to the actual water need. Some of the locations classified as "Possible" were, after further investigation, found to have a lower potential for groundwater supply than first expected. In any event, the results from the study concluded that 25 to 30 percent of Norway's population can be supplied with groundwater, in contrast to the 13 percent which was using groundwater during the project period.

The GIN project was completed after four years and produced:

- 261 municipality reports covering 301 municipalities
- 15 county reports (summaries of the municipality reports)
- 13 GiN instruction manuals with practical information about finding and using groundwater.

A brief summary of the GIN project and a full list of reports and publications produced during the project period is given in NGU (1992).

NVE – from rain to groundwater

In 1961, the Norwegian Water Resources and Energy Directorate (NVE) began to reorganise their work with groundwater. Earlier groundwater observations had already been recorded by NVE in 1949 at Groset in Telemark. The assignments in the first years were mainly concentrated on regulation investigations. These studies mapped the influence of rivers on the surrounding groundwater regimes in order to evaluate the effect of the imposed regulation in the river basins on local groundwater conditions. The methods developed by NVE to perform these studies included correlation analyses of observed river levels with measured groundwater levels at selected observation points, as well as mapping of groundwater in the affected areas.

The first surveys were carried out in Østerdalen (Stor-Elvdal and Rendalen) in 1961 and required assistance from both NLH and NGU. In 1963 a new survey was performed in Lærdal, and others followed as the result of new plans to utilise more of Norways' waterfalls for hydroelectric power production. The growing number of groundwater surveys soon required additional administration in the Hydrologic Department. The Groundwater Office (GWO) was formed in 1961, but it was only in 1963 that the necessary resources were in place for the office to begin functioning properly.

During 1971 the possibility of severe flooding after a period of large snow fall gave rise to the "Committee for run-off predictions" (UFT) with their secretariat placed at the GWO. The feared flooding did not occur because a large part of the flood waters went to recharge the groundwater reservoirs along the rivers. The hydroelectric producers were so fascinated by what happened that they funded new research around run-off predictions. UFT remained active until 1977 after which their activities were transferred to the GWO.

During their period of activity, UFT developed and published a method to calculate the size of available groundwater resources based on depletion curves constructed from water level observations at an observation point in the river basin. UFT equipped and ran a large number of observation fields over the entire country, where records were kept of groundwater levels, ground frost depths, snow depths, air temperature, and runoff in the lower river basin. In coordination with record keeping at the GWO, these records were entered into the shared Hydrologic Department's data archives.

When the IHD period came to an end in 1974, the Norwegian Hydrologic Committee suggested the establishment of a National Groundwater Network (LGN), combining the infrastructure established with IHD support with other existing activities. The national observation network was put together and completed by 1977, as they could use a number of the already established UFT stations. The observation programme was increased to include measurements of groundwater temperature together with analysis of chemical parameters. Administration and budgeting for the LGN was given to NGU. NVE provided an engineer for the field work, a position which in 1984 was permanently itemised in the NVE budget for the LGN. NVE was responsible for all of the LGN data processing since they already had a fully operational data processing centre for the storage and presentation of hydrologic data. Since then, important development and modernisation of the LGN stations has taken place. By the end of the 1980s the network measured groundwater levels at approximately 80 measuring points in 43 different localities.

Agricultural University of Norway (NLH)

The pressing need to improve the water supply situation in Norwegian farming communities in the middle of the 1960s was one of the reasons for the establishment of a new Professorship in geology at the Agricultural University of Norway (NLH) in 1965. It was an easy transition for Skjeseth from NGU to fill this position and take responsibility for hydrogeology at NLH, since he had already been teaching part-time classes on the subject for several semesters. Skjeseth's background from NGU contributed to the good working atmosphere that already existed between hydrogeologists at NGU and NLH.

Skjeseth was an inspiration for his colleagues and worked tirelessly at championing the use of groundwater in Norway. His hundreds of visits in the field gave him an intimate insight into the hydrogeology of southeast Norway. Beside his professional duties, he conducted courses and held lectures for well drillers and helped them in the field. The drillers saw Skjeseth as someone they could ask complicated questions from the field and receive good, practical advice. For the general public, Skjeseth published popular science publications and had a series on Norwegian TV about geology and groundwater in Norway.

Skjeseth's experience with hydrogeology was first incorporated into the basic geology curriculum at the college. Later, the first formal education in hydrogeology began with the creation of the first undergraduate course in hydrogeology in the spring semester of 1977. Graduate courses were also added and the first hydrogeologist in Norway graduated in 1978 after performing a detailed study of glacifluvial aquifers under marine clays. The first Norwegian doctorate in hydrogeology was granted in 1982, based on work in the Fetsund delta area (Ree, 1984). Life, however, was not easy for the new hydrogeology students working in the field. Drillers were not used to new techniques that involved pumping tests, wells were expensive and equipment had to be purchased or borrowed. One hydrogeologist recalls purchasing an old cable tool machine from the widow of an operator who had perished under the mast a few years earlier.

In the late 1970s, well stimulation of crystalline aquifers became popular. In practice, this meant using considerable amounts of water pressure tolerant explosives. The spectacular geysers, shooting out from boreholes loaded with 20 to 40 kg of explosives, were impressive sights. The stimulation would often stimulate the yield, sometimes reduce the yield, and invariably render the water from the well unfit for human consumption for months!

The field of hydrogeology at NLH expanded during the 1970s and the 1980s. The Norwegian Farming Science Research Council (NLVF) gave research grants for studies on water supply within small farming communities and later for studies of water in the unsaturated zone. A new position for a professor, working within several areas of hydrogeology and specialising in hydrochemistry, was added at NLH in the 1980s. In 1984, an assistant professorship was also added at NLH to carry out research in hydrogeology. The Norwegian Hydrologic Committee (NHK) was an important driving force behind granting research positions at NLH and received their funding from the state controlled Concession Fund. Employees were persistent in their efforts to obtain funding for hydrogeological research at NLH. NHK was later incorporated into the Norwegian Research Council (NFR), where interest in hydrogeologic research diminished after the 1980s.

In spite of the lower funding by NFR, cooperation between NLH and the University of Bergen (UiB) began in the middle 1980s. By 1987, a new national field course in hydrogeology was organized with practical support from NGU at Kaldvelladalen, south of Trondheim. The two-week field course soon became popular with aspiring hydrogeologists and eventually included the University of Oslo as well as the Norwegian Institute of Technology.

Jordforsk – georesources and pollution research

In 1977 NLVF was of the opinion that farming research was poorly organized and they created, after an initiative from Skjeseth, a "Steering committee for agricultural research". The committee began so much original farming research that by 1981 it received from NLVF status as an independent institute with the name of GEFO ("Institute for Georesources and Pollution Research"). In the late 1980s GEFO merged with the Norwegian Soil and Peat Company and was reorganized as a foundation. Jordforsk, as the new group was known, was run as an independent research institution in Ås with less funding from NFR.

Public consulting sector

Locating water supply wells in rural areas was always an important activity for the Hydrogeology Section at NGU. The number of such projects neared an all time high towards the late 1970s. By this time NGU had performed about 500 assessments. Most of these concerned water supplies for single family dwellings, mountain cabins and farms; the evaluations were mostly performed for well drillers and private persons, as well as some large municipal water supply projects.

Many of these groundwater projects were located in unconsolidated sediments, with the project leadership being provided by NGU. They were responsible for choosing the locations and performing the follow-up studies. This led to the establishment of a number of large public groundwater supplies, as in the case of Kongsvinger, Elverum, Lillehammer, Voss and Hønefoss. Some smaller ones were also established in Odda, Kautokeino and in most of the population centres in Gudbrandsdalen.

Over the years, the requests for help from the state hydrogeologists employed by NGU began to decrease for a number of reasons:

• water well drillers began using hydraulic fracturing to increase borehole yield
• the appearance of a new generation of water well drillers who often were better educated in their profession
• consulting firms with well qualified hydrogeologists began to offer their services in the market.

Noteby

In 1975, the Oslo based consulting engineering firm Noteby established a Hydrogeology Group to meet the growing demand for hydrogeological consulting skills. This was a daring move, in direct competition with the government employed hydrogeologists at NGU. The company employed some of the first students with dedicated training in hydrogeology. Most of these employees came from the new programme for hydrogeology at the Norwegian Agricultural College and from the geology department at the University of Oslo. Projects mostly involved development of groundwater sources and sanitary/industrial landfills in Norway.

By the late 1970s, Noteby had increased their activities so much in Africa that they transferred six hydrogeologist to Kenya. Groundwater studies in East Africa and especially Sudan were performed for the Norwegian Agency for Development Cooperation (NORAD), the Swedish International Development Cooperation Agency (SIDA) and the United Nations Development Program (UNDP) well into the 1980s.

NGI

Groundwater has historically created geotechnical challenges in Norway. During the 20th century, new techniques were developed to prevent leakage and subsidence problems near tunnels and open excavations. The Holmenkollen tunnel is a classical example of this type of subsidence (Holmsen, 1953). The construction was started in 1912 with sections still being completed in the 1970s. The Norwegian Geotechnical Insitute (NGI), a private foundation established by a government resolution in January 1, 1953, was engaged to determine the geologic conditions and to monitor pore pressures and follow up on the subsidence problems. With assistance from Noteby, they were responsible for schemes for artificial infiltration of water.

NGI has also worked extensively with the construction of dams for hydropower. With over 250 dams in Norway, there has been considerable research around the development of these structures. This research has led to new ideas about dam siting, construction material and requirements needed to control groundwater flow and seepage.

By the 1980s, hydrogeology was an integral part of NGI's consulting activities in working for national and international clients within the areas of natural hazards, environmental protection, oil and gas, building and construction, and transportation.

REFERENCES

Banks, D., Morland, G. & Frengstad, B. (2005) Use of non-parametric statistics as a tool for the hydraulic and hydrogeochemical characterization of hard rock aquifers. *Scottish Journal of Geology*, 41 (1), 69–79.

Ellingsen, K. & Banks, D. (1993) An introduction to groundwater in Norway – Promotion and reconnaissance mapping. In: Banks, S.B. & Banks, D. (eds.) *Hydrogeology of Hard Rocks, Mem. 24th Congress of International Association of Hydrogeologists, 28 June–2 July 1993, Ås (Oslo), Norway*. pp. 1031–1041.

Helland, A. (1898) Fiskeværenes forsyning med vand. *Norsk Fiskeritidende*, h. 4.

Holmsen, G. (1940) Grunnvannsbrønner. *Norges geologiske undersøkelse. Småskrifter Nr. 4.*

Holmsen, G. (1953) Regional settlements caused by a subsidence tunnel in Oslo. *International Conference on Soil Mechanics and Foundation Engineering, 3. Zurich 1953. Proceedings,* Vol. 1. pp. 381–383.

Jansen, O. (1995) Private conversation between T. Klemetsrud and O. Jansen as related to the editor.

Klemetsrud, T. (2003) Brønnboring i Norge. Historisk oversikt. Unpublished diary.

NGU. (1992) Grunnvann i Norge (GiN). Sluttrapport [Groundwater in Norway (GiN). Final report – In Norwegian] the publication series. NGU Skrifter 111, 23 p + appendices.

Olesen, O. & Rønning, J.S. (2008) Dypforvitring: fortidens klima gir tunnelproblemer. Gråsteinen 12, 100–110, Norges geologiske undersøkelse, Trondheim.

Rasmussen, W.C. & Haigler, L.B. (1953) Ground-water problems in highway construction and maintenance. *Delaware Geological Survey Bulletin No. 1.* 24 pp.

Ree, B. (1984) Recent delta deposits as groundwater resources: the Fetsund delta aquifer in Lake Øyern, SE Norway. Dr. Scient. dissertation 651, Department of Geography, University of Oslo. Rapportserie hydrologi 4, University of Oslo.

Rekstad, J. (1922) *Grunnvatnet. Norges Geologiske Undersøkelse Nr. 92.* H. Aschehoug & Co.

References in the series *"Meddelelser fra vannboringsarkivet"*:

Nr. 1: Holmsen, P. (1953) En orientering om arkivets arbeidsgrunnlag. Om Samarbeide med borefirmaene. Den viktigste fennoskandiske faglitteratur. NGU 184, 5–11.

Nr. 2: Skjeseth, S. (1953) Vannboringer utført i traktene omkring Mjøsa og Randsfjorden 1950–52. NGU 184, 12–22.

Nr. 3: Skjeseth, S. (1956) Kambro-silurbergartenes hydrogeologi i Mjøstraktene. NGU 195, 15–36.

Nr. 4: Holmsen, P. (1956) Oppsprekning, topografi og vannføring i massive dypbergarter. NGU 195, 37–42.

Nr. 5: Skjeseth, S. (1957) Kvaliteten av grunnvann. NGU 200, 55–67.

Nr. 6: Skjeseth, S. (1958). Vann i grus og sand. NGU 203, 80–87.

Nr. 7: Skjeseth, S. (1958). Norske kilder. NGU 203, 88–99.

Nr. 8: Skjeseth, S. (1959) Rørbrønner ved Rena og Elverum. NGU 205, 160–170.

Nr. 9: Hagemann, F. (1959) Vannboring i Øst- og Midt-Finnmark. NGU 205, 84–98.

Nr. 10: Bryn, K.Ø. (1961) Grunnvann øst for Oslofeltet. NGU 213, 5–19.

Nr. 11: Hagemann, F. (1961) Grunnvann i Vestfold. NGU 213, 29–48.

Nr. 12: Skjeseth, S. & Klemetsrud, T. (1962) Rørbrønner. NGU 215, 87–100.

Nr. 13: Hagemann, F. & Klemetsrud, T. (1965) Rørbrønnfiltere. NGU 234, 53–63.

Nr. 14: Englund, Jens-Olaf (1966) Grunnvann i Sparagmittgruppens bergarter i Syd-Norge. NGU 242, 5–18.

History of hydrogeology: Poland

Jan Dowgiałło[1] & Bogdan Kozerski[2]
[1] Polish Academy of Sciences, Institute of Geological Sciences, Warszawa, Poland
[2] Gdańsk University, Geography Department, Gdańsk, Poland

INTRODUCTORY REMARKS

Since the 15th century, when the first signs of interest in groundwater appear in written documents, the Polish territory has changed many times, both in size and location. During the period 1795–1918 the Polish independent state did not exist at all due to its partition between Austria, Russia and Prussia. After WW2, Poland lost one third of its territory in the East and expanded to the West. Therefore, to write about the history of hydrogeology in Poland, a science strictly linked to territory, one must first define its borders

We work on the premise that Polish hydrogeology was developed not only in the territories now belonging to Poland, but also in areas both lost and gained throughout the course of history. This methodological assumption results in considering areas now belonging to the Czech Republic, Slovakia, Hungary, Ukraine, Belarus, the Baltic republics and Russia as well as former parts of eastern Germany. Changes to polish borders only since 1795 are shown in Figure 1.

Another question to be considered is the fact that the Polish independent state practically did not exist for 123 years. During this period a considerable number of Polish hydrogeologists worked abroad, some of them with great success. Although achieved far from the Polish territory, this success, especially in the area of research methods, certainly forms a stage in the development of Polish hydrogeology and cannot be ignored.

The first part of this paper, covering the period between the 16th century and WW1, was prepared by J. Dowgiałło The second part, covering the period between 1918 and the late 1970s, was written by B. Kozerski. Both authors assume joint responsibility for the complete paper.

PREHISTORY AND EARLY HISTORY (15TH–18TH CENTURIES)

The times preceding the 19th century may certainly be called the prehistory of modern science. This obviously concerns hydrogeology, which at that time did not yet exist as a science although many, often quite correct, scientific observations had been made and published in this early period.

Figure 1   Map showing important border changes of Polish territory.

The first writers who recorded information about groundwater were early histori-
ans. Almost every comprehensive work concerning the history of Poland was equipped
with an introductory chapter containing a general description of the country, its borders, rivers, mountains and natural resources. The first and most eminent of them was
by Jan Długosz (1415–1480) [Figure 2]. In his "opus magnum": "*Annales seu Cronicae
incliti Regni Poloniae*" ("Annals or Chronicles of the famous Polish Kingdom" – first
printed 1883–1887) he gives a detailed description of Poland's hydrography (in the
introductory part entitled "*Chorographia*"). Writing about rivers he always mentions
the location of their fountain-heads, sometimes including interesting details as for
instance in the following passage: "... Nida R. the fountain spring of which is in the
Moskorzew village, at a certain section hides and flows underground and again in
the forest of the Dzierzkowa village ... reappears ...". In fact this is the first known
record of the Miocene gypsum karst phenomenon present in the Nida R valley. Długosz
also describes rivers which have their sources in lakes. Sometimes his information is
wrong, such as in the sentence quoted by Kowalenko (1975): "Aquae acidulae in
Polonia absunt" ("There are no carbonated waters in Poland").

Figure 2 Jan Długosz (1415–1480).

Matthew Karpiga of Miechów, called Miechowita, is the author of "*Chronica Polonorum*" ("Chronicle of Poles") (1519) in which several references to ordinary and mineral springs appear (Kowalenko, 1975). In another work, "*Conservatio Sanitatis*" ("Maintenance of Health") (1522), he included a part entitled "Tractatus de Aqua" ("Treatise on water"). Basing his reasoning on Aristotelian ideas he discusses the origin of heat and cold in water, and following Avicenna he mentions the healing effects of drinking and bathing in waters containing iron, copper, aluminum and sulphur.

Another historian providing some hydrogeological information was Marcin Kromer (1512–1589) [Figure 3]. His highly informative work known as "Polonia" (the complete Latin title is extremely long) was first published in 1575. In the introductory chapter the author mentions amongst other things the presence of saline springs in southern Poland (the Spisz province, now in Slovakia). In the same area there exist, according to Kromer, waters which become "petrified" and build channels within which they flow. This is certainly reminiscent of karstic caves appearing in the eastern part of the Tatra Mountains, as well as of travertine rocks accompanying some carbonated water warm springs. In the same area Kromer situates springs of water "harmful for the health", producing "disastrous reeks" which kill animals drinking it or inhaling its fumes. On the other hand he also writes about springs of "warm water having a fragrance of sulphur" which heal ulcers and lichens both human and animal. In general, these early works informing about groundwater show a mixture of superstition and some rational reasoning.

Figure 3 Marcin Kromer (1512–1589).

In the 16th and 17th centuries a somehow more scientific approach was represented by physicians and pharmacists who tried to apply mineral and thermal waters
to heal diseases. In their works they could not ignore the springs of water they studied
and noted sometimes important details concerning the shape and yield of the spring
as well as the smell, colour, taste and temperature of the water. Thus, these springs
became the first "hydrogeological objects" which appeared in scientific writings.

Wojciech Oczko (1537–1589) [Figure 4] was an outstanding physician, the official doctor of the royal dynasty, author of among other works two treatises published
in Polish: "*Cieplice*" ["Hot springs"] (1578) and "*Przymiot*" ["Syphilis"] (1581).
Although both works concern mainly medical treatments with the application of
numerous medicaments including thermal waters, considerable parts of them are
devoted to considerations of the origin of these waters, the possible sources of their
mineral content, the causes of their high temperature and their ways to the surface.
Oczko argues that even cold mineral waters may be called "thermal" as they lost their
heat during their way to the spring. This is the reasoning which substantiates the
common notion of "thermalism" or "thermal station" used in many European spas
which in fact have nothing to do with real natural thermal waters. In fact, the only

Figure 4 Wojciech Oczko (1537–1589).

"thermal" springs mentioned by Oczko are those producing cold, sulphurous mineral waters at Szkło near Lwów, Mikulińce near Trembowla, Swoszowice near Kraków and at Drużbak in the Spisz province. Only the last water can be called thermal by our standards, its temperature slightly exceeding 20°C.

As a humanistic erudite Oczko devotes much consideration to thermal waters in several European countries. Writing about some of them he wonders how coming from depths where coldness seems to dominate they become hot near the surface. Certainly, scientific consequence is in this case not his strong point.

Oczko is also discussing the ideas of Aristotle concerning the origin of "holy" (curative) waters. According to him: "... He (Aristotle) attributed their holiness to sulphur and thunder; we will attribute it not to their warming capacity, not to their sulphur smell, but to the fact that they greatly help to fight diseases which doctors and barber surgeons resigned to treat".

Another famous physician interested in mineral waters was Erasmus Sixtus (1570–1635). He lived and worked in Lwów, and his main work concerns mineral waters at Szkło situated about 50 km North West of this city. His work "*De Thermis in Pago Sklo libri tres*" ("About hot springs in the village Szkło three books") appeared first in Polish in 1617 (the second Latin edition was published in 1780). Szkło (now in Ukraine) was once a fashionable spa often visited by King Jan Sobieski and his court. Mineral water is linked to the presence of Miocene (Badenian) gypsum deposits and their reduction resulting in H_2S production. The work of Sixtus deals mostly with medical problems but Book II is almost entirely devoted to considerations on the origin of the mineral water and its chemical properties. All the then accessible methods of analysis (organoleptic proofs of water and its precipitate after evaporation) were used by Sixtus and he can justly be called "the first Polish balneochemist" (Kowalenko, 1984). He also gives a description of the then existing buildings and "technical" spa installations.

Jan Innocenty Petrycy (1592–1641) was the third member of the group of important Renaissance physicians interested in the application of mineral waters. In 1635 he published a book entitled "*About waters at Drużbak and Łęckova, their use and diseases they are able to heal*". Drużbak (now Rużbachy Vyżne in the Spisz province of Slovakia) is known for its thermal (up to 23°C) carbonated water producing travertine walls which were exploited for construction. Petrycy writes about the mineral water as "being petrified", but rightly attributes the origin of travertine to limestone "mixed" with water. His other interesting observation is that occasionally the clear spring water becomes "turbid like scum" and that this phenomenon is preceded by bad weather. Not knowing the notion of atmospheric pressure and its drop he, of course, could not explain what he observed.

Although the major part of Petrycy's work is devoted to medical problems and healing capacity of waters in both localities (Łęckova was a village close to Druzbak as indicated by Majer-1841), he considers in detail the "chemistry" of waters and devotes some attention to the shape and size of the timber lining of the springs. Although Petrycy's work was comprehensive and detailed, it has to be mentioned here that thermal waters at Rużbachy and at several other localities of the Spisz province were described much earlier by Wernher (1549). The question whether Petrycy knew the work of his predecessor remains open.

Kowalenko (1984) quotes two French authors who published papers about mineral waters. Both wrote about Iwonicz, a renowned spa in the Polish Carpathians based on saline waters with high methane content. One of these authors was supposedly K. L. Conradi, physician in ordinary of Queen Maria Sobieska (in fact the name of the actual author was Braun). In 1684 Conradi published in Latin an impressive volume (325 pages) entitled "*Descriptio curiosa fontis cuiusdam inflammabilis et medicinalis in Polonia*" ("Accurate description of an inflammable and medicinal spring in Poland"). Another supplementary paper was published in French by the physician J. Ch. Denis: "*Relation curieuse d'une fonteine decouverte en Pologne laquelle entre autres proprietes a celles de prolonger la vie jusqu'a cent cinquante ans*" ("Peculiar account about a fountain discovered in Poland which besides other properties has that of extending life till hundred fifty years") (Paris, 1687). Although the author calls the spring "mont merveille" ("miracle mountain"), thus demonstrating elements of exaggerated marketing, both pathetic and improbable, he tries to give a scientific commentary to the phenomena only noted in the previous book, such as the "combustibility" of water (due to the presence of methane) decrease and increase of the water level in the spring depending on moon phases and, last but not least, the healing potential of water.

Wojciech Tylkowski (1624?–1659), a philosopher and theologian, is the author of the 9 volume work "*Philosophia curiosa*" ("A thorough Philosophy"). In volume

9, entitled *"Meteorologia curiosa"* (1669), there is one chapter dealing with springs, rivers and shallow ground waters. Although not original and based mostly on the ideas of Aristotle, Tylkowski is the first Polish author who gives a synthesis of the then existing opinions regarding ground waters. According to him (fide. Piasecka, 1970) all rivers originate from the sea. Their sources are filled with condensed vapor produced from sea water evaporating underground. The condensation takes place close to the surface due to contact with cold rocks. The acceptance of this hypothesis after Aristotle enabled Tylkowski to explain the presence of springs in the mountains. He also did not exclude the existence of direct links of some springs with the sea. In some lakes, according to him, pieces of ships sunk in the sea are found on the water surface, pushed there by water.

Tylkowski was the first Polish author to explain the red color of water in some streams as originating from adjacent soil and rocks. However, his considerations about springs are much less convincing. He was interested only in peculiar cases, entirely ignoring ordinary springs. He thus writes about springs which are periodical, yield hot or unusually coloured water, about springs producing healing or harmful water, springs which react to thieves and illegitimate children, to perjury and ill manners of women.

In another work "Scholarly conversations" (1692) Tylkowski specifies once again that seawater penetrating the Earth's interior forms numerous water veins. Water is evaporating, afterwards condensing close to the surface and feeding the springs. A new idea regarding the disappearance of some springs is that dynamic processes underground may hinder the groundwater movement towards the spring. Answering the question why river waters are not saline he states that seawater from which rivers originate lose their salt content during migration through the Earth's interior.

The first half of the 18th century brings an important work of Gabriel Rzączyński (1664–1734), a Jesuit, one of the last Polish authors writing in Latin. In 1721 he published a big volume (488 pages): *"Historia naturalis curiosa Regni Pooniae, Magni Ducatus Lithuaniae annexarumque provinciarum."* ("Thorough natural history of Poland, the Grand Duchy of Lithuania and adjacent provinces"), a comprehensive description of the inanimate and animate nature in the territory defined by the title.

A spring according to the definition of Rzączyński is the beginning of a river. Spring water originates directly from the sea. It is fresh, being desalinated during its flow through the Earth strata. The author excluded the role of meteoric precipitations in the formation of ground waters. He distinguished three types of springs: 1. flowing out like boiling water although it is cold, 2. producing "trembling" water and 3. not discharging in the shape of a river but appearing in wells. He could not answer the question whether there are periodical springs in Poland (Piasecka, 1970).

Rzączyński worked until his death on appendices to and corrections of his work. The addendum to "Historia naturalis ..." appeared in print in 1742, after his death as: "Auctuarium historiae naturalis Regni Poloniae Magni Ducatus Lithuaniae" ("Enrichment of the natural history"). This volume of 504 pages was an object of a detailed analysis of Balińska-Wuttke (1976) who has put special emphasis on Rzączynski's considerations about ground waters. They are dealt with in Chapter 5 of *"Auctuarium"* entitled: "About mountains, springs, bituminous waters, sulphurous waters, waters changing into stone, boiling, healing, harmful, saline waters" A part of this chapter is entitled "*De Aquis qualitatum variarum*" ("About waters of various qualities").

Among springs Rzączyński distinguishes burning, stormy and seasonal springs. Burning springs occur according to him in the vicinity of the town Krosno (Carpathians) between Turaszówka and Potok which is the area where saline ground waters linked to oil and gas deposits contain methane. In fact one of these springs is that at Iwonicz, and was studied by Conradi and Denis (op. cit.). According to Rzączyński stormy springs when set in motion produce rain. One of such springs is supposed to exist in Huszcza Wielka village in the Polesie province (now Belarus). Among seasonal springs the weather forecasting ones are known at Szkło near Lwów. During rains the two springs there are stagnant, but in periods of good weather they seethe and eject water.

The classification of spring waters proposed by Rzączyński does not necessarily follow the rules of logic. He distinguishes the following water types: bituminous, sulphuric, nitrogenous, aluminous, petrifying, frothing, medicinal, acidulous, noxious, blood-red, saline spring- and lake waters.

This water typology is chaotic and the author realizes it, writing for example: "Burning springs near Krosno contain also frothing waters called noxious; they froth in the cold state and never flow out (from the spring)".

Rzączynski must have collected his data very meticulously, although not always from credible sources. He quotes many localities in Poland and abroad, where springs and waters he writes about, appear. At Białobrzegi south of Warszawa there is according to him a spring producing small fish, in particular during summer and autumn. An underground stream supposedly brings them to the spring. Sulphurous waters appear close to the salt mines at Wieliczka and Bochnia (which is true). Nitrogenous water appears at Kamieniec Podolski (now Ukraine) and in Lithuania, petrifying water occurs in the Spisz province (the author probably meant Drużbak) and in Supruszkowce (Podole, now Ukraine). Frothing waters occur in many places including in the Polish Lowlands (Reda, Kleszczewo among others) and in Volhynia (now Ukraine) "between Międzyrzecz and Ostróg close to the villages Hulcza and Andruszówka, where frothing water sprays and throws out not only sand but gives back gravels thrown-in".

Other types of water also appear in "Auctuarium ... " as linked to particular regions and localities, like carbonate waters of the Carpathians (e.g. in the Poprad R. catchment area: Krynica, Muszyna etc.), Drużbak, Lubownia, etc., (now in Slovakia), saline waters in Kałusz (now Ukraine), in the vicinity of Toruń and Szubin, at Stokliszki (Lithuania) etc.

It has to be concluded that although Rzączyński, being a type of a bookworm rather than a field-worker, did not verify all information collected and his work constitutes a summary of whatever was known (or imagined) about hydrogeology at his time.

Medical literature with only rudimentary information regarding the mode of appearance and data on the quality of curative waters and sometimes also some theoretical-philosophical considerations gets in the second half of the 18th century also a more hydrogeology-oriented trend and several papers devoted to the mere description of mineral springs and waters are published. Among them one of the first is an anonymous paper (probably by A. Krupiński) concerning "*The description of mineral waters in general and in particular of that at Kozin located in the Polish Kingdom, district of Krzemieniec, Voivodship of Wolhynia*" (1773).

Figure 5 Krzysztof Kluk (1739–1796).

A treatise by Krupiński (1750), although devoted to the "Description of diseases" includes a chapter: "*About waters endowed with healing strength*" with numerous details concerning mineral water springs.

H. J. Crantz (1777) publishes a book concerning mineral springs of the Austrian Monarchy concerning among others the province of Galicia, tweaked away from Poland in 1772. The book contains descriptions of the already mentioned springs at Szkło and Rabka (now a well-known health resort) as well as those of Nahojowice and Tustanowice. Also in 1773 an anonymous booklet appears (as reported by Czarniecki, 1977 the name of the author, J. G. Morgenbesser, was determined only in 1775) concerning mineral waters in Silesia (the present names of spas in which waters have been described are: Kudowa, Duszniki, Stary Zdrói, Jedlina, Szczawno and Świeradów). Detailed descriptions of the intakes including their size, shape and material used are accompanied by the characteristics of "chemical" procedures applied to determine the water quality.

The first volume of the great work of Krzysztof Kluk (1739–1796) [Figure 5] in which he considers groundwater, appeared in 1781. Part II of this volume is devoted to water both common and mineral. On 66 pages he presents physical characteristics of water adequate to the knowledge of the epoch ("… water particles must be round, which causes its ability to flow like grains of beans which contact each other only in one point ...". "Although water is one of the four elements it never appears in its elemental form, always including 'alien' admixtures", etc.). Kluk's classification of waters is very detailed and amazingly logical. Common waters are divided into "living" and "stagnant" with further subdivisions based on the form of their appearance (springs, lakes, marshes etc.). Mineral waters are divided into cold (superior and more paltry), acidulous and thermal waters with many further detailed subdivisions. This classification deserves a comprehensive discussion which, however, would go beyond the framework of the present paper.

Kluk shares the opinion pronounced by the majority of his contemporaries that all groundwater which appears in springs and wells is of marine origin, but quotes also the rare and timidly expressed views underlining the role of atmospheric precipitations in groundwater forming. Interesting is his opinion concerning the origin of thermal waters as often connected with exothermic chemical reactions taking place underground. These reactions according to him occur mostly in the presence of sulphur and its compounds.

Gałkiewicz (1955) calls Kluk's book the first work in applied geology. This concerns the part of the letter devoted to water, in particular to groundwater. Kluk may be rightly qualified as the first Polish expert in practical hydrogeology, although the notion of "hydrogeology" appeared in the literature almost 120 years later (Krisztafowicz, 1899). Kluk's work contains a precise description of qualitative and semi-quantitative analysis of water. This analytical part of the work is of course obsolete, but shows details of the analytical "kitchen" in the second half of the 18th century. His suggestions regarding methods of water treatment, water prospection and execution of dug wells, ways of spring protection etc. are of value also for a groundwater practitioner of modern days.

In 1789 the royal physician L. Lafontaine publishes a comprehensive book about thermal and cold mineral waters at Krzeszowice near Kraków; in 1790 the first modern chemical analysis of the "Main Spring" mineral water appears at Krynica (carried out by Hacquet). The paper by P. B. Hoppen about karst mineral waters at Birże (Lithuania) appears in Wilno in 1791, while in 1792 S.B. Jundziłł writes about saline springs and salt exploitation at Stokliszki and Birsztany (Lithuania).

The work of Michał Hube (1737–1807) published in 1791 was a textbook for military high schools. As pointed out by Piasecka (1970), his great contribution to science was the refutation of the long-living, commonly propagated and accepted theory that springs are recharged by seawater. Hube was the first Polish writer to state that ground- and spring waters originate as a whole from atmospheric precipitations which sink into permeable soil and rock fissures. An aquifer has an impermeable bottom and in places where it intersects the surface water flows out in the form of springs. Therefore most springs occur on mountain and hill slopes and in valleys where the lowest rock strata appear. "Springs therefore should be considered as groundwater mouths which supply water even if there was no rain for a long time" (op. cit., p. 153). The groundwater table oscillates depending on the precipitation amount and this dependence decides about the number and yield of springs. Both these figures are also dependent on the altitude of mountains which are always recharged by rains as well as melting snow and ice.

It is worth mentioning that by the end of the 18th century the famous scholar Alexander v. Humboldt prepared expert evidence concerning the saline springs at

Kołobrzeg (now Western Pomerania, Poland). According to Cramer (1892) who made a comprehensive analysis of Humboldt's work, it included the inventory of all wells at Kołobrzeg (40 or so) producing saline water, their yield and the salt concentration in the water. Humboldt's conclusions concerning the possibility of salt production increase in the Kołobrzeg salt manufactory were optimistic enough to encourage the Prussian Treasury to buy it from the salt guild. Investigations of Dowgiałło (1965) who had at his disposal drilling data, only partly corroborated this optimism.

The period of partition (1795–1918)

In 1795 Poland lost its independence, becoming partitioned between Austria, Prussia and Russia. The lack of political freedom did not hinder the development of hydrogeology, although it was often exercised by foreign specialists.

Stanislaw Staszic (1755–1826) [Figure 6] is often called "the father of Polish geology". His main work "About the Earth products ... etc." (1815) is a collection of 12 dissertations published earlier, concerning various regions of Poland. He was not only an outstanding scholar of the Enlightenment era, but also a gifted organiser and administrator as well as a man of great diligence. Hydrogeology, however, doesn't seem

Figure 6 Stanislaw Staszic (1755–1826).

to have been his beloved child and unlike other problems of both basic and practical geology, information about ground waters is rather scarce in his writings. Nevertheless, he is an author of some ideas concerning mining hydrogeology (proposals for dewatering of coal as well as zinc and lead mines in the Kraków-Silesian area). The description of mineral springs by Staszic doesn't give much more information as compared with data published by previous authors, but he proposes for instance a kind of generalised chemical characteristics of Carpathian carbonated waters. The analysis defined as generally typical "with small variations" is presented in the form of a table where particular gases (CO_2, H_2S) and salts contents are expressed in grains per 20 pounds of water.

Systematic temperature measurements of common water springs were done for the Warszawa area by G.G. Pusch (1844) and for the Tatra Mountains by L. Zejszner (1844). The interpretation of the results obtained by Zejszner was an object of an animated polemics among both tourists and scientists.

Georg Gottlieb Pusch (1790-1846), an outstanding German geologist, worked in the Mining Department in Warszawa and was named professor of the Mining School at Kielce. In his fundamental work "*Geognostische Beschreibung von Polen*" ("Geognostic Description of Poland"), he paid attention to springs of both ordinary and mineral waters. Many chapters describing particular geological regions include a separate part headed "*Quellenfuehrung*" ("Occurrence of Springs"), which enables us to call Pusch a pioneer of Polish regional hydrogeology. As far as the origin of groundwater is concerned he states clearly that there is no other possibility of its forming than by infiltration of meteoric water into the ground. Discussing the origin of mineral waters composition he stresses among other things the role of living organisms which are able to produce many components of these waters. Thus, rock leaching, by which we may explain the origin of simple saline (sodium chloride) waters (in case salt rock is present in the underground), is not sufficient to understand how waters of more complex chemical composition are formed. Pusch's mostly intuitive reasoning shows directions leading to the development of modern hydro geochemistry.

The development of spa medicine typical of Central Europe in the 19th and at the beginning of the 20th century (until WW1) on one hand and the progress in analytical chemistry on the other resulted in several hundred published chemical analyses of mineral waters. They were performed both by chemistry professors working in the universities of Kraków, Lwów, Wrocław, and Wilno and by practicing pharmacists. Waters in renowned spas like Krynica, Iwonicz, Rabka (Carpathian Mountains), Szkło near Lwów, Cieplice (Warmbrunn), Szczawno (Salzbrunn), Duszniki (Reinerz) (Sudetes Mountains) and so on were analysed several times and the results published mostly in professional medical periodicals. Also ground waters in areas in which there were no spas were analysed. Particularly "productive" analysts were: A. Aleksandrowicz, J. Celiński, N. W. Fischer, A.M. Kitajewski, K. Olszewski, I. Radziszewki, K. Trochanowski, I. Radziszewski a. o. A detailed description of their achievements would go beyond the scope of this paper.

The interest in spa medicine resulted also in synthetic regional papers concerning the hydrogeology and chemistry of mineral waters. Works concerning: Lower Silesia by G. Mogalla (1802), Pomerania by C.J.B. Karsten (1847), Lithuania by A. F. Adamowicz (1851), Silesia by A. Hoennicke (1857) and Galicia by W. Szajnocha (1892) may be quoted in this respect.

Figure 7 Ignacy Domeyko (1802–1889).

In the last decade of the 19th century, great importance has been attached to the water supply of big towns and industrial districts. Opinions concerning this question were worked out among others for Kraków by W. Szajnocha (1892), for Lwów by S. Olszewski (1895) for Gdańsk by A. Jentsch (1899) and for Lublin by N. J. Krisztafowicz (1899). Problems of water supply of the Upper Silesian industrial district were considered by Kunitz (1894) and Gellhorn (1894).

During the period of partition two national uprisings against Russia took place in 1830–1831 and in 1863. Both failed and their participants were trodden down by the tsar's oppressive system. Many educated young persons had to emigrate and make their life abroad.

Ignacy Domeyko (1802–1899) [Figure 7] took part in the conspiracy at the Wilno University, preceding the first of the above-mentioned uprisings and in the uprising itself. After studies in the Ecole des Mines in Paris he set out for Chile where he became university professor of chemistry and mineralogy. He ended his career as president of the Santiago de Chile University.

As an extensively educated person Domeyko worked also as hydrogeologist. In 1847 he published a study concerning the hydrogeology of Santiago de Chile and its surroundings He is also the author of ten or so papers containing hydrogeological descriptions and chemical analyses of Chilean mineral waters summarised in a monograph published in 1871.

Karol Bohdanowicz (1864–1947) was not a political emigree. He studied in Russia and became there professor of the St Petersburg Mining Institute. He did extensive geological and hydrogeological work in Russia (mainly in its Asian part) as well as in other Asian countries. Except many important hydrogeological observations included in his excellent geological papers he published at least two separate papers devoted exclusively to hydrogeological questions (Bohdanowicz, 1890, 1915). Within the period 1915–1917 he was director of the Geological Committee (Geological Survey) of Russia.

After WW1 he worked in Poland as professor of the Kraków Academy of Mining and Metallurgy as well as director general of the Polish Geological Survey (State Geological Institute) in the years 1938–1939 and 1946–1947 (Wójcik, 1997).

Between the two World Wars (1918–1939)

After regaining its independence in 1918 Poland's territory was much smaller than before losing it. The biggest changes were in the East where vast areas which belonged to Poland before the partition took place had now been transferred to the Soviet Union. Also the northern border was changed due to the establishing of the Baltic countries, out of which Lithuania and Latvia became Poland's immediate neighbours.

In the new state structures the State Geological Institute was created, whose activity will have a bearing on the development of geological sciences in Poland. In 1923 a new hydrological department was formed within this institute. This department was later converted into a hydrogeological department. The leadership of this department was assumed by Romuald Rosłoński (1880–1956) (Figure 8) He carried out groundwater research since the beginning of the 20th century, publishing works on groundwater movement, discharge and the effect of well interaction (1908). In the period between the world wars he performed research on groundwater within the San river valley and its basin, as well as within the Jasiołda river basin (1948a). He focused his attention mainly on the groundwater balance. Based on his research on the Jasiołda river basin he presented the "law on increase of groundwater resources retention". The novelty of this law was in checking the amount of annual evaporation by monitoring the levels of groundwater in the river basin at the beginning and at the end of the hydrological year, and by out of these calculating the changes in retention. He linked his research activity with projects dedicated to utilizing groundwater for municipal water works for Lwów, Przemysl and cities of Upper Silesia. His work on the groundwater intakes in Pralkowice (Rosłoński, 1924) is a brilliant example of combining theoretical research and the practical needs to estimate the groundwater capacity. In his vast scientific and engineering heritage we also find works devoted to mineral waters and cartography. However special recognition should be given to his activity in teaching hydrogeology. As professor of the Lwów Technical University he developed a chapter entitled "*Hydrogeology as Part of the Groundwater Science*" (Rosłoński, 1928) in the widely known engineering handbook edited by S. Bryła. The most distinguished Polish professors and practitioners in civil and water engineering were co-authors of the book. The fact that the subject of groundwater was singled out in this handbook indicates that the technicians understood and appreciated its role in water management. Since 1937 R. Rosłoński was professor at the department of Hydro-Engineering at the Technical

Figure 8 Romuald Rosłoński (1880–1956).

University of Lwów and after 1945 he became professor at the Mining and Metallurgy Academy (AGH) in Kraków. His lectures given at the abovementioned Academy
were subsequently published in under the title "Course on Hydrogeology" (Rosłoński,
1948b).

Alongside R. Rosłoński who was undisputedly the pioneer of modern hydrogeological research in Poland, groundwater was the subject of interest to many known Polish
hydrologists and hydro technicians. In 1934 a second volume of the hydrology handbook was published by K. Pomianowski *et al*. This volume was devoted exclusively to
groundwater and contains a review of the conditions in which groundwater is present
in different areas of Poland. Contained in the manual are also a theory of groundwater
movement, the issue of groundwater quality and examples of calculations.

Groundwater was also in the geologists' scope of interest. J. Lewiński, professor
at Warszawa University, wrote a comprehensive study on groundwater in the vicinity
of Warszawa (Lewiński, 1921). The area of research covered over 3000 sq. km. Apart
from thorough geological characteristics hydrogeology was also presented in great
detail. A complete analysis of groundwater occurrence in the Warszawa artesian basin
is also presented indicating areas of recharge, the piezometric surface and describing
the chemical composition of groundwater. The study also contained a hydrogeological

Figure 9 Jan Samsonowicz (1888–1959).

map on the scale 1:100 000 and constituted the first comprehensive regional hydroge-ological synthesis in Poland. Its content and form are stunningly modern and for many years it was a benchmark for hydrogeological research in the Warszawa area, also that performed many decades thereafter. As a curiosity one may add that the geological cross – section of the Warszawa artesian basin contained in the study had been, until very recently, included in school textbooks.

J. Lewiński undertook many expertises and studies associated with water supply of towns, among others Wloclawek, Kalisz and Rowne (Pomianowski *et al.*, 1934). He also worked on groundwater in the vicinity of Tomaszow Mazowiecki (Lewiński, 1933).

Another distinguished geologist, a former student of J. Lewiński, Jan Samsonowicz (1888–1959) [Figure 9] also carried out important hydrogeological work. He inves-tigated groundwater of the Lodz artesian basin (Samsonowicz, 1928, 1930), mineral and thermal waters of Ciechocinek and groundwater of the Hel Spit. He included hydrogeological data in some geological maps (Samsonowicz, 1934). The Mazovian area groundwater was also a research subject of S.Z. Różycki, Z., Sujkowski (1937), J. Kirkor (1938) and J. Samsonowicz, who worked for the Warszawa water supply company during WW2.

One also has to mention J. Czarnocki who investigated groundwater of the Kielce area and brines at Busko (Czarnocki, 1923). In the period between the two wars an important role in investigating groundwater utilisation as well as collecting and disseminating data on this subject was played by drilling companies founded and operating already in the partitioned Poland.

The most important of them was the "Technical Office Eng. Rychłowski, Wehr and Co" in Warszawa which in 1917 published *"Materials on the Hydrology of the Polish Kingdom and Surrounding Territories"*. In 1930 B. Rychłowski updated this work and published it as "Materials on Hydrology of the Republic of Poland" containing numerous geological profiles of drilled wells along with hydrogeological data. This work gave a beginning to the *"Materials of the Drillings Archives"* which were continuously published by the State Geological Institute after World War II.

In the area formerly governed by Prussia the "Jan Kopczyński Well Company" continued its activity. In 1935 Jan Kopczyński published a handbook entitled *"Drilled and Dug Wells"*. Beside the technical and organisational advice a number of geological profiles of the wells of the Wielkopolska region were also included in the handbook.

In summary one may conclude that during the 20 years between the two wars the development of hydrogeology was associated mainly with the growing utilisation of groundwater. Drilling new wells and their groups was preceded by comprehensive hydrogeological research which led to significant progress in the knowledge of groundwater at regional level and to enriching groundwater research methods. Among the people who made it happen were R. Rosłoński i K. Pomianowski, J. Lewiński and J. Samsonowicz. Distinguished "pure" geologists' interest in groundwater helped the development of hydrogeology in Poland and turned it into an important geological discipline. This direction will be preserved and solidified in the following years.

The post-war period

The period of the Second World War (1939–45) brought not only enormous human and material losses to Poland. Due to the war and its outcome the geographic position and the political system of Poland changed. Poland lost its eastern territories, which represented about 30% of its pre-war area. This loss was partially compensated by gaining the former German lands east of the Neisse and Oder rivers and in the north the southern parts of East Prussia. These territorial changes were accompanied by a people migration on an unprecedented scale. 6 million Polish citizens had to leave the eastern lands, which were taken by the Soviet Union, and 4 million Germans were forced to leave the land granted to Poland in the west. It is to be added that in this period of great instability the armed opposition against the new political system forced on Poland was continuing.

In 1945–50, in this very unstable political situation the reconstruction of the country began. Universities started to operate again also on post–German territories, the State Geological Institute has resumed its activity. The state took control over drilling and geological companies. There was an acute shortage of specialists in all fields of the country economy, including geology and water supply. The development of natural resources became one of the main economic priorities. This resulted in a decree on Polish Geological Survey being issued, which in turn led to the establishment of the Central Geological Board in 1952. Within this office a department of hydrogeology and engineering geology was soon createand in 1955 the Commission on Hydrogeological

Figure 10   Zdzisław Pazdro (1903–1998).

Documentations was established. The geological administration played a significant role in the development of hydrogeology through its legislative and supervising activity and through financing hydrogeological research and projects.

Once the geological administration was established, in 1951 a re-organisation of the higher geological education was undertaken (Bolewski, 1976). The geologists' education was concentrated in the Universities of Warszawa and Wrocław as well as in the Academy ofMining and Metallurgy (AGH) in Kraków. At the Gdańsk Technical University's Hydro-Engineering Faculty a new section of Technical Geology was formed. This section specialised in engineering geology and hydrogeology. The organiser and head of this section was Professor Zdzisław Pazdro (1903–1998) [Figure 10], Head of the Engineering Geology Department. In 1955 the first diplomas for geologists in the field of hydrogeology were awarded. In total 180 students graduated from this section in the years 1955–60. Later the education in this field was transferred to the Geology Faculty at Warszawa University where Professor Pazdro had also moved. In the years 1945–50 the lectures in hydrogeology were conducted there by F.Z. Rutkowski. In 1954 the Section of Hydrogeology was established – converted in 1957 into a Chair headed by Professor Jozef Gołąb (1904–1968). It became an important centre for education and research in the field of hydrogeology.

The second important centre was established in Kraków where since 1945 Professor Roman Krajewski (1906–1993) was lecturing on hydrogeology at the Geological-Mining Faculty of the AGH. The scope of this subject was extended after establishing the Department of Mining Geology (Kleczkowski, Sadurski, 1999). In the field of education and research the emphasis was put on issues associated with mining hydrogeology. This scope would be significantly widened in the future.

In the State Geological Institute the hydrogeology department continued its activity. Since 1948 this department was directed by F.Z. Rutkowski. Many experts' opinions for the purposes of country economy were performed in this department. Also work on the artesian basin of Warszawa was carried out there (Olendski, 1960). The Institute started collecting and publishing the "*Materials from Drilling Archives*" which included profiles and data from drillings carried out within the new Polish borders. The majority of them were water well drillings. In 1957 the Geological Institute (new name) started publishing hydrogeological maps on the scale 1:300 000. The map (composed of two sheets) showed the first and the deeper aquifers, taking into account water quality and yield. By 1964 maps covering the whole territory of Poland were developed. The map was edited by Cyryl Kolago, the creator of its concept, who later in 1970 would present its synthesis on the scale 1:1 000 000 and in the years 1981–87 would become the co-coordinator of the development of a hydrogeological map of Poland on the scale 1:200 000. Kolago devoted to hydrogeological cartography 30 years of his work and contributed greatly to its development.

The decree on the state geological survey mentioned earlier resulted in the reorganisation of drilling and geological companies. Important here was the fact that individual ministries, when needed, were establishing their own companies. A few of those companies were concentrated mainly or exclusively on prospecting and utilising groundwater for municipal, agricultural and industrial use. Companies to mention here are those of the agriculture, construction, municipal and health ministries. In the latter a company named "Balneoprojekt" was created with the purpose to prospect and investigate the therapeutic waters. Companies under the mining and chemical ministries jurisdiction investigated hydrogeological issues associated with mines construction and operation.

The demand for groundwater specialists was, therefore, growing. One can estimate that at the end of the 1950s there were around 300 specialists with a university degree who were involved in hydrogeology, mainly graduates from the Technical University of Gdańsk, the Academy of Mining and Metallurgy in Kraków (AGH) as well as Warszawa and Wrocław Universities. At these universities either hydrogeology specialisations were established or master theses on these subjects were written.

The development of hydrogeology was aided by further legislative steps. In 1957 the President of the Central Board of Geology issued a regulation on obligatory elaboration and approval of plans concerning geological works such as water well drilling and groundwater prospecting. In 1960 an obligation to determine and approve the safe yield of groundwater for any new well deeper than 30 m was introduced.

A new Geological Law passed in 1960 strengthened the role of hydrogeology and tied the probability of locating groundwater resources with the decision on financing the investment. Once this law was passed, the formation period of a solid legislative base for hydrogeology in Poland was over.

The years 1960–80 bring significant development of hydrogeology which was mainly the result of drilling thousands of new wells and carrying out numerous exploratory boreholes. These boreholes systematically increased the knowledge of regional hydrogeology and helped the development of new research methods. This made it possible to perform both hydrogeological syntheses and research on the origin of groundwater. The continued development of raw materials mining both underground and open-pit required comprehensive hydrogeological studies taking into account also consequences of deep excavations dewatering. Due to extensive use of groundwater the necessity arose to protect its quality and resources. The danger of groundwater pollution coming from the surface is an associated issue and this threat would grow systematically. After 1980 the issues of groundwater protection would dominate Polish hydrogeology.

During the course of work on a variety of sometimes complex projects, teams of experienced geologists were formed at the companies operating in Gdańsk, Kraków, Poznań and Warszawa. Special recognition has to be given to the Poznań centre where the use of analysis of test pumping data in unsteady filtration flow was propagated (Dąbrowski., Przybyłek, 1980) and where specialists worked on methods of renewable water resources evaluating (Pleczyński, Przybyłek, 1974).

Academics from universities and research institutes were actively co-operating with these companies. The growing amount of hydrogeological information and monitoring allowed the undertaking of research in a number of areas of hydrogeology.

This variety of subjects considered can be seen in the Hydrogeology Chair of the Warszawa University. J. Gołąb who was its head turned his attention mainly to the hydrogeology of the Podhale region (1947). He also indicates the necessity to determine circulation paths and to estimate the age of groundwater. Z. Pazdro, who takes the leadership of the chair in 1968, published in 1964 the first hydrogeology handbook in Poland (Pazdro, 1964). He also conducted research on salinity of groundwater in the Polish Lowlands (Agopsowicz, Pazdro, 1964) and on hydrogeology of the Gdańsk region (Pazdro, 1958). In 1956–1974 Z. Pazdro was the Chairman of the Commission on Hydrogeological Documentations at the Central Board of Geology. His teaching as well as academic and organisational activity accelerated the development of hydrogeology and he is rightly considered the founder of modern Polish hydrogeology in the post-war period.

Alongside the professors their students also performed research. T. Macioszczyk moved from regional issues to the dynamics of groundwater (Macioszczyk, 1964, 1973b), and then to analogue and digital modelling (Macioszczyk, 1973a). S. Krajewski investigated the hydrogeology of the Lublin Plateau, the characteristics of fissured aquifers (Krajewski 1970, 1972) and methods of estimating groundwater resources. B. Kozerski analyzed the interrelation of the filtration parameters of Quaternary aquifers (1972).

D. Małecka devoted her work to groundwater of the Inner Carpathians, in particular the Podhale Basin (1981). A. Macioszczyk investigated groundwater chemistry in Tertiary and Cretaceous aquifers of the Polish Lowlands as well as the groundwater mineralization background and its anomalies (Macioszczyk, 1973, 1979). In 1987 she published a textbook on hydrogeochemistry. J. Szymanko worked on geophysical methods and on digital modelling in hydrogeology (Szymanko, 1980). At his initiative the first Polish programmes for digital modelling of groundwater were developed. The programme was named "Hydrolib" and was used to estimate the regional resources of groundwater.

In the Warszawa center J. Dowgiałło carried out research on mineral and thermal waters covering a variety of regions in Poland. He started the research while working in "Balneoprojekt" and continued it in the hydrogeology section at the Polish Academy of Sciences. In the following years he would work on the origin and salinity of water in the Mesozoic of the Polish Lowlands (Dowgiałło, 1971) as well as on mineralized waters in the Mesozoic and Cenozoic of the Carpathians, the Carpathian Foredeep and the Sudetes Mountains (Dowgiałło, 1973). It was the first ever work in Poland in which the analysis of oxygen and hydrogen isotope composition was used to determine groundwater origin. An important part of his research concerned thermal waters. He co-authored and edited the book *"Geology of balneological resources"* (Dowgiałło *et al.*, 1969). It is worth pointing out that J. Dowgiałło was one of the first Polish members of the International Association of Hydrogeologists and it is thanks to his endeavors that the Polish National Committee of I.A.H. was established in 1973.

The hydrogeological department of the State Geological Institute was headed – in various periods – by S. Turek and C. Kolago. Work on hydrogeological maps and regional research was performed in this department. In 1959–1972 work on the estimation of groundwater resources in the whole of Poland has begun. This work was later used to publish the Atlas of Fresh Groundwater in Poland on the scale 1:500 000. In the years 1960–1980 nine different hydrogeological maps on the scales from 1:1 000 000 to 1:2 000 000 were published, including the Map of Mineral Waters of Poland (Kolago *et al.*, 1966; Dowgiałło *et al.*, 1974). In 1977 a hydro geochemical atlas of Poland on the scale 1:2 000 000 was initiated and edited by S. Turek. B. Paczyński worked on the hydrogeological regionalisation of Poland (1977a) and on methodology of groundwater resources evaluation (1977b). The chemistry of groundwater is the subject of Z. Płochniewski's work, he devoted his publications to both fresh and mineral waters (Płochniewski, 1973). The groundwater of the Roztocze area was investigated by J. Malinowski (1974).

A great achievement was the establishment of the network of groundwater monitoring stations in Poland (Pich, 1979) consisting of around 800 monitoring sites. 100 of them are hydrogeological observation stations where in the boreholes carried out specifically for that purpose, the water table is being monitored daily in all aquifers pierced through. Twice a year the chemical composition of the water is determined. Other observation points are wells and springs. The results of monitoring are compiled continuously and published. Based on this observation network a country groundwater quality monitoring system would be established in the 1980s.

Hydrogeology teams are also working in the regional branches of the State Geological Institute. The hydrogeologists from the Upper Silesian Branch who work on groundwater in the mining areas could boast significant achievements. The work of A. Różkowski who led the team between 1958 and 1986 is of special importance here. His work includes mainly the hydrogeological conditions in the Carboniferous of the Upper Silesian and Lublin regions. The results of his work are included in many publications among which is the work concerning the hydrogeological characteristics of the upper Carboniferous Silesian Basin (Różkowski, 1965) and the chemistry of waters in the Tertiary of this area (Różkowski, 1971). In the Lublin coal basin research was carried out on hydrogeological conditions and natural gas occurrence (Różkowski, Derdzinska, 1969). In the work of Różkowski an important scientific issue was karst waters in the Triassic of the Silesian-Kraków area. He also worked on the application of isotopes in hydrogeology (Różkowski, Przewłocki, 1974). In 1975 Różkowski becomes Professor at the Silesian University (Sosnowiec) where he heads the Faculty of Hydrogeology and Engineering Geology and organizes a team of younger hydro geologists. Also in Upper Silesia at the Central Mining Institute (Katowice) and the Silesian Technical University (Gliwice) hydrogeologists worked on the hydrogeology of mines and water threats therein (Sztelak, 1975) as well as on hydrogeological mining damages and their origin (Rogoż, Ryłko, 1970).

In other branches of the State Geological Institute the hydrogeological activity is limited to map drawing. A somewhat wider scope of activity is pursued in the Swietokrzyski branch where over many years C. Zak and E. Maszonski carried out groundwater investigations of the Holy Cross Mountains and their Mesozoic cover. In the Carpathian region A. Michalik pursued research on Outer Carpathian ground waters in particular on saline waters in Silesian Beskid Mts. (Michalik, 1973).

Another important hydrogeological research center was developing at the Mining and Metallurgy Academy (AGH) in Kraków. The AGH Department of Applied Geology was headed by Professor Roman Krajewski. He was a distinguished expert on mining geology and laid the groundwork for the research development on hydrogeology of mineral deposits. These subjects were later taken over by Krajewski's student, Zbigniew Wilk. He worked on forecasting water inflows to mines, and investigated the impact of hydrogeological conditions on coal mining (Wilk, 1965, 1967), the impact of mining on the water environment and worked on the methodology of hydrogeological investigations. Most of his publications are devoted to hydrogeology of the Upper Silesian Mining Area. However, one can also find among his work studies concerning water threats to brown coal open-pit mines as well as copper, zinc and lead mines (Różkowski, Wilk, 1980). Professor Z. Wilk was internationally acclaimed for his work on mining hydrogeology.

In 1961 A.S. Kleczkowski joins the Faculty of Applied Academy of Mining Metallurgy and takes up the hydrogeological issues. He carried out research on hydrogeology of the Hopei Plain, China (Kleczkowski, 1963). In his later work he pays special attention to groundwater of Kraków area and the changes taking place therein (Kleczkowski, 1964, 1967). He investigated the hydrogeology of the Opole and Kraków regions and even the Baltic coast area. In 1979 he published the "Hydrogeology of the territories surrounding Poland". The fact that the Polish areas were omitted in this work was the result of censorship constraints. In this work he presented the principles of groundwater regionalisation and its resources evaluation. Kleczkowski's main research object was, however, the protection of groundwater against the impact of man's stress. In the 1980s he developed groundwater protection strategy distinguishing and demarcating major groundwater basins in Poland (Kleczkowski, ed. 1990). This work was subsequently taken into account in legislation and regional planning. In the period of 1967–1969 Kleczkowski headed the AGH Hydrogeology Chair and in 1976–1982 he was director of the Institute of Hydrogeology and Engineering Geology of the AGH. In the years 1982–1988 he was rector of the Kraków Academy of Mining and Metallurgy.

The hydrogeological team at the Academy of Mining and Metallurgy worked on many hydrogeological issues. Z. Śmietański (1969) applied hydraulic and electric analogy methods to estimate water inflows to mines and groundwater intakes. A. Szczepański works on groundwater dynamics (1977) and methods of estimating groundwater resources (Szczepański, 1979). S. Witczak investigated hydrogeological characteristics of the Carboniferous rocks and mineral waters of Upper Silesia. He works also on issues of groundwater chemistry and protection (Maloszewski et al., 1980). J. Motyka. investigated hydrogeology of mineral deposits paying special attention to zinc and lead ores and groundwater flow in karst aquifers. An important role in the application of isotopes in hydrogeology was played by the Institute of Nuclear Physics and Technics at the AGH in Kraków. Most of the research on isotope composition of groundwater was performed there (Rozanski, Florkowski, 1979). Also A. Zuber and his associates' publications were devoted to the origin and age of groundwater and the tracer dispersion problem (Kreft, Zuber, 1978). Worth mentioning is also the activity of A. Wieczysty from the Kraków Technical University who worked on protection of groundwater intakes and published a handbook on applied hydrogeology (Wieczysty, 1970).

The Kraków hydrogeological centre is one of the most active in Poland and is developing continuously.

When describing the history of hydrogeology in Poland the important role of geological and hydrogeological companies was already emphasized. This can be very clearly seen in Wrocław. The companies working here take part in solving many important issues associated with planning and construction of brown coal and copper mines. Special planning offices and companies with separate hydrogeological teams are formed specifically for that purpose. The Hydrogeology Department at Wrocław University was formally established in 1970, however, hydrogeological research started there even before that time. It was carried out mainly by M. Różycki, who worked on ground waters of the Opole, Wrocław and Sudetes Mts. areas.

J. Bieniewski carried out research on groundwater dynamics. He combined his work at the Mining Faculty of the Wrocław Technical University with his activity in the brown coal mining industry. Late in his life he was also Professor at Wrocław University where for a short time he directed the Hydrogeology Department. This position was taken over in 1980 by T. Bocheńska. Important research objects of the Wrocław centre were mineral waters of the Sudety Mountains. Since the 1950s they had been investigated by J. Fistek (1977).

In the Gdańsk center, formed by Z. Pazdro, studies were published on river basins hydrogeology (Wróbel, 1960; Piekarek-Jankowska, 1979). After the Department of Hydrogeology at the Gdańsk Technical University had been reactivated in 1972, the research was concentrated on groundwater of the Baltic coastal area and the Vistula River delta. This work was conducted by B. Kozerski (1981). Also the Gdańsk Cretaceous artesian basin was investigated by A. Sadurski.

In Poznań T. Błaszyk (1968) investigated groundwater of the Wielkopolska Plain. He also worked on changes of groundwater quality during its exploitation (Błaszyk, Górski, 1978). An important part of the Poznań hydrogeological centre is evaluating groundwater resources, their protection and bank infiltration intakes. They investigated also the hydraulic properties of aquifers (Górski, 1979).

For the last three decades Polish hydrogeology was developing both in the field of regional investigations and in research methodology. Polish hydrogeologists also carried out research and applied work abroad, in particular in developing countries. This period, however, is beyond the scope of the present chapter.

ACKNOWLEDGMENTS

The authors are indebted to Prof. Zbigniew Wójcik and Dr Stanislaw Czarniecki for their helpful remarks and kind assistance in finding some publications which were difficult to obtain.

REFERENCES

Adamowicz, A. (1851) About mineral waters in the Kowno province. *Pam. Tow. Lek. Warsz.*, 26 (1), 75–82 (in Polish).

Agopsowicz, T. & Pazdro, Z. (1964) The salinity of waters in the Cretaceus of the Polish Lowlands. *Zesz. Nauk. Polit. Gdanskiej*, 49a, 34 p. (in Polish).

Balińska-Wuttke (1976) "De aquis qualitatum variarum" of Gabriel Rzaczynski or about various qualities of water (News from the XVIII century). *Bulletin of Geology*, 21, 237–348 (in Polish).

Bilikiewicz, T. (1932) Outline of balneological bibliography of Polish territories, XVIII century inclusive. *Mem. Pol. Baln. Sdoc.*, XI, 237 p. (in Polish).

Błaszyk, T. (1968) Groundwater of the Quaternary aquifers in relation to the present land forms of the Wielkopolska Plain. *Zesz. Nauk. Inst. Gosp. Kom.*, 25, 1–70 (in Polish).

Błaszyk, T. & Górski, J. (1978) Changes in groundwater quality in condition of intense exploitation. *Warszawa*, 64 p. (in Polish).

Bohdanowicz, K. (1890) Demarcation of the protection perimeter of Staraja Russa mineral springs in the Novgorod province Gorn. *Zurn.*, 4–5, 699–701 (in Russian).

Bohdanowicz, K. (1915) Saline springs in the vicinity of the Irkutsk salt-works. *Izv. Gieol. Kom.*, 34, 348–351 (in Russian).

Bolewski, A. (1976) Reorganization of the higher geological education in Poland. *Przegl. Geol.*, 9, 510–517 (in Polish).

Cramer, H. (1892) Zur Geschichte der Saline zu Colberg und ein Gutachten gegen Ende des XVIII Jahrhunderts nebst Mitteilungen ueber Soolquellen in Pommern. *Ber. Sitz. Naturforsch. Ges. Halle*, 11–104 (in German).

Czarniecki, S. & Martini, Z. (1972) Retrospective geological bibliography of Poland 1750–1950 Supplement, 332 p. (in Polish).

Czarniecki, S. (1977) Epilogue to Morgenbesser's "Public announcement ... etc". *Opole-Kraków*, 4 p. (in Polish).

Czarnocki, J. (1923) About geological structure of the Busko environs in connection with the problem of saline springs. *Pos. Nauk. PIG* 5, 2–4 (in Polish).

Dąbrowski, S. & Przybyłek, J. (1980) Methodology of test pumping. *Warszawa*, 197 p. (in Polish).

Domeyko, I. (1847) Sobre las aguas de Santiago e de sus imediaciones. 16 p. (in Spanish)

Domeyko, I. (1871) Estudio sobre las aguas minerales de Chile. *Ann. Universid. Chile*, 39, 221–283 (in Spanish).

Dowgiałło, J. (1965) Saline waters of Western Pomerania. *Szczecin*, 123 p. (in Polish).

Dowgiałło, J. (1971) Study on the origin of mineralized groundwater in Mesozoik formations of North Poland. *Biul. Geol. Wydz. Geol. UW* 13, 319–336 (in Polish).

Dowgiałło, J. (1973) Results of measurements of oxygen and hydrogen isotopic composition of ground waters in South Poland. *Biul. Inst. Geol. Z badan hydrogeologicznych w Polsce*, 3, 319–338 (in Polish).

Dowgiałło, J., Karski, A. & Potocki, I. (1969) Geology of balneological raw materials. *Warszawa*, 295 p. (in Polish).

Dowgiałło, J. (ed.), Plochniewski, Z. & Szpakiewicz, M. (1974) Map of mineral water in Poland 1: 500,000. *Panstw. Inst. Geol.* (in Polish).

Fistek, J. (1977) Acidulous springs of the Klodzko Basin and Bystrzyckie Mountains. *Biul. Geol. UW*, 22, 61–115 (in Polish).

Fleszarowa, R. (1966) Retrospective geological bibliography of Poland. 2. 1750–1900 1. 573 p. (in Polish).

Gałkiewicz, T. (1955) First Polish work on applied geology. *Przegl. Geol.* 11, 520–523 (in Polish).

Gellhorn (1893) To the question of water supply of the Upper-Silesian industrial district. *Zeitschr. Oberschl. Berg-u. Huettenm.*, Ver. 32, 421–423 (in German).

Gołąb, J. (1947) Hydrogeological characteristics of the Gubalowka Range near Zakopane. *Biuletyn PIG*, 32, 39–46 (in Polish).

Górski, J. (1979) Filtration features of incoherent, clastic water-bearing deposits. Panstw. Wyd. Nauk, 59 p. (in Polish).

Hacquet, B. (1790) Neueste physikalisch-politische Reisen durch die Dazischen und Sarmatischen oder noerdlichen Karpathen, *Nuernberg*, 1, 232 p. (in German).

Hoennicke, J.A. (1857) Mineral springs of the Silesia Province in respect of physics-chemistry, geology, and medical practice. Wolow. p. 174 (in German).

Hopp (1791) About waters at Birze.Wilno. 28 p. (in Polish).

Hube, J.M. (1791) Physical letters or natural science adapted to common understanding: 1–467 (in Polish).

Jundzill, S.B. (1792) About saline springs and salt at Stokliszki. Wilno. 18 p.

Karsten, C.J.B. (1847) Saline springs in Pomerania. *Beitr. Kunde Pommerns*, 1, 24–27 (in German).

Kirkor, J. (1938) Groundwater of the capital city of Warszawa and its vicinity. *Gaz, Woda,Techn. Sanit.*, 18 (6), 162–172 (in Polish).

Kleczkowski, A.S. (1963) Hydrogeology of the Hopei Plain. North China. Pr. Geol. PAN 15. Warszawa. 158 p. (in Polish).

Kleczkowski, A.S. (1964) The geological structure and groundwater of the high terrace of the Vistula R. to the East of Kraków. *Rocz. PTG*, 34 (z.1–2), 191–224 (in Polish).

Kleczkowski, A.S. (1967) Hydrochemical anomalies and their relations to the structure of the bedrock of the Kraków Old Town. *Rocz. PTG*, 34 (z.1–2), 161–169 (in Polish).

Kleczkowski, A.S. (1979) Hydrogeological conditions on territories surrounding Poland. *Wyd. Geol.*, 184 p. (in Polish).

Kleczkowski, A.S. (ed.) (1990) Map of critical protection areas of major groundwater basins in Poland 1:500,000. 33 p. [English translation].

Kleczkowski, A.S. & Sadurski, A. (1999) Genesis and evolution of Polish hydrogeology. *Bull. Pol.Geol. Inst.*, 388, 7–14 (in Polish).

Kluk, K. (1781) Prospection, study and use of mineral objects, in particular those more useful. Vol. 1 About mineral objects in general, about waters, salts, earth greases and earths. P. II, 78–144 (in Polish).

Kolago, C., Pich, J. & Płochniewski, Z. (1966) Map of mineral waters in Poland 1: 1,000,000. *Panstw. Inst. Geol.* (in Polish).

Kopczyński, J. (1935) Drilled and dug wells. *Poznan*, 167 p. (in Polish).

Kowalenko, H. (1975) The history of Polish balneology in the 16 and 17th C. *Studies and Materials from the History of Polish Science. S*, B26, 33–84 (in Polish).

Kowalenko, H. (1984) Erazm Sixtus – The first Polish balneochemist. *Studies and Materials from the History of Polish Science S*, B31, 5–27 (in Polish).

Kozerski, B. (1972) Interrelations among specific yield, effective porosity and filtration coefficient in the light of laboratory investigations of t water-bearing deposits in the Suwalki Lake region. *Biul. Geol. UW*, 14, 115–174 pp. (in Polish).

Kozerski, B. (1981) Salt-water intrusion into the coastal aquifer of the Gdansk region. *Sverige Geologiska Undersokning. Rapporter och neddelanden*, 27, 83–87 (in English).

Krajewski, S. (1970) The type of groundwater circulation in fissured Upper Cretaceus rocks of the Lublin Upland. *Przegl. Geol.*, 8–9, 367–370 (in Polish).

Krajewski, S. (1972) Zonality of water content in the Upper Cretaceous of the Lublin Coal Basin. *Inst. Geol. Prace hydrogeologiczne. Ser. Spec.*, 3, 60 p. (in Polish).

Kreft, A. & Zuber, A. (1978) On the physical meaning of the dispersion equation and its solutions for different initial and boundary conditions. *Chemical Engineering Science*, 33, 1471–1480 (in English).

Krisztafowicz, N. (1899) Hydro-geology and topography of the Lublin waterworks. *Warszawa*, 15 (in Russian).

Krupinski, A.J. (1775) About waters endowed with healing strength. In: *Description of diseases V*, 13. p. 209–233 (in Polish).

Kunitz (1893) To the question of water supply of the Upper-Silesian industrial district. *Z. Oberschles. Berg- u. Huettenm. Ver.*, 32, 348–349 (in German).

Lafontaine, L. (1889) Description of warm sulphurous and cold ferruginous baths at Krzeszowice. *Kraków*, 230 p. (in Polish).

Lewiński, J. (1921) Hydrogeological investigations in the Warszawa environs. *Roboty Publ.*, 3, 121–144 (in Polish).

Lewiński, J. (1933) Geological structure and morfology of the Tomaszow Mazowiecki and its environs. *Spraw. PIG*, 7 (3), 399–420 (in Polish).

Macioszczyk, A. (1973) Chemistry of groundwater in the Polish Lowlands Miocene. *Biuletyn Inst. Geol.*, 227, 293–317 (in Polish).

Macioszczyk, A. (1979) Chemistry of waters in the Tertiary and Cretaceus formations of the western part of the Mazowian Basin. *Prace hydrogeologiczne. Seria specjalna*, 11, 227 p. (in Polish).

Macioszczyk, A. (1987) Hydrogeochemistry. *Warszawa*, 475 p. (in Polish).

Macioszczyk, T. (1964) Flow into wells under mixed and turbulent filtration. *Biul. Geol. UW*, 4, 173–205 (in Polish).

Macioszczyk, T. (1973a) Model methods of evaluating groundwater resources and balances. *Przegl. Geol.*, 10, 555–558 (in Polish).

Macioszczyk, T. (1973b) Varability of parameters of non-linear groundwater filtration. *Biul. Geol. Wyd. Geol. UW*, 15, 5–58 (in Polish).

Majer, J. (1841) Communication about the life and scientific work of Jan Innocenty Petrycy. *Ann. Fac. Med. Jagiell. Univ.*, IV, 1–30 (in Polish).

Malinowski, J. (1974) Hydrogeology of the Western Roztocze area. *Inst. Geol. Prace hydrogeologiczne. Seria specjalna*, 2, 6–91 (in Polish).

Malecka, D. (1981) Hydrogeology of the Podhale region. *Inst. Geol. Prace Hydrogeologiczne. Seria specjalna*, 14, 181 p. (in Polish).

Maloszewski, P., Witczak, S. & Zuber, A. (1980) Prediction of pollutant movement in ground waters. In: Nuclear Techniques in Groundwater Pollut. Proc. Advisory Group Meeting, Cracow 1976. IAEA Vienna, 61–81 (in English).

Michalik, A. (1973) Mineral waters in the Polish Western Carpathians. *Z badan hydrogeologicznych w Polsce*, 3, 279–289 (in Polish).

Mogalla, G. (1887) Mineral springs in Silesia and the Klodzko District. *Wroclaw*, 102 (in German).

Morgenbesser, J.M. (1777) Public announcement on healthy baths or healing mineral waters in Silesia at Kodowa etc … situated. Wroclaw, G. B. Korn. 20 p. (in Polish).

Olendski, W. (1960) On the origin of artesian water mineralization in the Mazowian Basin. *Przegl. Geol.* 8, 355–360 (in Polish).

Olszewski, S. (1895) About advantages of deep Canadian drilling in Lwów and about water supply of Lwów. *Kosmos*, 20, 924–929 (in Polish).

Paczyński, B. (1977a) General regional division of groundwater in Poland. *Kwart. Geol.*, 21(4), 831–851 (in Polish).

Paczyński, B. (1977b) Some problems of systematics and estimation of groundwater resources. *Kwart. Geol.*, 21 (3), 619–631 (in Polish).

Pazdro, Z. (1958) Groundwater of the Gdansk region. *Przegl. Geol.*, 6, 241–244 (in Polish).

Pazdro, Z. (1964) General hydrogeology. *Wyd. Geol.*, 575 p. (in Polish).

Piasecka, J.E. (1970) History of Polish hydrography till 1850. *Monographs in history of science and technics LXV*, 1–196 (in Polish).

Pich, J. (1979) The network of groundwater monitoring stations in Poland. *Przegl. Geol.*, 4, 229–232 (in Polish).

Piekarek-Jankowska, H. (1979) Groundwaters of the upper Radunia river basin and their importance for the groove-lakes recharge. In: *Materials of the scientific session on groundwater resources of Kashubian Lake District GTN.* pp. 69–86 (in Polish).

Pleczyński, J. & Przybyłek, J. (1974) Groundwater resources in river valleys and their documenting Wyd. Geol. Warszawa. 196 p. (in Polish).

Płochniewski, Z. (1973) The occurnce of iron and manganese in groundwater of the Quaternary. (Selected examples from North and central Poland.) *Biul. Inst. Geol.*, 227, 221–278 (in Polish).

Pomianowski, K., Rybczyński, M.M. & Wóycicki, K. (1934) Hydrology, part II: Groundwater. 314 p. (in Polish).

Pusch, G.G. (1833) (P. I, 359 p.), 1836 (P. II, 695 p.). Geognostische Beschreibung von Polen so wie der uebrigen Nordkarpathen-Laender. Stuttgart u. Tuebingen (in German).

Pusch, J.B. (1844) About the temperature of springs in the Warszawa area. *Bibl. Warsz.*, 3, 1–36 (in Polish).

Rogoż, M. & Rylko, L. (1970) Hydrogeological mining damages and their origin. *Przegl. Gorniczy*, 7–8 (in Polish).

Rosłoński, R. (1908) About the wells' yield and their interaction. *Czas. Techn. Lwów* (in Polish).

Rosłoński, R. (1924) The underflow of the San river valley near Przemysl (Pralkowce). A hydrogeological study. *Pr. PIG*, 1 z.2-5, 237–272 (in Polish).

Rosłoński, R. (1928) Hydrology pertaining to groundwater science for the needs of settlements. In: Podr. Inzyn. S.Bryla. Warszawa Lwów, 1567–1583 (in Polish).

Rosłoński, R. (1948a) Water balance of the river basin and the method of its calculation. *Wiad. Sluzby Hydrol.-Meteor.*, 2, 157–169 (in Polish).

Rosłoński, R. (1948b) A course of hydrogeology. *Wydz. Polit. AGH Kraków*, 85 p. (in Polish).

Różański, K. & Florkowski, T. (1979) Krypton-85 dating of groundwater. In: *Isotope Hydrology 1978.* IAEA, Vienna, II: 949–961 (in English).

Różkowski, A. (1965) Hydrogeological characteristics of the Upper Carboniferous in the Upper Silesian Basin. *Biul. Inst. Geol.*, 1.249, 7–63 (in Polish).

Różkowski, A. (1971) The chemistry of groundwater in the Tertiary of the Upper Silesian Basin. *Biul. Inst. Geol.*, 1.249, 7–63 (in Polish).

Różkowski, A. & Derdzinska, X. (1969) Hydrogeological and gaseous conditions othe Lublin Coal Basin. *Inst. Geol. Seria spe.1.* 45 p. (in Polish).

Różkowski, A. & Przewłocki, K. (1974) Application of stable environmental isotopes in mine hydrogeology taking Polish coal basins as an example. In: *Isotope techniques in groundwater hydrogeology*. IAEA 1, 481–502 (in English).

Różkowski, A. & Wilk, Z. (1980) Hydrogeological conditions of zinc and lead ore deposits in the Silesian- Kraków region. *Studies of the Geol. Inst. 1.* 319 p. (in Polish).

Rychłowski, B. (1917) Materials to the hydrology of the Polish Kingdom and adjacent territories Tow. *Nauk. Warsz.* 738 p. (in Polish).

Rychłowski, B. (1930) Materials to the hydrology of the Republic of Poland PIG 1/3. 1413 p. (in Polish).

Samsonowicz. 1928 On saline springs in Leczyca area and their connection with the Quaternary bedrock structure *Wszechswiat*, II.1/34, 141–147 (in Polish).

Samsonowicz, J. (1930) About artesian water at Ozorków. *Czas. Przyr.*, 4, 84–89 (in Polish).

Samsonowicz, J. (1934) Explanation to the geological map of Poland 1:100,000, sheet Opatow. *Panstw. Inst. Geol.*, 117 p. (in Polish).

Staszic, S. (1815) About Earth- production of the Carpathians and other mountains and plains of Poland. *Warszawa*, 390 (in Polish).

Sujkowski, Z. & Różycki, Sz (1937) Geology of Warszawa. *Zarzad Miejski*, 44 p. (in Polish).

Szajnocha, W. (1889) Pronouncement about the durability and constancy of the Regulice springs. In: *Report and conclusions concerning the construction oh the Regulice waterworks.* Kraków. 19 p. (in Polish).

Szajnocha, W. (1892) Mineral springs of Galicia, view on their situation, chemical composition and origin. *Rozpr. AU 2.2*, 30–140 (in Polish).

Rylko, L. & Szczepański, A. (1977) The dynamics of groundwater. *AGH Kraków.* 151 p. (in Polish).

Rylko, L. & Szczepański, A. (1979) Groundwater exploitation resources in view of schematization of calculations conditions. *Zesz. Nauk. AGH Geologia*, 5, 5–69 (in Polish).

Sztelak, J. (1975) Mining hydrogeology and water threats in mines. *Polit. Slaska, Gliwice*, 318 p. (in Polish).

Szymanko, J. (1980) Concept of the water-bearing system and methods of its modelling. *Wyd. Geol.*, 263 p. (in Polish).

Śmietański, Z. (1969) The method of hydraulic analogy in the assessment of unsteady state filtration processes during de-watering of open- pit mine. *Pr. geol. Komisji Nauk Geol. PAN Kraków*, 56, 95 p. (in Polish).

Turek, S. (ed.) (1977) Hydrogeochemical atlas of Poland 1:2,000,000. *Inst. Geol.* (in Polish).

Wernher, J. (1549) De admirandis Hungariae aquis hypomnemation [Slovak translation from Latin, 1974].

Wieczysty, A. (1970) Engineering hydrogeology. *Warszawa*, 1068 p. (in Polish).

Wilk, Z. (1965) Relationship between mine water inflow, and mines size and depth in the eastern part of the Upper Silesian Coal Basin. *Prace Geol. PAN, Oddz. w Krakowie*, 117 p. (in Polish).

Wilk, Z. (1967) Development trends and variations in the quality of groundwater in Polish coal mines. *Zesz. Nauk. AGH 179, Geologia 9.* 160 p. (in Polish).

Wójcik, Z. (1997) Karol Bohdanowicz. Geologist and traveller–explorer of Asia. *Warszawa.* 410 p. (in Polish).

Wróbel, B. (1960) Hydrogeology of the Radunia river basin. *Biul. Inst. Geol. Z bad. hydrogeol. w Polsce*, 1–35 (in Polish).

Zejszner, L. (1836) About sour or acidulous waters in the Carpathians. *Pam. Farm. Krak*, 3, 265–290 (in Polish).

Zejszner (1844) About the temperature of springs in the Tatra Mts. and adjacent (mountain) ranges. *Bibl. Warsz.*, 2, 257–281 (in Polish).

Zieleniewski, M. (1891) Bibliographic-balneological dictionary of domestic and some foreign spas ... etc. 177 p. (in Polish).

A brief history of Romanian hydrogeology

P. Enciu[1], A. Feru[2], H. Mitrofan[3], I. Oraseanu[4], M. Palcu[5] & A. Tenu[6]

[1] Institute of Geography, Dimitri Racouita Street, Bucharest, Romania
[2] National Company of Mineral Waters, Bucharest, Romania
[3] Institute of Geodynamics of the Romanian Academy, JK Calderon Street Bucharest, Romania
[4] Prospectiuni, Caransebes Street, Bucharest, Romania
[5] Geo Agua Consult, Bucharest, Romania
[6] National Institute of Hydrology and Water Management, Bucharest, Romania

INTRODUCTION

The information for this chapter is taken mainly from the Proceedings of the National Symposium "*One Century of Hydrogeology in Romania*" held in Bucharest, 24–26 May 2000. It is more than 100 years since the first paper on the hydrogeology of Romania was published in 1895. This was a monograph on the structural features and hydrogeological potential of the Sub-Carpathians Hills and Romanian Plain by Mateiu Draghiceanu, a mining engineer and geologist who studied at the Paris School of Mines and is considered the father of modern hydrogeology in Romania (Bretotean 2000).

This chapter starts with a review of groundwater exploitation up to the 19th century, followed by summaries of some of the main themes of the Symposium: hydrogeological mapping, groundwater resources, mineral and thermal groundwaters, karst hydrogeology and mining hydrogeology, followed by the development of education and science up to the 1980s.

OVERVIEW OF GROUNDWATER EXPLOITATION IN ROMANIA

The first record of groundwater exploitation in Romania is from the 6th century BC, when cisterns were built to store spring water to serve Costesti, a Getic fort in south Romania. In the 2nd to 1st centuries BC, a spring with a flow of 35 m³/hr was tapped and the water conducted 7 km to the city of Istros, in the 2nd century AD, the occupying Romans built collection systems for thermal springs in Ad Mediam (now Baile Herculane). The Romans believed that the waters had miraculous qualities, and dedicated the waters to healing divinities such as Hercules and Hygeia. The Roman legion XIII Gemina used groundwater for balneotherapy in Germisara (now Baile Geoagiu), and developed the thermal springs of Aquae (Ad Aquas in the Tabula Peutingeriana, a mediaeval copy of a map of the Roman road network, and Hydata by Ptolemy – c. 90–168 AD. During the 2nd and 3rd centuries, groundwater collection systems were built to tap springs for several towns, including Histria (old Istros) and Callatis (now Mangalia) in the area of Dobrogea. To supply Tomis (now Constantza) with drinking water, drainage pits, galleries and storage tanks were constructed inland, with an aqueduct to take the water to the town. From the 4th to 9th centuries, water works were constructed to supply many fortresses, works carried out in difficult times due to successive incursions by Goths, Huns, Gepids, Avars, Slavs, Bulgarians, Hungarians and others.

Between the Middle Ages and modern times, groundwater activities developed in three directions: drinking water supply, mineral waters and thermal waters. There is reference to an early centralised groundwater supply at Tismana Monastery, founded by the Orthodox monk Nicodim by the mid-14th century. In the 17th century, the public fountains in Bucharest were constructed, taking water from springs in the Crevedia Valley, and the water supply system for the town of Iasi was installed. During the reign of Prince Constantin Brancoveanu (around 1700), new water supply works were installed for Bucharest, and in the 1800s, major works were carried out in the town of Craiova, with the installation of several public supply wells.

In the second part of 19th century, studies for drinking water supply multiplied and a number of French and German experts, such as Lindley, Guilloux, Ziegler and Thieme, designed large urban water supply systems. Urban water supply also concerned Romanian specialists: Elie Radu was notable in this field, coordinating water supply works in many towns, including Bucharest, Sinaia, and Braila.

The concern with groundwater for potable supply was dictated by urban development and town planning in the 19th century. However, studies of mineral water started long before this. The earliest information regarding mineral waters is mention in the 16th century by the Italian doctor Brucella of Borsec water. The waters of Calimanesti in Wallachia and the therapeutic qualities of Harghita waters were formally recognised in the 1600s. The priest Istvan Lakatos described the therapeutic effect of sulphurous–carbonic gas waters from Harghita baths in 1702. In 1749 the quality of saline waters from Bazna was extolled in documents such as "*Memorialae Europae*", while in 1760, the mineral waters from Olanesti were attested in Wallachia. The quality of Odorheiul Secuiesc waters was recognized by Istvan Matyus in "*Dietetica*" (1764). In 1773 Heinrich Krantz published "*De aquis medicalis Transylvaniae – On medicinal waters of Transylvania*" in Vienna, and in the same period Lucas Wagner from Brasov defended his doctoral thesis on medical-chemical aspects of these waters. By the end of the 18th century, groundwaters from Vatra Dornei (in the Dorna area of Bukovina region) were used in balneal treatment, and Richard Haquet mentioned the quality of Dorna waters in his 1790 "*Note de voyage*". In 1802, Kauski published "*Die Heilquellen der Bukovina – The mineral springs of Bukovina*" in Poland, and in 1808, Osan published the article "*Izvoarele minerale de la Dorna Candreni din Bucovina – Dorna Candreni Mineral Springs from Bukovina*" in Vienna. Due to Lindenmayer, a physician, new mineral water sources were developed in Buzias in the period 1809–1911. In 1821, Vasile Popp, published a paper "*On the mineral waters of Arpatac, Bodoc and Covasna*", the first document in Romanian about Romanian medicinal waters. The first mineral waters from the Slanic Valley in Moldavia were identified in 1800, and in 1824 building started on the spas here, with the same quality of water as the famous spas of Vichy, Karlovy Vary (Karlsbad), Marienbad and Bad Kreuznach.

With regard to mineral waters, those of Calimanesti were described in 1827 by Marsil in Curierul Romanesc magazine. Stefan Starostescu wrote "*Apele metalice ale Romaniei Mari – Metallic Waters of Greater Romania*" in 1837. This book is considered the first major text in the field of water resources with specific reference to drinking water. Three years later, Eduard Siller published "*Die Mineralquellen der Walachei* – The Mineral Waters of Wallachia" in Hannover, and "*Descrierea curei de apa rece* – Description of Cold Water Cures" by Manole was published in Bucharest. Steege published "*Les eaux minerals de Slanic en Moldavia*" in Iasi in 1856, and in 1871, Bernath's article "*Apele minerale din tara* – Minerals Waters from the Country" appeared in Columna lui Traian magazine. The first properly scientific works were by the geologist Grigore Cobalcescu (1831–1892), in the article "Sorgintile minerale de la Calimanesti – Mineral Springs from Calimanesti" (Figure 1).

Figure 1   Grigore Cobalcescu (1831–1892).

In the same period, the quality of these mineral waters were investigated. Chemical analyses are first mentioned in Valcele (SE Transylvania) in 1761. In the first half of the 19th century, systematic physico-chemical analyses and references to them (in Buzias, Vatra Dornei, Poiana Negri and other places) increased.

The first attestation of Oradea's thermal sources was in 1221, and in 1405 Pope Innocent VII sent a letter to his bishop there, mentioning the use of the water for warm baths. In 1734 the first analyses of the thermal springs in the Cerna Valley (old Ad Mediam) were carried out, and in 1736 works to modernise and rearrange the spa were started. In 1777, Stephan Hatvany gave full details of treatment with thermal waters in Oradea in his work "*Thermae Varadiensis*". At the beginning of the 19th century, Constantin Golescu, a scholar and nobleman, described a method of treatment using scalding waters from Mehadia. In 1863, Arune Pumnul, philologist and professor of Romanian language and literature, gave details about treatments at Mehadia Baths (Herculane) and elsewhere in the Cerna Valley, where the spa is located.

THE HYDROGEOLOGICAL MAPPING OF ROMANIA

The hydrogeological mapping of Romania was carried out in three stages by the Geological Institute of Romania (GIR), founded in 1906. In the first stage, before World War II, the first hydrogeological maps of the main arable regions were prepared by the Agro-geology Section. This included mapping the groundwater dynamics in the eastern part of the Romanian Plain (over 15 000 sq. km) by Murgoci, Protopopescu and Enculescu (1908). To the east, between the Danube River and the Black Sea, the main hydrogeological features of the South Dobrogea hydrogeological province were mapped by Murgoci and Macovei (1915).

The second stage was from the end of World War II to the 1990s, when the GIR undertook a programme of mapping the major hydrogeological units, the Dacic and Pannonian Basins, at the scale of 1:100 000. The Dacic Basin occupies the south and southeast of the country, an area of about 22 000 sq. km, and comprises thick (between 300 and 5000 m) Miocene-Quaternary strata that include productive alluvial deposits. Thirty six maps were prepared of the phreatic aquifers between 1968 and 1987, showing the morphology and hydrology, geological boundaries and hydrogeological characteristics (contour lines of potentiometric surface, depth of water table, location of investigation wells, hydrochemical data etc).

The Pannonian Basin is located in eastern Hungary and western Romania, west of the Apuseni Mountains. During the Middle Miocene to Quaternary, up to 4500 m of sediments were deposited, including Upper Quaternary fluvio-lacustrine and proluvial deposits. Between 1970 and 1989, the hydrogeological conditions of the phreatic aquifers of this great basin were mapped (on 11 sheets, scale 1:100 000), giving the main characteristics of the Quaternary formations and locally providing information on the deep thermal aquifers.

In addition to the 1:100 000 scale mapping, the GIR collated hydrogeological data in the Atlas of Geology, which includes the Hydrogeological Map of Romania (Ghenea et al., 1969), the Thermal and the Mineral Waters Map (Bandrabur et al., 1981) and the Geothermal Map (Veliciu et al., 1985), all at the scale 1:1 000 000. In the period 1960–1970, Elizabeta Frugina from INMH drew up a map of recharge and groundwater flow for Romania as part of Central Europe map, prepared by the CAER (Council for Mutual Economic Assistance).

GROUNDWATER FOR POTABLE AND AGRICULTURAL PURPOSES

Due to their extent and productivity, the Dacic and Pannonian Basins are the most important aquifers in Romania, followed by South Dobrogea and intermontane basins in the Carpathian Mountains. In third place are restricted portions of the Transylvanian Depression and East-European Platform (EEP). The EEP has Middle-Upper Miocene mainly clayey-marly sediments, with highly mineralised water in sandy horizons (Liteanu et al., 1963). As result, during the 20th century researchers focused on the phreatic groundwaters in the Quaternary alluvium, mainly in the floodplains of the Siret (Cadere et al., 1964; Frugina et al., 1975), Suceava (Constantinescu et al., 1973) and Moldova (Cadere et al., 1967) rivers.

The Transylvanian Depression (DT), was formed in the Upper Cretaceous, and in the Paleogene-Miocene it became an inland sea. As result, the huge pile (over 10 500 m thick) of cyclical sediments locally contains aquifers with saline waters. Due to the active transit of infiltrating meteoric water in the central part of the DT, the Miocene formations (0–200 m depth) contain locally potable waters (Marosi, 1980). Another area with potable water was identified on the southern edge of the DT, within the thin alluvial-proluvial deposits, near the foot of the South Carpathian mountains.

The main intermontane basins were formed during the Cenozoic Alpine orogeny, and in places the sediments contain locally important aquifers. The main contributors to hydrogeological research on these aquifers were Bandrabur (1964) and Tenu et al. (1985) for parts of the Sf. Gheorghe Basin; Vasilescu (1967) and, Liteanu et al. (1968) for the Baraolt Basin; and Pascu (1973) and Tomescu (1978) for the Barsei Basin.

The South Dobrogea aquifer extends over an area of approximately 4000 sq. km. It is part of the Moesian Platform, and is bounded by the Capidava-Ovidiu crustal fault in north, the Danube River in the west, and the Black Sea in the east. Here, highly productive fissured and karstified aquifers are found in the Upper Jurassic-Lower Cretaceous sandstones and carbonate rocks. The main data about the upper, unconfined, sequence were published by Avramescu (1973). The confined aquifers were studied by a generous programme of hydrogeological investigation wells (Vasilescu and Dragomirescu, 1983). Due to the economic importance of the confined aquifers, for agriculture and urban supply, subsequent efforts focused on its hydrodynamic background (Zamfirescu et al., 1987), recharge area, the age of deep waters etc (Tenu et al., 1975; Tenu and Davidescu, 1984).

That part of the Pannonian Basin (PB) lying in Romania covers an area of around of 28 000 sq. km, including the Crisana Hills and Western Plain. The first hydrogeological papers from the 1950s deal with the phreatic aquifers in Quaternary alluvial fans of the middle section of the Western Plain. Later, the highly productive phreatic aquifers of the southern and northern sections were studied by Ghenea (1962), Feru and Mihaila (1963, 1971), Mihaila and Giurgea (1985), Vancea (1972), Bandrabur (1974) and others. For the regional confined aquifer, containing potable water in Upper Miocene sands, the main contributions encompassed the southern and northern parts of the PB (Tenu, 1975, 1981, 1984).

The most important hydrogeologic unit, the Dacic Basin covers around 80,000 sq. km between the Carpathians in the north and the west, the Balkan plateau in south and Dobrogea in the east. Due to the development of the city of Bucharest, the first works dealt with available potable groundwaters in the proximity of this settlement (Draghiceanu, 1895; Radu, 1902). Much later, based upon a few hundreds of deep wells, the medium to highly productive confined formation (Upper Pliocene-Lower Pleistocene in age), containing potable water, was the subject of a number of monographs (Liteanu, 1952; Constantinescu, 1963; Bretotean and Reich, 1983 and others).

Due to the sparse network of perennial watercourses, low to medium precipitation and high evapotranspiration, surface water does not provide a reliable or sufficient resource in the southern half of the Dacic Basin. The aquifers were, therefore, intensively studied for water supply and agricultural requirements. Information on the hydrostratigraphy and hydrochemical features of the Upper Neogene-Quaternary sedimentary formations (150–750 m thickness) were published, on the basis of division into the following interfluves: Arges-Ialomita (Liteanu, 1956; Avramescu *et al.*, 1963; Cadere *et al.*, 1971), Ialomita-Mostistea (Bandrabur, 1961), Ialomita-Buzau (Pricajan, 1961), Jiu-Olt (Liteanu *et al.*, 1961; Bandrabur, 1971), Topolnita-Desnatui-Jiu (Ghenea *et al.*, 1963), Olt-Arges (Slavoaca and Opran, 1963) and Ialomita-Buzau-Siret (Florea, 1971; Gastescu *et al.*, 1979).

The middle and northern parts of the Dacic Basin correspond to the Carpathian Foredeep, containing mainly permeable formations of Pliocene-Quaternary age up to 2000 m thick. From west to east, the main contributors to knowledge of this multi-aquifer system were Mihaila and Giurgea (1980), Craciun (1984 – Danube-Motru-Drincea interfluve), Liteanu *et al.*, (1971 – Motru-Jiu and Olt-Arges interfluves) and Enciu and Grigorescu (1986 – Oltet-Olt interfluve). Large parts of the Arges River basin were studied by Constantinescu (1971). To the east, the contributions of Tenu and Tenu Sanziana (1968 – Contesti-Titu High Plain), Avramescu *et al.*, (1964) and Constantinescu (1973) and others were significant.

MINERAL WATERS

As a result of its complex geology, Romania has an enormous number and variety of mineral water springs. The use of mineral waters for therapeutic purposes dates back at least to Roman times, while medicinal and table waters have been bottled and distributed in Romania for more than 200 years.

Between the 17th and 19th centuries, subject to the historical and geo-political situation in Central and Eastern Europe, the first descriptions with geological character concerning mineral waters in Transylvania were published by scientists from the Austrian Empire. The first paper in Romanian was by Vasile Popp (1824). The investigation of mineral waters was continued by many physicians and geologists, the Romanian school being increasingly well represented. In addition to sources given above, the synthesis "*The mineral metallic waters of Great Romania*" (Episcopescu, 1837), the activity of the naturalist Hanko, who published an important monograph addressing the mineral waters in Transylvania at Cluj (1900), and that of geologist Saabner Tuduri, author of "*The mineral waters and the climatic resorts in Romania*" published in Bucharest (1900 and 1906) are worthy of mention.

Starting in the 20th century, two distinct approaches to the investigation of the mineral waters developed: therapeutic and hydrogeological. The therapeutic approach addressed curative effects and the development of the therapeutic resorts, while the hydrogeological approach addressed the chemical character, origin of the waters and storage and other characteristics of the aquifers. Both approaches led to an improvement of analytical methods, chemical analyses being repeated and published for most mineral springs in Romania. The hydrogeological objectives became increasingly diverse, by addressing additionally oil field brines and saline springs in the proximity of salt diapirs (Mrazec, Macovei, Popescu-Voitesti) and thermal-mineral waters (Popescu-Voitesti, Cantuniari, etc.). In 1939, Atanasiu and Lobontiu provided the first regional description of all mineral waters in the country and their genesis.

In the second half of the 20th century, mineral waters were of interest for both balneo-therapy and hydrogeology. Until the Romanian hydrogeological school was established, mineral water investigations were carried out by eminent researchers in the field of geology (Ciocardel, Filipescu, Papiu, Oncescu, Pauca, and others).

Starting in the 1960s, there was a strong development of interdisciplinary and systematic research on Romanian mineral waters under the auspices of the Geological Committee and then the Ministry of Geology. A true school of applied hydrogeology in the field of mineral waters developed in this period, based in several specialist institutes such as Prospectiuni S.A. (Prospecting Company), Balneology and Physical Medicine Institute, Institute for Studies and Design for Land Reclamation, National Institute of Meteorology and Hydrology (INMH), etc. A few of the experts (mostly hydrogeologists, but also drilling engineers, chemists, etc.) who contributed to this development were Liteanu, Pricajan, Bandrabur, Dinculescu and Pascu.

Four national and international symposiums held before 1985 followed the progress in the field of mineral water investigation: Brasov-Borsec (1967), Baile Herculane (1971), Eforie Nord (1974), Calimanesti-Caciulata (1978), each marking a distinct stage in terms of ideas about research, conservation and exploitation. The results of the investigations have been published in more than 500 papers and books, including the reference books of Pricajan *"Apele minerale si termale din Romania – Mineral and Thermal Waters of Romania"* 1972; *"Substantele minerale terapeutice din Romania* – Mineral Therapeutic Substances from Romania" 1985, and that of Dinculescu *et al.*, *"Mineral Waters and Therapeutic Muds in the SRR"* (1960–1980).

The most extensive investigations and measurements of groundwater radioactivity were carried out by Athanasiu over four decades (1926–1966). Also, studies based on environmental isotope techniques carried out on mineral or thermomineral waters were performed and published by Tenu and his team.

THERMAL WATER

The first Romanian hydrogeologists to address issues related to thermal waters were Popescu-Voitesti for outlets at Herculane (1921) and Mangalia (1933), followed by Nicolescu (1965), Constantinescu and Croitoru (1968), Berbeleac (1968) for the area of Tusnad, and Feru (1971) for Harsova-Vadul Oii. They tried to explain the genesis of the waters, although they had little evidence to go on. Hence it is not surprising that Slavoaca and Avramescu (1956, 1971), for instance, thought that there was no connection between the thermal springs at Tusnad and nearby recent volcanic activity. This connection was later proven through the use of techniques such as geothermometry (Mitrofan 2000).

As for the part of the Pannonian Depression located within Romania, only thermal springs discharging along the basin rim were studied initially (Pauca, 1958; Cohut, 1961). However, towards the end of the 1960s, deep wells started to be drilled in order to tap geothermal water, from aquifers whose existence had been detected by wells drilled in Hungary, across the border.

Starting from that reasonable – and to a large extent successful approach (works co-coordinated by FORADEX hydrogeologists Cohut, Paal, Tonko, Plavita, Sinka, Mircescu, Vasilescu, Nechiti and Vamvu, the geothermal well frenzy took hold in other areas of Romania. While the Caciulata – Calimanesti field in the Southern Carpathians was another success story, with more than 70% of its artesian discharge of 40° to

>90°C water yet to be utilised, the hinterland of the country's capital – Bucharest – failed to yield such spectacular results. Wells drilled to 3 km depth should have warranted withdrawal of water with temperatures in excess of 70°C, but in many cases not even 30°C was recorded at well bottom. Some people tentatively ascribed this seemingly erratic distribution of temperature to a vigorous natural convection regime, probably induced by the interplay of both temperature and salinity gradients. Such unfavorable findings discouraged exploration drilling of further prospects identified by certain hydrogeologists (Craciun, Bandrabur, Ghenea, Albu, Opran) in the same thick carbonate aquifer. However, even though the geothermal heating project had to be abandoned, the investigations had at least one valuable outcome: now $NaCl – H_2S$, 36°C thermal water pumped from a 3 km deep well is used for curative purposes in the very City of Bucharest!

KARST HYDROGEOLOGY

In 1863, the Austrian geographer Adolf Schmidl published the book "*Das Bihar Gebirge and der Grenze von Ungar und Siebenburgen* – The Bihor Mountains and the Boundary of Hungary and Transylvania", the first major geographical study addressing the karstology and speleology of part of Romania. He provided a detailed description of the karst in Bihor Vladeasa and Codru Moma mountains, the main karst springs, and the mineral and thermal springs in the mountains and in Beius basin. In 1901, the Romanian geologist Mihutia performed the first tracer test of an underground stream course in Romania. By tracing the Tarina stream with charcoal powder, he outlined the hydraulic connection between Campeneasca cave, on the Vascau karst plateau, and Boiu spring at Vascau (Mihutia, 1904).

Protopopescu-Pache, as part of his research in the Mangalia area, used uranite to trace the water that discharges from Kara-Oban lake into a swallet and found that it flows to sulphur springs on the Black Sea shore (Ciocardel, Protopopescu-Pache, 1955). The world's first Speleological Institute was founded in Cluj in 1920 by Emil Racovita, a biologist and speleologist, and later it was transferred to Bucharest.

The first systematic geomorphologic studies of Romanian karst areas were carried out by the Emil Racovita (Figure 2) Institute of Speleology and the Institute of Geography. The papers published from these studies (e.g. Serban, Coman and Viehman, 1957; Rusu, Racovita and Coman, 1964; Viehman, 1966) reflect the interest in defining groundwater flow paths, largely by means of fluorescein tracer tests.

Sencu (1970, 1978, 1986) is known for his detailed investigation of karst in the central part of the Resita-Moldova Noua synclinorium. His work with the Institute of Geography, dedicated to geomorphologic studies, continued for over 30 years, and through the roughly 25 fluorescein tracer tests he carried out, he made a major contribution to the knowledge of groundwater flow paths and the hydrologic relationship between the karst aquifers and the coalfields in the Anina area of south Banat.

The karst in the Padurea Craiului mountains was subject of systematic investigations of karst morphology and hydrology under the leadership of Rusu (1988), who carried out many fluorescein tracer tests between 1960 and 1988, by means of which he outlined about 40 flow paths.

Figure 2 Emil Racovita 1868–1947.

In 1967, FORADEX investigators Bisir and Pascu performed the first tracer tests to delineate the recharge area of the Hercules spring at Baile Herculane. They performed further fluorescein tracer tests in 1969, in the upper catchment of the Western Jiu river.

From 1970, the Hydrogeological Survey department of Prospecting Company (Simion, Oraseanu, Iurkiewicz) carried out extensive investigations in most of the carbonate rocks areas in Romania. The work included completion of hydrogeological maps, drawing up water balances, tracing the main karst flow paths (with more than 200 tracer tests), groundwater resource assessments, and investigation of groundwater chemistry. The investigations were performed jointly with specialists from the Institute of Physics and Nuclear Engineering (Gaspar, Tanase), the National Institute of Meteorology and Hydrology (INMH, Hotoleanu), the Institute of Speleology (Rust, Viehman) and the Higher Education Institute in Baia Mare (Pop). The INMH performed hydrological investigations, combined with tracer tests in the Piatra Craiului Mountains (1971) and in the Cerna – Western Jiu catchment (Bulgar). The Institute of Speleology carried out a long-term investigation of the karst hydrology of the Cerna river catchment, the Mehedinti mountains and the Western Jiu catchment (Povara) of the Piatra Craiului and Bucegi mountains.

In 1974, Marcian Bleahu of the Geological Institute published the book *"Karst Morphology"*, an encyclopaedic work dealing with karst and its genesis and which has had a strong, beneficial impact on the methodology of karst investigations in Romania.

Research on the karst aquifers of south Dobrogea was the focus of a large group of investigators, with major contributions by Ciocardel and Protopopescu-Pache (1955), Dragomirescu (1971), Nainer (1971–1973), Tenu *et al.* (1975) and Pitu (1980).

In the same period, starting in 1970, the tracer department of the Atomic Physics Institute (now the Institute of Physics and Nuclear Engineering), led by Emilian Gaspar, made a major contribution to the development and the implementation of tracer technology in karst aquifers using radioactive and inert tracers. Gaspar and Oncescu published the volume *"Radioactive Tracers in Hydrology"* (1972) and Gaspar is the author of *"Modern Trends in Tracer Hydrology"* published by CRC Press, Florida (1987).

The journal Theoretical and Applied Karstology started in 1983, publishing the proceedings of symposia jointly convened by the Institute of Speleology and Prospecting Company.

MINING HYDROGEOLOGY

Mining hydrogeology is a branch of applied hydrogeology with specific research methodologies (Gheorghe and Radulescu, 2000). In Romania, this discipline developed as a consequence of lignite mining (1955–1980) and open pit coal and ore exploitation (1980–2000).

In Romania, the coal beds for coal-fired power stations are found in Pliocene sands, clays or gravels. Major problems encountered in exploiting the coal include water ingress, flooding with water and sand, and instability. As a function of the lithology, groundwater pressure and aquifer hydraulic properties, hydrogeological conditions for coal mining are classified as "soft", "hard" and "very hard".

Hydrogeological research was carried out by Prospecting Company from 1959 to 1984 (Maieru et al., 2000) for sulphur ores (at Rosia Poieni, Baia de Aries, Herja, Baia Sprie, Cavnic, Tolovanu-Suceava, Macarlau – Gura Baii – Toroioaga, Baita-Bihor), iron ores at Teliuc, bauxite in the Padurea Craiului mountains, gold at Suior, potassium and boron salts in the Carpathian Fore-Syncline, and Transylvanian Basin, phosphate in the Resita-Moldova Noua Synclinorium and at Anina-Oravita.

In the period 1955–1985, detailed hydrogeological investigations were carried out in over 80 coalfields in Oltenia, Wallachia, Transylvania and Banat. Hydrogeological investigation wells and mining investigations were carried out by specialised units, including the Carboniferous Exploring Trust (1955–1956). Open New Mining and Prospecting and Exploring Trust (1957–1970), FORADEX (1955–1994), Prospecting Company (1957–1964, 1981–1983) and other exploration companies (IPEGs) in Oltenia, Arges, Harghita, Suceava, Cluj, Maramures, and Banat.

In addition to the physical methods of investigation mentioned above, "black box" techniques introduced by Albu were applied in many areas, especially in karstic, fractured hard rock and faulted aquifers. Modelling techniques were also used: analogue models predominated from 1965 to 1975, but numerical finite difference and finite element models were developed by Enachescu and Albu in the early 1980s (*Procese nestationare de redistribuire a energiei in crusta terestra* – Nonsteady Processes of Energy Re-distribution in the Terrestrial Crust 1985). Tracers (Orasanu et al., 1978) were used to determine at least the local direction and real velocity of groundwater flow and, later, other hydrogeological parameters. Laboratory technical – scientific and hydrogeological experimental research was carried out by the Research Mining Institutes of Bucharest and Craiova in collaboration with field hydrogeologists from the mining areas.

Experimental dewatering of coal mines was applied by FORADEX and several IPEGs in many places. Dewatering methods developed for open-pit coal mining consisted of preliminary, parallel and combination dewatering (Gheorghe and Bomboe 1963). The design of dewatering schemes is a function of geological and hydrogeological conditions, access, underground working, economic efficiency etc. In the case of deep coal beds (over 200–350 m in parts of Transylvania and Oltenia) in difficult hydrogeological conditions, with great groundwater pressure, access to the coal was achieved by drilling large (3.5–6 m) diameter wells. Preliminary dewatering by drainage wells was followed by combination dewatering methods. For open pits, the surface, underground and combination drainage systems were designed to create safe working conditions and to reduce groundwater pressure on the pit bottom. Dewatering for open pit trenches was achieved by one or more deep well dewatering systems, placed in parallel, followed by a well point system. All open pit dewatering from the surface with deep wells were applied using surrounding (north and south Pesteana, Rosia de Jiu), parallel (north and south Pesteana) or network well systems. In the case of open pits with a great thickness of overlying aquifer(s), the drainage wells were installed in stages and finally the dewatering continued by horizontal drains. In the case of fine-grained aquifers with a small storage capacity, vacuum pumps were used following gravity dewatering, and sometimes, after reaching steady state conditions, additional pumping (west Prunisor, Negomir Darova).

Romanian mining hydrogeology was represented by numerous papers in specialist journals (e.g. *Mining Journal, Journal of Mining, Petroleum and Gases, "Hidrotehnica" Journal*) or in yearbooks (e.g. of the Geology Survey, Bucharest University). In addition, several specialist books with direct reference to mining hydrogeology were published, including *"Mining Hydrogeology"* (1963) by Alexandru Gheorghe and Petre Bomboe and *"Ore Dewatering"* (1985) by Constantin Enache.

EDUCATION AND SCIENCE

Although several chapters of hydrogeology were included in different textbooks between the two World Wars, specialisation in hydrogeology at university level did not start until the 1950s. The first hydrogeological course was in the Bucharest Mining Institute Faculty of Geology, under Ciocardel (1952–1957), who published the text book *"Hydrogeology"* in 1957, following the first Romanian book on hydrogeology, *"Applied Hydrogeology"* by Liteanu (1953). Liteanu, in the Geological Committee of Romania further advised the graduates from Ciorcadel's course in their hydrogeological work in the Romanian Geological Survey. After reorganisation, the centre for hydrogeological studies transferred from the Bucharest Mining Institute to the Institute of Petroleum, Gas and Geology (1957–1974), and after 1974 to Bucharest University (Faculty of Geology and Geophysics). The course started by Ciocardel was developed here by Gheorghe (1960–1985) and continued by Zamfirescu (1975–1985). Gheorghe's text book *"Prelucrarea si sinteza datelor hidrogeologice* – Processing and Synthesis of Hydrogeological Data" was published in Romania in 1973 and translated and published in England in 1978. Before 1985, beside those mentioned above, an important contribution to the education of undergraduates and postgraduates in hydrogeology was made by Albu, Scradeanu, Pascu, Vasilescu and others. An important contribution to the literature at this time was the first synthesis of the hydrogeology of Romania: *"Apele subterane din Romania* – Groundwaters from Romania" (Pascu, 1983). Hydrogeology was also taught on geological engineering courses by Preda in Bucharest University, Marosi in Cluj University and Olaru in Iasi University.

At the same time as hydrogeology, text books were written and courses were developed in underground hydraulics by Cristea: *"Hidraulica subterana* – Underground

Hydraulics" (v1 1956, v2 1958); Oroveanu; Cretu: "*Hidraulica generala si aplicata –* General and Applied Hydraulics" (1971), and Soare from the Institute of Petroleum and Gas in Ploiesti, following its transfer from Bucharest in c. 1974; Iacob and Gheorghita from the fluid mechanics department of the Mathematical Institute; Harnaj, and many others. A significant contribution to the science was also made by Ene and Gogonea from the fluid mechanics department of the Mathematics Institute, e.g. "*Probleme in teoria flltratiei* – Problems in Filtration Theory" (1973), and by Cioc, Iamandi, Pietraru, Manescu, Petrescu, Trofin and Drobot from Bucharest University of Civil Engineering among others.

Applying the concept of potential movement to groundwater studies, the first analogue and physical models were developed after 1960. In the Hydrotechnical Research Institute (ISCH) Bucharest laboratory from 1960 to 1970 different types of analogue models were used to solve practical problems: Pietraru (1960, 1967, 1968), Zaharescu (1960), Ivan (1965), Juster (1968), Bobeica (1968). Numerical simulation of groundwater flow was made possible by the advent of third generation electronic computers after 1970, and was applied extensively in water management and the assessment of water resources. Numerical models for hydraulics were first developed in the ISCH and Hydrotechnical Faculty in the Civil Engineering Institute: Constantinescu (e.g. 1972, 1985), Danchiv (e.g. 1972, 1985), Capritza (e.g. 1972, 1978) Zaharescu (1977), Pietraru (1977), Drobot (1981–1985).

Following 1970, several important books were published in this field, including "*Alimentari cu apa* – Water Supply" (Trofin, 1972), "*Modele de calcul analogic in hydraulica subterana* – Analogue calculus models in underground hydraulics" (Ivan, 1975), "*Calculul infiltratiilor* – Water Seepage Calculus" (Pietraru, 1977), "*Mecanica apelor subterane* – Mechanics of Groundwaters" (Albu, 1981), "*Calculul sistemelor de drenaj* – Drainage Systems Calculus" (Ivan, 1985), "*Bazele hidrologiei technice* – Basis of Technical Hydrology" (Vladimirescu, 1984) and "*Hidraulica aplicata* – Applied Hydraulics" (Hancu *et al.*, 1985).

Staring in 1960, tracer techniques using environmental and artificial isotopes, dyes and other tracers were widely applied in groundwater studies, e.g. by Gaspar, Constantinescu and Tenu. These techniques proved to be an efficient instrument in both the hydrochemical and hydrodynamic study of groundwater. During the 1970s, the first modelling of groundwater chemical composition appeared in its thermodynamic and mineralogical context. This was an important step in the development of coupled groundwater chemistry and flow models, which were developed mainly by Aqua – Project Institute, Prospecting Company, ISCH, Environment Research Engineering Institute and the Institute of Speleology.

FINAL REMARKS

Although groundwater in Romania was exploited at least from the 2nd century BC, the first major works for abstraction and supply (e.g. Herculane Spa, Geoagiu Spa, Ulpia Traiana) are attributed to the Romans when they occupied the country in the early 2nd century AD. Then, from the 2nd to 16th centuries, works were constructed to supply the water needs for forts, monasteries and, later on, for large settlements such as Constantza, Bucharest, Craiova.

The first manuscripts about the hydrothermal water of the East Carpathians (an area of 8000 sq. km of carbonated waters) and the highly mineralized connate waters around the Flysch and Molasse Units appeared during the 16th–19th centuries.

Beginning with the 20th century, a major concern of hydrogeology, potable groundwaters, saw an impressive development, including the installation of more than two hundred deep wells for urban water supply by the Enterprise for Water Supply Bucharest. After World War II, the National Council of Waters and the Ministry of Geology each funded a long-term programme of hydrogeological investigation. The first comprised an observation network of about 4500 wells for unconfined and 3,500 for confined aquifers, while under the second, Prospecting Company and the Romanian Geological Institute in particular had the task of advancing knowledge of geological structure, and stratigraphy, hydrogeology and hydrochemistry.

Based on data from the large number of boreholes for hydrogeological and geological investigation, the Geological Institute published the Hydrogeological Maps of the main basins at 1:100 000 scale, followed by the Hydrogeological, Thermal and Mineral, and Geothermal Maps for the whole country, at the scale of 1:1 000 000.

Hydrogeological investigations of karst areas, including mapping, water balance studies, flow path tracing and resource assessment, have been carried out mainly by Prospecting Company and the Institute of Speleology. Tracing technology with radioactive and stable tracers was developed by the Atomic Physics Institute, starting in the 1970s.

In the field of the mining hydrogeology, dewatering of sedimentary deposits containing coal, iron, sulphur, bauxite and phosphate, and fissured aquifers with metals and metalliferous ores was the main focus. For the Pliocene lignite-bearing formations, more than 80 coalfields in the Pannonian and Dacic Basins were studied. Investigations of mineralized bodies in fissured aquifers concentrated mainly on the South Apuseni mountains.

In the period 1960 to 1970 many problems of applied hydrogeology concerning: water supply for drinking and industrial water, dams and reservoirs canals, harbours and dry docks, mining, tunnels and water pipelines, mineral water research and exploitation, irrigation and drainage systems were solved. The practice of applied hydrogeology, together with investigations for potable and irrigation water, stimulated advances in the basic science and in education, and with the development of computer simulation set the scene for the "modern stage" of Romanian Hydrogeology after 1970 (Zamfirescu, 2000).

MAIN SOURCES OF INFORMATION

Proceedings of the National Symposium "100 de ani de hidrogeologie moderna in Romania – 100 years of Modern Hydrogeology in Romania" 24–26 May 2000, Hydrogeological Association of Romania (AHR):

- Bretoteanu, M. Mathei Draghiceanu – precursorul hidrogeologiei moderne din Romania. pp. 9–14.
- Craciun, P. Elaborarea hartilor hidrogeologice nationale. pp. 15–19.

- Danchiv, A. Modelarea matematică în hidrogeologie in Romania. Scurt istoric. p. 97–108.
- Feru, A. & Simion, G. Istoricul cercetărilor si evolutia cunoasterii zacamintelor de ape minerale din Romania. pp. 23–33.
- Gheorghe, A. & Radulescu, C. Aportul hidrogeologiei miniere romanesti in dezvoltarea stiintifica si tehnologica a hidrogeologiei în Romania. pp. 53–62.
- Oraseanu, I. Consideratii privind cercetarea hidrogeologică a arealelor carstice din Romania. pp. 34–52.
- Tenu, A. & Davidescu, F. Treizeci de ani de hidrogeologie izotopica in Romania. pp. 82–95.
- Zamfirescu, F. Un secol de hidrogeologie in Romania. pp. 1–8.

REFERENCES

Albu, M. (1992) Quelues moments de 1'exploitation et de la recherche des eaux souterraines du territoire roumain. Premiere partie. *De 1'antiquite a la fin du XVIII siecle. AHR Bulletin, no. 1*, vol. 1 Bucharest.

Albu, M. (1993) Quelues moments de 1'exploitation et de la recherche des eaux souterraines du territoire roumain. *Deuxieme partie. lere moitie du XIXe siecle. AHR Bulletin*, no. 1, vol. 11. Bucharest.

Pricajan, A. (1985) Substantele minerale terapeutice din Romania. Ed. St. si Enciclopedica, Bucharest.

250 years of Russian hydrogeology (1730–1980)

E. Zaltsberg

Interenvironment Torresdale Avenue, Toronto, Ontario, Canada

ABSTRACT

Development of hydrogeology in Russia between 1730 and 1980 is briefly described. The contributions from many distinguished researchers in various fields of hydrogeology are highlighted and comparison of the achievements of the Russian hydrogeologists is made with their respective Western colleagues.

INTRODUCTION

The history of Russian hydrogeology goes back to the 18th century. During this period hundreds of Russian scientists and engineers worked in various fields of hydrogeology and contributed to its development and achievements. Their findings were published in thousands of monographs, papers and reports, but the limited scope of this paper allows only some of the names and publications to be mentioned.

1730–1790

Peter the Great established the Russian Academy of Sciences in 1724. During the 18th century the Academy organised and directed several expeditions to various regions of the country to collect scientific information on natural resources and their potential use. These expeditions explored the Ural Mountains, Siberia, the Volga basin, the Kol'sky and Kamchatka peninsulas, the Dnepr basin and the Caucasus Region. The expeditions were led by prominent Russian scientists, such as academicians V. Zuev, N. Richkov, S. Krasheninnikov, I. Lepekhin, and N. Ozeretskovsky. Each expedition spent several years in the field and collected vast amounts of data including information on the occurrence and distribution of fresh and mineral springs, water table elevations, the relationship between groundwater chemistry and chemical composition of water-bearing formations, and the role of groundwater in karst processes. The first of the expeditions was led by S. Krasheninnikov and carried out in 1733–1743; its results were published in 1765 in the monograph entitled *Description of the Land of Kamchatka*. It contained information on the rivers, lakes and the fresh, mineral and thermal springs in Kamchatka.

Hydrogeological information collected by the expeditions enabled some preliminary regional groundwater classification. For example, it was found that within the European part of Russia hardness in shallow groundwater increases in a southerly direction. It was also shown that the distribution of mineral springs is less dense in the interior areas of the Russian Plain than in the surrounding mountainous areas. The widespread occurrence of brackish and salty groundwater in Siberia was also recorded.

1790–1917

As early as in 1791 V. Severgin compiled the inventory of thermal springs in Russia. In 1800 he published the manual entitled *The Method of Mineral Groundwater Testing*. In 1809 he also prepared the first classification of mineralised groundwater based on its chemical composition and temperature, and published a summary on mineral groundwater occurrences in Russia.

G. Abikh studied the geology and hydrogeology of the Caucasus Region in the 1840s through the 1860s. He delineated the main groundwater recharge and discharge areas, found artesian groundwater in Armenia and studied and calculated the baseflow contribution into the Lake Sevan. He also studied the saline lakes in Armenia and numerous mineral springs in the Caucasus Region. He was the first scientist to relate the occurrence of mineral springs with tectonic lineaments.

Ilia Tchaikovsky, the father of the composer Pyotr Tchaikovsky, was a mining engineer who worked in the Ural Region in the 1830s and 1840s. He investigated saline springs and how to extract the salt from the spring waters. Reports were compiled by him for the salt plant owners and the Mining Department in Saint Petersburg.

In the 1850s and 1860s G. Gel'mersen studied artesian groundwater resources as a potential source of water supply for big cities. In 1864 he published the paper entitled *On Artesian Wells*. Based on his results and conclusions, the first deep flowing water well was drilled in Saint Petersburg between 1861 and 1864 under the supervision of A. Inostrantsev. Over the next two decades numerous other deep artesian wells were installed in the capital. In 1884 Inostrantsev published an inventory of these wells which contained all the borehole logs, the hydrochemistry data and some suggestions on the future use of artesian water for drinking and industrial water supply.

G. Gel'mersen and G. Romanovsky were the first geologists to identify artesian groundwater as supply to Moscow. Based on their findings, the first deep artesian well was drilled in Moscow in 1863. The total depth of this well was 459 m and it encountered a thick Devonian water-bearing deposit at a depth of 325 m.

The Geological Committee established in 1882 within the Ministry of State Properties and headed by G. Gel'mersen, initiated systematic large-scale hydrogeological investigations. At the end on the 19th century these investigations were managed by S. Nikitin, now acknowledged as one of the founders of Russian hydrogeology. He defined hydrogeology as the science of "underground waters, their origin, occurrence, distribution, movement and spring discharge". He stressed that the hydrological cycle is the main principle behind groundwater movement. In another definition of hydrogeology S. Nikitin pointed out that this is the science of "the underground component of the overall hydrological cycle."

I. Sintsov and S. Nikitin compiled the first water well inventory for the whole of Russia. S. Nikitin also conducted extensive studies of geological and hydrogeological conditions in various regions of the European part of Russia, focusing particularly on the Moscow Region. From 1884 until 1890 he collected detailed information on the geological setting of Moscow and the Moscow Region in order to define the Moscow artesian basin and its recharge area, evaluate the groundwater potential of the main water-bearing formations and describe the chemical composition of the groundwater bodies. These investigations were summarized in two papers published in 1890: *A Geological Map of Russia, Sheet 57*, and *Carboniferous Deposits and Artesian Waters in the Moscow Region*.

In 1900 S. Nikitin published the monograph *Unconfined and Confined Groundwater in the Russian Plain*. He identified that the hardness of the confined aquifers increases with depth and distance from the recharge area. He also delineated the main artesian basins within this huge territory and proposed its hydrogeological zoning.

After Gel'mersen's death in 1885 the Geological Committee was headed by A. Karpinsky who initiated further systematic geological surveys and hydrogeological investigations within the European part of Russia. Under the supervision of the Committee, a wide range of groundwater investigations were conducted in the early 1900s in Ekaterinoslav, Kherson and Samara Provinces in European Russia and in some Asian Provinces as well.

In 1905 the Geological Committee commenced an extensive study of mineral springs in the Caucasus Region led by A. Gerasimov, A. Ogil'vi, and Y. Langvagen who published a fundamental work on the origin of mineral springs and their piping in 1911.

In the early 1900s the Committee conducted a regional study of the main components of streamflow in the European part of Russia. The work focused on the baseflow component of the big rivers and was conducted under the guidance of S. Nikitin and N. Pogrebov.

The Geological Committee supervised hydrogeological investigations for water supply to big cities and the railway network that so depended on steam generation. Exploration of numerous artesian aquifers eventually satisfied demand for drinking water at Saint Petersburg, Moscow, Baku, Taganrog, Voronezh, Novorossiysk, and Sebastopol. In addition, in 1905 N. Pogrebov carried out investigation of the main springs at the Izhorskoye Plateau in order to supplement the water supply of Saint Petersburg. In 1909 he also investigated Shollar springs near Baku in order to evaluate their supply potential.

In 1914 an extensive hydrogeological study was conducted by N. Sokolov and A. Ivanov to find out the reason behind the steady deterioration of the tap water quality in Moscow. This led to detailed hydrogeological survey and mapping of the hydrochemical composition of the main aquifers both laterally and in depth, i.e. in 3D.

Based on ideas of V. Dokuchaev who studied zoning of soils and the role of the forest in water balance, original works on groundwater zoning and groundwater regimes in various climatic areas have been published by P. Ototsky and E. Oppokov. Of special interest is the investigation of A. Lebedev on the role of vapour in groundwater formation (1912). Based on the highly accurate measurements he found that groundwater recharge due to vapour condensation is much less than those due to infiltration of precipitation.

Significant achievements were made in hydrodynamics and well hydraulics. In a paper entitled *Theoretical Study of Groundwater Movement* (1889) N. Zhukovsky derived differential equations describing groundwater movement in porous media.

Investigating groundwater inflow into wells installed into the fractured media, A. Krasnopol'sky found that Darcy's law is inapplicable when the groundwater flow velocity exceeds some specific limit. He published his findings on the validity of Darcy's law and turbulent groundwater flow in *Mining Journal* (1912).

In the early 20th century systematic investigations of groundwater in the permafrost areas of Siberia began. In 1916 A. L'vov, who worked in Irkutsk, published a comprehensive summary on groundwater in the permafrost areas. At the same time the first works on groundwater within oil fields were presented by D. Golubiatnikov and others.

1917–1941

After the 1917 Revolution and the end of the Civil War in 1922, the Geological Committee conducted and coordinated hydrogeological surveys in central Russia, the Volga basin, the Komi area and the Caucasus. The Committee continued extensive investigations of the Moscow artesian basin and V. Khimenkov, A. Ivanov, A. Danshin and V. Zhukov prepared a summary of the geological and hydrogeological data that had been gathered on the basin.

The Committee continued investigations of mineral waters and mineral springs in the Caucasus Region (A. Gerasimov, N. Slavianov, A. Ogil'vi, N. Ignatovich). At the same time N. Tolstihin started a study of mineral waters in Eastern Siberia. The Siberian scientist M. Kurlov proposed a formula for expressing the chemical composition of groundwater (1928), which is still widely used in the hydrogeological literature.

In the 1920s the Institute of Meteorology established the first regional groundwater monitoring network in the European part of Russia, and the Federal Government funded an extensive groundwater monitoring programme over the huge irrigated areas in Central Asia.

Special courses on hydrogeology were available at several Russian Universities from the 1900s. The first Russian textbook on hydrogeology for University students was published in 1922 by P. Chirvinsky, and by the end of the 1920s Bachelor and Master degrees in hydrogeology could be obtained from the Universities of Moscow, Leningrad (now Saint Petersburg), Kiev, Odessa, Tomsk, and Tashkent.

Construction of new industrial complexes in the northern and eastern parts of Russia promoted studies of geological and hydrogeological conditions in these remote areas. During the 1920s M. Sumgin, A. Leverovsky, N. Tsitovich, and V. Sokolovsky studied the hydrogeology of permafrost. In 1922 V. Il'in published a summary of all the hydrogeological investigations conducted in the European part of Russia. At the same time three maps edited by V. Il'in were published: *A Map of Hydrogeological Studies in the European Part of the USSR*, *A Schematic Groundwater Map of the European Part of the USSR* (scale1:2,800,000) and *A Groundwater Map of the Central Industrial Zone* (scale 1:1,000,000). Il'in came to the conclusion that sustainable groundwater regimes depend on climate, topography and geological setting.

In 1925 A. Semikhatov published *Artesian and Deep Unconfined Aquifers in the European Part of the USSR* in which he described the 15 artesian basins in this territory.

In 1922 N. Pavlovsky published the monograph *Theory of Groundwater Movement Beneath Hydrotechnical Constructions and Its Main Applications* and concluded that groundwater flow velocity depends on the properties of the fluid as well as on characteristics of the porous medium. He demonstrated that groundwater flow velocity is proportional to the pressure gradient, the square of the effective diameter of the particles, and the square of the pore diameter and is inverse proportional to the kinematic viscosity of water. For solving problems associated with groundwater movement through and beneath hydrotechnical structures, N. Pavlovsky developed the so called conductive-sheet analogy method.

The First All-Union Congress of Hydrogeologists was held in Leningrad in the end of 1931. 570 delegates from all over the USSR attended this unique scientific forum and discussed 230 technical presentations. Significant attention at the Congress was given to papers by B. Terletsky, D. Sokolov, V. Zhukov and G. Kamensky dealt with the methodologies of both site specific and regional hydrogeological mapping.

As a result of the Congress, the comprehensive summary *Materials for Characterization of Groundwater Potentials in Various Regions of the USSR* edited by F. Savarensky, M. Vasil'evsky and D. Shegolev, was published in 1933.

Savarensky was a leading figure in Russian hydrogeology during the 1930s and 1940s and was one of the first lecturers to teach hydrogeology in Russia. He started teaching in 1920 at the Institute of Agriculture in Saratov, then at Saratov University and later in various Institutes in Moscow. Savarensky was author of one of the best Russian textbooks on hydrogeology published in 1933 and 1939 which remained "the Bible" for several generations of Soviet hydrogeologists thereafter. He also possessed a tremendous managerial skill for organising research, investigation and study. Savarensky presided over numerous expert Committees and Commissions and was chair of the Hydrogeological Commission of the All-Union Scientific and Technical Society and a member of various Editorial Boards. He was also the author of numerous publications on groundwater resources evaluation. He was one of the first hydrogeologists to stress the need to study man-made factors influencing the environment in general and groundwater in particular, see *The influence of the Kuybishev Reservoir on Geological Processes in the Surrounding Area* (1940, a) and *The Influence of Dams on the Transformation of River Banks* (1940, b).

Many other Soviet hydrogeologists worked successfully in various fields of hydrogeology before World War II. In 1934 A. Semikhatov published a monograph on groundwater in the European Part of the USSR. In 1937 A. Dzens-Litovsky and N. Tolstihin published a comprehensive summary of the occurrence and distribution of mineral waters in the USSR, and monographs and papers on hydrogeology in irrigated areas were published by V. Priklonsky and O. Lange.

F. Savarensky, M. Vasil'evsky, N. Ignatovich, I. Zaitsev defined hydrogeological zones of the USSR in the 1930s. In 1938 M. Vasil'evsky completed the groundwater zoning of the USSR, delineating hydrogeological basins, provinces and regions. This was the first time this was attempted for the Asian part of the USSR.

Significant achievements were made in further understanding well hydraulics and groundwater dynamics (G. Kamensky and G. Bogomolov, 1932; G. Kamensky, 1933; M. Al'tovsky, 1934). Corrections were introduced into Dupuit's equations, and limits for their practical applications identified. It was found that groundwater withdrawal from a fully penetrating artesian borehole depends not only on the confined aquifer thickness, its hydraulic conductivity and drawdown, but also on the borehole diameter. Factors that define the radius of influence of a borehole (duration of pumping, yield, drawdown, groundwater flow gradient and velocity) were identified and modified equations for calculating the radius of influence developed.

M. Al'tovsky (1934) derived equations for well yield calculation based on the results of a short term pumping test. Meanwhile N. Zhukovsky and A. Kostiakov developed the original theory of groundwater inflow into horizontal wells and galleries and the main achievements in hydrodynamics were summarised in the monograph by G. Kamensky *Principles of Groundwater Dynamics* (1933 and 1935).

Advances in groundwater chemistry were associated with the establishment of the Institute of Hydrochemistry in Moscow in 1921. It was led for many years by O. Alekhin. The Institute published the journal entitled *Materials on Hydrochemistry* in which seminal works by N. Slavianov, N. Tolstihin and M. Kurlov were presented.

Significant funding was provided for studying mineral waters which could be used for medical purposes. In 1936 V. Ivanov, L. Yarotsky, V. Shtil'mark and A. Ovchinnikov compiled a series of hydrochemical maps of mineral waters in the USSR and a summary map of mineral waters in the USSR (scale 1:20 000 000) was compiled by N. Tolstihin in 1937.

In 1933–1936 V. Vernadsky published the monograph entitled *History of Natural Waters*. He considered the various phases of natural waters as components of the global underground hydrosphere and evaluated the role of each phase in the geological and chemical evolution of the Earth and transformations of the Earth's crust. V. Vernadsky subsequently demonstrated the role of natural water in the geological and geochemical evolution of the Earth and highlighted the cyclic development of geological processes and natural water evolution. V. Vernadsky's ideas were fruitful for further development in various fields of hydrogeology including regional hydrogeology, paleohydrogeology, hydrochemistry and others.

Significant achievements were made in studying groundwater movement in fractured rocks. In 1939 D. Shegolev and N. Tolstihin published the monograph *Groundwater in Fractured Rocks* in which they summarized the findings of Russian and Western hydrogeologists. They pointed out the need for studying various geological processes, which influence the occurrence, size and distribution of fractures in rocks as well as the need to study the chemical interaction between the rock and water.

1941–1961

Understanding groundwater flow was a key theme for this period in which the solution of critical differential equations was central. Various scientists were involved, but the work of P. Polubarinova-Kochina, one of the founders of modern hydrodynamics and hydromechanics is perhaps the most famous. Her major work entitled *Theory of Groundwater Movement* was published in 1952 and in a few years several new editions followed along with translations into many languages. This monograph was and still is a handbook for many hydrogeologists throughout the world.

The application of finite difference method allowed site-specific characteristics of an aquifer to be considered. G. Kamensky initially developed this method for one dimensional groundwater flow and later modified it for the two dimensional flow. V. Lukianov constructed and used conductive-solid analogs for simulating groundwater flow within multi-layered water-bearing complexes.

Special emphasis was placed on studying deep aquifers and the role of riverbeds, sea and lake depressions as their discharge areas (K. Makov, F. Makarenko, A. Silin-Bekchurin, and B. Arkhangel'sky). Vertical movement of groundwater through thick clayey deposits was also investigated (N. Girinsky and A. Myatiev).

The extensive development of hydrotechnical works in the European part of Russia made the prediction of groundwater flow retention caused by dam and reservoir constructions of great practical importance. G. Kamensky developed the theory of steady (maximum) groundwater flow retention. Various forecasting equations for retention calculations were developed by G. Kamensky (1943) and N. Bindeman (1951).

The need to provide water supply for numerous cities, towns, and newly built industrial complexes initiated new studies in the evaluation of groundwater resources, their mapping, extraction and utilization. M. Al'tovsky developed methods of evaluating the total discharge from several interacted water supply wells and N. Plotnikov introduced a new method of calculating resource potential based on the size of the regional cone of depression.

In 1948–1955 N. Girinsky published on seawater intrusion into confined and unconfined coastal aquifers.

In 1944 the Laboratory of Hydrogeological Problems was created within the USSR Academy of Sciences. F. Savarensky was appointed as the first Director of this newly established scientific body. Under the directorship of F. Savarensky and then N. Slavianov, G. Kamensky, and V. Priklonsky, a team of hydrogeologists including O. Lange, F. Makarenko, A. Silin-Bekchurin, I. Garmonov, and others worked on fundamental theoretical problems of modern hydrogeology. The work included but was not limited to the following subjects: formation of groundwater under various climatic conditions; formation and regularities of regional baseflow; regional and global groundwater zoning; vegetative indicators of groundwater; the role of groundwater in the formation of ore deposits; and formation and utilization of mineral and thermal groundwater. Between 1944 and 1961 the Laboratory played a significant role in development of Russian hydrogeology.

1961–1980

Restructuring of the Academy of Sciences created a new Laboratory for studying geothermal energy and deep hydrogeochemistry under the leadership of F. Makarenko. The distribution of geothermal anomalies and their influence on groundwater chemistry was recorded. The role of groundwater in heat flow was evaluated and areas of geothermal groundwater identified, see *Thermal Waters in the USSR and Their Prospective Use* by F. Makarenko (1963), and other documents from this era.

Other Laboratory works included the chemical evolution of groundwater, chemical loading and the hydrochemical balance within deep aquifers and artesian basins. The Laboratory produced various regional hydrochemical maps and monographs on the hydrochemistry of thermal groundwater, physical and chemical equilibriums in groundwater under various natural conditions, and the regional distribution of groundwater resources.

Works of the Geological Institute of the Soviet Academy of Sciences included investigation of submarine hydrogeological systems; the chemical composition of modern sediments; hydrochemical conditions associated with modern sedimentation; and the interrelation between modern sediments, seawater and thermal springs on the sea bed.

Significant achievements in regional hydrogeology especially in studying and forecasting groundwater regimes and the water balance were made in the All-Union Institute of Hydrogeology and Engineering Geology (VSEGINGEO) established in 1939. In 1957 A. Konoplyantsev headed the groundwater resources, regimes and dynamics department of the VSEGINGEO. Konoplyantsev advocated the need to establish a regional groundwater monitoring network within the vast territory of the USSR and by the end of the 1960s several regional groundwater monitoring agencies had been established.

By 1970, the federal monitoring network consisted of about 30 000 monitoring sites to measure groundwater level, temperature, and chemical composition. All groundwater-monitoring wells were classified as first class (or basic observation wells) and second class (or auxiliary observation wells). The main purpose of carrying out observations in the 20 000 first class monitoring wells was establishing the regional patterns of groundwater regimes and obtaining baseline groundwater data.

The second class monitoring sites, consisting of more than 10 000 observation wells, provided information on groundwater regimes under natural and stressed conditions for site-specific groundwater problems in various environments.

The monitoring agency's responsibilities included not only data collection from first class and second class monitoring sites but also preparation of annual monitoring reports, evaluation of groundwater resources, groundwater mapping and forecasting. All regional monitoring reports were forwarded to the VSEGINGEO where groundwater predictions and maps were compiled for the entire USSR territory. This unique groundwater forecast information service is still operational. In order to process and summarise this huge amount of information, a new approach was needed, and A. Konoplyantsev became one of the pioneers in applying statistical methods to groundwater monitoring results.

The comprehensive hydrogeological map of the former USSR (scale 1:5 000 000) was compiled in 1966 by the team of hydrogeologists from the All-Union Institute of Geology (the VSEGEI) led by I. Zaitsev. The following features were depicted on this map: main hydrogeological structures (artesian basins and hydrogeological massifs), the groundwater type, the groundwater runoff within the active circulation zone, the permafrost areas, icings, old river valleys, volcanoes, and geysers. In addition, information on groundwater withdrawal within main hydrogeological regions was given. It included the stratigraphical indexes of main aquifers and typical pumping rates of water supply wells installed in these aquifers. At several specific locations within each main aquifer/complex borehole logs were shown. On each borehole log the groundwater type, the age of waterbearing rocks, and groundwater chemical composition were indicated.

In order to show such a variety of features, the compilers used numerous symbols including colors, shadings, lines, lettering, numbering and others.

The unique hydrochemical map of the former USSR (scale 1:5 000 000) edited by I. Zaitsev and N. Tolstihin was compiled in 1966. It was based on hydrochemical zoning and depicted three main zones: the zone of fresh water with total dissolved solid (TDS) values not more than 1 g/l; the zone of saline water (TDS = 1–35 g/l); and the zone of brines (TDS more that 35 g/l). Within cross-sections a certain combination of hydrochemical zones set up hydrochemical belts. Within artesian basins the depth of belts was limited by the bottom of the sedimentary cover or the top of the basement. Within hydrogeological massifs it was limited by the bottom of the active circulation zone.

At specific points typical borehole logs were presented and contained the following information: stratigraphical indexes of waterbearing deposits, the presence of halogenated deposits, the bottom of waterbearing deposit(s), the hydrogeochemical zones, and the mineral water's number.

Within the artesian basins and hydrogeological massifs (folded areas) mineral water provinces were shown. Main mineral water deposits, mineral lakes and mineral mud deposits were marked with special symbols. In addition, the groundwater temperature at the bottom of artesian basins was shown with isolines.

This map was state-of-the art compilation and presentation of the enormous amount of information on groundwater hydrochemistry collected in the former USSR by this time.

Detailed groundwater balance studies in the different climatic zones of the USSR were conducted by A. Lebedev at the VSEGINGEO. He developed methods for groundwater balance calculations based on analysis of water table fluctuations. He also compiled numerous maps of groundwater balance components for several basins located in various parts of the USSR. Study of the changes in groundwater regimes and water budgets caused by water infiltration from irrigation systems was made by D. Katz (1967).

B. Kudelin of the Moscow State University was able to organize a wide-scale study of groundwater runoff and groundwater resources in the USSR. Many University staff members (V. Vsevolozhskii, I. Zektser, R. Djamalov, I. Fidelli and others) as well as representatives from other scientific institutions and Regional Geological Surveys participated in this project. During the period 1964–1967 maps of groundwater runoff for the entire USSR territory (scales 1:5 000 000) were completed and published. Every map provided the following information: the average annual value of groundwater runoff or the modulus of groundwater runoff (in $l/s \times km^2$); the ratio between groundwater runoff and streamflow (in percentage); and the ratio between groundwater runoff and precipitation (in percentage). The maps show the quantitative distribution of groundwater flow and enable the evaluation of the groundwater resource potential.

Prior to compiling these maps, hydrogeological regionalisation for the whole USSR was conducted. It was based on consideration of such factors as the occurrence, age, thickness and lithological composition of the main waterbearing complexes/aquifers within the active circulation zone. Depending on the regime of groundwater runoff within these complexes/aquifers and the degree of their hydraulic connection with the rivers, the appropriate scheme for hydrograph separation was chosen for each region. For compiling the map for average annual groundwater runoff the streamflow data at 2128 streamflow gauges with the long observation periods had been analyzed and 25 317 streamflow hydrographs had been separated using one of Kudelin's techniques. The set of maps was awarded several prizes from the Moscow State University, The Academy of Sciences, and Exhibition of Achievements in the Economy. The summary monograph *Groundwater Flow in the USSR* was published in 1966.

In 1967 the Water Problems Institute (IWP) of the USSR Academy of Sciences was established in Moscow. The Institute consisted of several Branches including the Groundwater Resources and Subsurface Runoff Branch. It was headed by B. Kudelin and after his death in 1972, by I. Zektser. Hydrogeologists R. Dzhamalov, V. Kovalevsky, N. Lebedev, and M. Nikitin worked at this Branch for many years and made a sufficient input into Branch's scientific achievements. The main topics which investigated included: distribution of groundwater resources; regional evaluation of groundwater resources; groundwater quality and its dependence on various natural and man made factors; and groundwater protection and conservation.

In accordance with the UNESCO International Hydrological Programme and in collaboration with many Russian and European Institutions, the IWP played a leading role in compiling a map of groundwater flow in the Central and Eastern Europe (scale 1:1 500 000).

In the 1930s through the 1950s new concepts of aquifer compressibility and elasticity were developed in the USA by Theis (1935), Jacob (1940, 1946) and Boulton (1954) and eventually introduced into the Russian hydrogeological literature. They have since been successfully applied by Russian hydrogeologists for problem solving. The significant input into development of modern groundwater dynamics and its practical application in evaluating groundwater resources have been made by V. Babushkin, G. Barenblatt, N. Bindeman, F. Bochever, I. Gavich, V. Gold'berg, E. Kerkis, A. Lebedev, N. Ogil'vi, I. Zhernov, and others.

V. Shestakov was one of Russian hydrogeologists who made a smooth transition from Darcy-Dupuit-Boussinesq-based classical groundwater dynamics into modern hydrogeodynamics and modelling that opened new possibilities and means for solving the wide range of complicated practical problems.

He developed, verified and successfully applied various mathematical methods in many areas of hydrogeology such as the design of lateral drainage systems, calculation of the regional drawdown caused by pumping in interacting wells, pumping from multi-aquifer groundwater systems, pumping from wells located in river valleys, interpretation of pumping test results.

In the numerous textbooks and articles on hydrogeodynamics Shestakov highlighted the importance of the following crucial issues: schematization of hydrogeological conditions as the first step in transition from field investigation and description of the site to its mathematical simulation; heterogeneity within the aquifer, aquitard and vadoze zone and its incorporation into hydrogeological models; and groundwater modelling as the necessary component and tool in hydrogeological research and studies.

In 1974, V. Shestakov and V. Mironenko published the monograph *Principles of Hydrogeomechanics* which was one of the most important contributions to understanding of interaction between water and the water-bearing media of that era.

V. Mironenko led the development of mining hydrogeology and groundwater dynamics. His early research involved evaluation of the efficiency of horizontal drainage wells, various aspects of dewatering of mineral deposits, an assessment of the stability of quarry and pit slopes with regard to dewatering operations, and rock deformation due to the influence of extensive dewatering. Gradually his interest switched to contaminant hydrogeology, groundwater protection, and environmental hydrogeology. In the early 1970s Mironenko pioneered the use of modern computer techniques, especially numerical modelling, in solving complex hydrogeological and environmental problems.

In the 1960s and 1970s many monographs and papers on hydrochemistry were published. As in the past, the scientific interest was focused on groundwater origin and formation (A. Silin-Bekchurin, A. Khod'kov, E. Posokhov, S. Smirnov, and others), and on regional hydrochemistry (K. Pit'eva, S. Shvartsev, E. Pinneker). Fundamental works were published on geochemical methods for ore deposits exploration (A. Brodsky, A. Germanov, A. Ovchinnikov, V. Borovitsky, V. Kiryukhin, and S. Shvartsev). Special attention was given to the groundwater geochemical features which could be used for oil fields exploration (V. Sulin, V. Krotova, and others). Hydrochemistry of mineral waters in various regions of the USSR continued to be studied and summarised in works by A. Perel'man (1968), A. Ovchinnikov (1970), N. Tolstihin and E. Posokhov (1977).

In the 1960s and 1970s many groundwater exploration studies were carried out. For example, in the mid-1960s about 1.5 million groundwater exploration boreholes were drilled annually. Between 1971 and 1975 groundwater abstraction was developed to supply more than 650 big cities including Tbilisi, Erevan, Tallinn, Vilnius, Irkutsk, Frunze; significant achievements were made in developing the principles and methods of groundwater resources evaluation and groundwater safe yield calculations. Among numerous publications in this field, the comprehensive monographs by F. Bochever and N. Verigin (1961), F. Bochever (1968), and L. Yazvin (1972) should be mentioned. In the early 1970s first detailed studies on groundwater protection were published by V. Gold'berg, F. Bochever and A. Oradovskaya (1972).

The 50-volume monograph *Hydrogeology of the USSR* was published between 1960–1976. Each volume describes in full the hydrogeology of a specific region of the USSR. It contains descriptions of climatic and geological settings, areal and vertical distribution of main aquifers and water-bearing complexes, the hydraulic character-istics, hydrochemistry and groundwater temperature data, sources of groundwater contamination and their influence on groundwater quality, evaluation of groundwater potentials, and current and future groundwater use and management. The summary volume consists of five books and an atlas of various regional hydrogeologic maps. The current use of different types of groundwater is thoroughly evaluated and the long term forecast of their future utilization is given.

CONCLUSIONS

1   The development and achievements of hydrogeology in Russia were always based on and closely associated with geology, and with geological surveying and map-ping. One of the most prominent Russian hydrogeologists, Professor N. Tolstihin repeatedly told his students "Nobody can pretend to be a good hydrogeologist without being a good geologist". Geology was and still is the main component of hydrogeological training, study and research in Russia.

2   A multi-theme approach was typical for hydrogeological investigations in Russia. The groundwater origin, occurrence, distribution and chemical composition were studied alongside geological, hydrogeological, hydrological, climatic, topo-graphic, soil and other natural conditions.

3   Regional investigations and studies were a significant component of the hydro-geological development in Russia. These were conducted by the Academy of

Sciences, the Geological Committee, the Federal and Republican Ministries of Geology, the Water Problems Laboratory, the All-Union Institute of Hydrogeology and Engineering Geology, the All-Union Geological Institute and other Institutions. As a result, such problems as regional evaluation of groundwater potentials, regional mapping, regional characteristics of groundwater regimes and balance, and regional groundwater forecasting were studied in depth, and Russian achievements in these fields were impressive.

4   Solving fundamental scientific problems was always a significant component and driving force behind Russian hydrogeology. For the period from 1960 to 1980 only, several hundred monographs were published in Russia dealing with a variety of topics including planetary and regional hydrogeology, hydrodynamics, hydrochemistry, hydrogeomechanics, groundwater in the permafrost areas and groundwater modelling.

5   Russia claims to host the largest number of Universities and Institutions teaching hydrogeology. As a result, the Russian hydrogeological community consists of several thousand professional hydrogeologists most of whom have a MSc degree.

6   The isolation of Russian hydrogeologists from their Western colleagues from the early 1930s to the late 1980s stimulated original investigations and studies to satisfy immediate needs of the industry, agriculture and urban and rural developments. Many achievements of Russian hydrogeologists working on in such fields as regional evaluation of groundwater resources, regional groundwater mapping and forecasting, evaluation of groundwater regimes and balances, regional and planetary hydrogeochemistry remained unknown for many years in the West. The political barriers created by the Cold War were counterproductive for the development of both Russian and Western hydrogeology. Nowadays the personal contacts and exchange of ideas and literature is common practice.

7   Considering 250 years of Russian hydrogeology in its broadest context, for much of the time Russia was more advanced than Western countries in large scale investigations and studies of evaluation and distribution of groundwater resources, groundwater mapping and forecasting, groundwater regimes and balance, and interaction between groundwater and surface water. Russia was also a pioneering country in establishing the national groundwater monitoring network and development of regional groundwater forecasts. The typical Russian approach was always first to get the big hydrogeological picture and use this to help solving site specific problems, rather than to collect piece meal site specific information to compile a big picture. The numerous regional studies and investigations resulted in a large amount of fundamental scientific monographs and papers being produced on various aspects of hydrogeology. Unfortunately, many of them remain unknown to the Western hydrogeological community to this day.

8   In the 1930s, 1940s and 1950s, Russia was behind the West in the field of modern groundwater dynamics. In the 1960s and 1970s Russian hydrogeologists were also behind their Western colleagues in applying numerical methods and modern computer techniques for solving complicated groundwater problems including those in the field of contaminant hydrogeology. However, starting from the late 1970s, Russian groundwater specialists were again in the front line of the international hydrogeological community with regards to their scientific achievements and its practical applications.

ACKNOWLEDGMENTS

The author would like to thank Drs. S. Pozdniakov of Moscow, Russia and M. Gogolev of Waterloo, Canada for providing valuable information on the history of Russian hydrogeology.

REFERENCES

Al'tovsky, M. (1934) Water well yield calculations based on the pumping test results. *Proceedings of the Fifth Water Supply System Congress, Moscow* (in Russian).

Bindeman, N. (1951) *Calculations of Groundwater Flow Retention and Filtration from Reservoirs.* Moscow, Ugletechizdat (in Russian).

Bochever, F. & Verigin, N. (1961) *Manual for Evaluating Groundwater Potential for Water Supply Purposes.* Moscow, Gosstroyizdat (in Russian).

Bochever, F. (1968) *Theory and Practical Methods for Calculating Groundwater Exploitation Resources.* Moscow, Nedra (in Russian).

Chirvinsky, P. (1922) *Hydrogeology.* Rostov-on-Don (in Russian).

Dzens-Litovsky, A. & Tolstihin, N. (1937) *Natural Mineral Waters in the USSR. Priroda*, No. 10 (in Russian).

Gel'mersen, G. (1864) On artesian wells. *Mesyatseslov.* Saint- Petersburg (in Russian).

Gerasimov, A., Ogil'vi, A. & Landvagen, Y. (1911) Caucasian mineral waters. In: *Materials for Studying Geological Conditions in Imperial Russia, vol. 3.* Moscow (in Russian).

Hydrogeology of the USSR (1960–1976) in 50 volumes. Nauka, Moscow (in Russian).

Il'in V (1925) A map of hydrogeological studies in the European part of the USSR; A schematic groundwater map of the European part of the USSR (scale 1:2,800,000); and a groundwater map of the central industrial zone (scale 1:1,000,000). *Proceedings of the First Russian Hydrology Congress, Leningrad, 1925* (in Russian).

Kamensky, G. & Bogomolov, G. (1932) Calculation of the hydraulic conductivity value for the case of the asymmetric radius of influence. *Hydrotechnical Construction*, 2–3 (in Russian).

Kamensky, G. (1933, 1935) *Principles of Groundwater Dynamics. Part I (1933), Part II (1935).* Moscow (in Russian).

Kamensky, G. (1943) *Principles of Groundwater Dynamics.* Moscow, Gosgeoltechizdat (in Russian).

Katz, D. (1967) The groundwater regime in the irrigated areas and its control. Moscow, Kolos (in Russian).

Konoplyantsev, A., Kovalevsky, V. & Semenov, S. (1963) *The Natural Groundwater Regime and Its Regularities.* Moscow, Gosgeoltechizdat (in Russian).

Krasheninnikov, S. (1765) *Description of the Land of Kamchatka.* Saint-Petersburg (in Russian).

Krasnopol'sky, A. (1912) Water wells in unconfined and confined aquifers. *The Mining Journal*, 3 (4–7) (in Russian).

Kudelin, B. (ed.) (1966) *Groundwater Flow in the USSR.* MGU.

Kurlov, M. (1928) *Classification of Curable Mineral Springs in Siberia.* Tomsk (in Russian).

Lebedev, A. (1912) The role of vapor in the regime of soil water and groundwater. *Proceedings of the Second South Russian Land Reclamation Congress.* Odessa.

Lebedev, A. (1976) *Methods of Studying Groundwater Balance.* Moscow, Nedra (in Russian).

L'vov, A. (1916) *Exploration and Testing of Water Supply in the Permafrost Area for the Amur Railway.* Irkutsk, The Ministry of Transportation (in Russian).

Luckner, L. & Shestakov, V. (1976) *Modelling of Geofiltration.* Moscow, Nedra (in Russian).

Makarenko, F. (ed.) (1963) *Thermal Waters in the USSR and Their Prospective Use*. Moscow, The USSR Academy of Sciences (in Russian).

Makarenko, F. & Kononov, V. (1973) *Hydrothermal Regions in the USSR and Hydrothermal Energy Use*. Moscow, Nauka (in Russian).

Mironenko, V. & Shestakov, V. (1974) *Principles of Hydrogeomechanics*. Moscow, Nedra (in Russian).

Mironenko, V. & Shestakov, V. (1978) *Theory and Methods for Test Result Interpretations*. Moscow, Nedra (in Russian).

Nikitin, S. (1890a) A Geological Map of Russia, Sheet 57. *Proceedings of the Geological Committee, vol. 5, Saint-Petersburg* (in Russian).

Nikitin, S. (1890b) Carboniferous Deposits and Artesian Waters in the Moscow Region. *Proceedings of the Geological Committee, vol.5, Saint-Petersburg* (in Russian).

Nikitin, S. (1900) *Unconfined and Confined Groundwater in the Russian Plain*. Saint-Petersburg (in Russian).

Ovchinnikov, A. (1970) *Hydrogeochemistry*. Moscow, Nedra (in Russian).

Pavlovsky, N. (1922) *Theory of Groundwater Movement Beneath Hydrotechnical Constructions and Its Main Applications*. Petrograd (in Russian).

Perel'man, A. (1968) *Geochemistry of Epigenetic Processes*. Moscow, Nedra (in Russian).

Pinneker, E. (1966) *Brines in the Angara-Lena Artesian Basin*. Moscow, Nauka (in Russian).

Polubarinova-Kochina, P. (1952) *Theory of Groundwater Movement*. Moscow, GTTI (in Russian).

Savarensky, F., Vasil'evsky, M. & Shegolev, D. (eds.) (1933) *Materials for Characterization of Groundwater Potentials in Various Regions of the USSR*. Moscow (in Russian).

Savarensky, F. (1939) *Hydrogeology*. Moscow-Leningrad, GONTI (in Russian).

Savarensky, F. (1940a) The influence of the Kuybishev Reservoir on geological processes in the surrounding area. *The Bulletin of the Academy of Sciences*, 8–9 (in Russian).

Savarensky, F. (1940b) The influence of dams on river bank transformations. *Papers of the Academy of Sciences*, 27 (9) (in Russian).

Semikhatov, A. (1925) *Artesian and Deep Unconfined Aquifers in the European Part of the USSR*. Moscow (in Russian).

Semikhatov, A. (1934) *Groundwater in the USSR, Part 1 – Groundwater in the European Part of the USSR*. Moscow, ONTI (in Russian).

Severgin, V. (1800) *The Method of Mineral Groundwater Testing*. Saint-Petersburg (in Russian).

Shegolev, D. & Tolstihin, N. (1939) *Groundwater in Fractured Rocks*. Leningrad, Spetsgeo (in Russian).

Shestakov, V. (1973) *Groundwater Dynamics*. MGU (in Russian).

Silin-Bekchurin, A. (1961) *Conditions of Formations of Saline Waters in Arid Zones*. Paris, UNESCO.

Tolstihin, N. (1937) *The Map of Mineral Waters and Mud for the USSR (Scale 1:20,000,000)*. *The Great Soviet Atlas of the World*. Moscow, TSNIGRI (in Russian).

Tolstihin, N. & Posokhov, E. (1977) *Mineral Curative, Industrial and Thermal Waters*.Leningrad, Nedra.

Vernadsky, V. (1933) *History of Natural Waters, Part I (1933), Part II (1934), Part III (1936)*. Moscow (in Russian).

Yazvin, L. (1972) *Reliability of Hydrogeological Forecasts in Relation to Evaluation of Groundwater Exploitation Resources*. Moscow, VSEGINGEO (in Russian).

Zaitsev, I. (ed.) (1966) *Hydrogeological Map for the USSR (Scale 1:5,000,000)*. Leningrad, VSEGEI.

Zaitsev, I. & Tolstihin, N. (eds.) (1966) *Hydrogeochemical Map for the USSR (Scale 1:5,000,000)*. Leningrad, VSEGEI.

Zhukovsky, N. (1889) Theoretical Study of Groundwater Movement. *The Journal of the Russian Physical and Chemical Society*, vol. XVI (Section 1, Issue 1) (in Russian).

The history of hydrogeology in Serbia

Zoran Stevanovic
Head of Department of Hydrogeology Faculty of Mining & Geology, University of Belgrade
Belgrade, Serbia

INTRODUCTION

In 1997, Serbia (which was at that time still with Montenegro in the Yugoslav Federation) celebrated one hundred years of its hydrogeological history. The jubilee was dedicated to Svetolik Radovanovic and his famous book "Groundwater", published by the Serbian Literary Guild in 1897, which represents the first analytical hydrogeological book in the former Yugoslavia as a whole. In 2005 Belgrade hosted a large international conference sponsored by IAH and UNESCO, dedicated to Jovan Cvijic and the 110 years since his major and influential work "Karst" was published. In fact, these two events commemorated two crucial scientists and their works which were milestones in the national story of hydrogeology.

However, the history of hydrogeology in this part of the Balkans does not actually begin in 1897; the tapping of mineral or fresh water was a very ancient technique in the area. Ruins which remain from Roman times are evidence of extensive water use. The first organized investigations of mineral water began during the 1830s in Prince Milosh Obrenovich's Serbia. After a long period of rule under the Turks, the country's rulers wanted to be at the same level as the rest of Europe, and the efforts of the prince and his son Prince Mihailo brought more water investigations, the construction of drinking-water wells and the creation of the first water protection law. At that time, numerous Austro-Hungarian and Serbian chemists were beginning research on thermal and mineral water, primarily as a basis for its intensive exploration and medicinal use.

During the second half of the 19th century, the Serbian geological school was established and its founders Jovan Zujovic, Jovan Cvijic and Svetolik Radovanovic were the first scientists to initiate and conduct groundwater research. The volume of surveys in hydrogeology was certainly modest, but sufficient to state that theoretical and practical knowledge in some fields of hydrogeology was fairly high (mineral and thermal waters, water supply, karst water, etc.). After the two Balkan wars, followed by the First World War and the suspension of all activities between 1912 and 1918, the development of hydrogeology proceeded more rapidly. Assisted by a new generation of specialists, some very experienced geologists continued to work and made significant progress in this field.

After the Second World War, the growing needs for national reconstruction led first to the reorganisation of the geological science and profession, and then to practical geological tasks, including systematic hydrogeological investigation. In contrast to the initially mainly descriptive hydrogeology, with its long history around the world, including in Yugoslavia, quantitative hydrogeology had its maximum development from 1945 onwards.

HYDROGEOLOGICAL CHARACTERISTICS OF SERBIA

The hydrogeological provinces of Serbia are equivalent to the geotectonic units of the national territory (Figure 1). These are: the Dacian basin (NE part); the Carpatho-Balkanides of Eastern Serbia; the Serbian crystalline core (central part); the Sumadija-Kopaonik-Kosovo zone; the Dinarides of Western Serbia and the Pannonian basin (north). The units are characterised by complex geology and non-uniform distribution of water-bearing rocks and water reserves. Due to the prevalence of highly karstified Mesozoic limestones, karst aquifers were formed in the Dinarides as well as in the Carpatho – Balkanides. The central part is relatively poor in groundwater with the exception of the thick alluvium deposited along major rivers. Thus, the Sava, Danube, Drina, and Velika Morava valleys are the most important and most valuable groundwater resources of the country.

1 - Dakijski basen
2 - Karpato-balkanidi istočne Srbije
3 - Srpsko kristalasto jezgro
4 - Šumadijsko-kopaoničko-kosovska zona
5 - Dinaridi zapadne Srbije
6 - Panonski basen

Figure 1   Schematic Map of the hydrogeological provinces of Serbia – *Dacian basin (1), Carpathian-Balkan arch of Eastern Serbia (2), Serbian crystalline core (Serbo-Macedonian massif, 3), Sumadija-Kopaonik-Kosovo zone (4), Dinarides of Western Serbia and Montenegro (5), Pannonian basin (6).*

These prevailing hydrogeological conditions directed groundwater investigations along two major lines: a) for practical purposes such as projects for municipal and industrial water supply, mineral water exploration, and dewatering of mines and irrigation, and b) for regional hydrogeological studies and hydrogeological mapping.

At present, Serbia's water supply systems provide groundwater for drinking purposes for around 80% of the population. According to some estimates, the total groundwater reserves in Serbia are around $100\,m^3/s$.

Serbia has about 230 identified mineral and thermo-mineral springs, while about 20 modern balneo-therapeutic or recreation centres are currently operating. However, there are significant potential reserves (approx. $1\,m^3/s$) that have not been sufficiently explored to provide a definite estimate. Depending on depth, temperatures vary between the usual 20–60°C to, in a few instances, a maximum of 90°C. Numerous mineral springs and the attempt to maximize their potential have made hydrogeochemistry and hydrogeology for medical application (balneology) well-developed disciplines since the beginning of national hydrogeology.

Serbia and its neighbours of ex-Yugoslavia represent one of the typical karst "strato-type" regions. The characteristic rough Karst landscape, large distribution of land surface and subsurface karstic occurrences always provoke interest in their study and description. This is why the contribution of national scientists to international hydrogeology can be assessed as greatest in terms of karst hydrogeology. Cvijic's explanation of water circulation in karst phenomena, the theory of vertically-positioned dynamic hydrographic zones, and terminology are well woven into the fabric of this scientific discipline. Indeed, karst explorers world-wide use the local Yugoslav terms polje, uvala, doline, ponor, which is both an honour and a commitment to the work of its hydrogeologists and scientists in general.

THE PERIOD UP TO THE END OF THE NINETEENTH CENTURY

Archeological traces of water structures in the country are numerous. The earliest known water well dates back to the Neolith, c. 4000 B.C. (Pancevo, Vojvodina). Later, the Sarmatian tribes introduced the digging of wells lined and supported by oak timbers. Many waterworks from Roman times have also been discovered. The Roman Empire's flourishing economy was destroyed when barbarian tribes overran its boundaries, their invasion causing earlier settlers to return to a nomadic way of life. From the 5th to the 10th century, due to frequent wars and great migrations, water was rarely drawn from artificial sources. Since the 10th century, the trend of building permanent settlements associated with the draining of marshland, sinking of wells and protecting and tapping springs to provide usable water continued in the northern part of Serbia until the Turkish invasion in the 15th and 16th century. A long period of stagnation in all activities occurred under the Turks (Milosavljevic et al., 1997).

After liberation from the Turks, prosperity was restored. Very soon, almost every household had its own dug well, while the organized drilling of artesian wells was first started in Vojvodina in the middle of the 19th century. Water supply from deep wells brought many improvements and soon expanded throughout the areas underlain by intergranular aquifers, especially in the Tertiary basins (Vojvodina, Smederevo, Paracin, Zajecar etc.).

The earliest known investigations of mineral water date back to 1834, with the first records made in Vienna by F. Hruschaner and Baron Hake. A year later, invited by Prince Milos, Z.A. Herder visited Serbia and made a tour of spas. Based on the results of chemical analyses, he wrote his comments on their value and usability, mainly through analogy with known mineral waters from European spas. Some time later, in 1846, mineral water was first analysed in the "state pharmacy" of Kragujevac, and later also in the High School of Belgrade. In 1856 E.P. Lindermayer published the book *"Description of mineral waters and their use in general, and medicinal water in particular, in the Princedom of Serbia"*. This is the first detailed description of water from some of the mineral springs of Serbia (Filipovic, 1997). The largest number of chemical analyses of mineral water was made from 1874 to 1886 by Professor S. Lozanic, one of the pioneers of scientific thought, particularly in natural sciences. Valuable information concerning mineral waters was published from 1884 to 1890 by M.T. Leko, M. Nikolic, A. Zego and others.

Serbian Prince Mihailo wrote in 1859: "There should be a description of both surface and groundwaters, since even groundwaters can be dangerous if their direction is not known, while their strength could be put to use. All this has yet to be investigated: Baba (Prince Mihailo's father) has been paying attention to healing waters in Serbia since long ago, and he says that they represent an invaluable treasure. This is a country rich in waters, land and people, but its potential is still vastly unused. Work must be begun from the very basis, so that we may reap some rewards in several decades at the earliest ..."

The first legal act, which contains five paragraphs for the protection of the environment and of water, had been written in 1850 and introduced severe punishment for those who polluted natural resources.

The beginning of the teaching of geology in Serbia is associated with the name of Josif Pancic, a prominent natural scientist who delivered his first lectures in the Nature – Engineering Department of the Lyceum in 1853. The first scientific contribution by national geologists was published in Vienna in 1854. In 1880, Jovan Zujovic became the newly founded Chair of Mineralogy and Geology in the High School. Three years later the Geological Survey of the High School was also founded in Belgrade. The first scientific journal *"Annales Geologiques de la Peninsule Balkanique"* was published in 1888. It is one of the oldest geological journals, which at present can be found in about seven hundred libraries over the world. Jovan Cvijic and Svetolik Radovanovic belong to the first generation of Zujovic's students and they took the role of chair in the further development of the national geological school.

Jovan Cvijic's role and contribution to karst hydrogeology

During his studies of Geographic Science at the University of Belgrade, Cvijic (Figure 2) was particularly interested in the geological disciplines. As an undergraduate, he published a paper on Serbian geographical terminology with explanations of many terms for surface and subsurface karst features. From 1889 to 1895 Cvijic carried out the first surveys in karst in the Kucaj massif of the Carpatho-Balkan mountain arc in eastern Serbia, and published five extensive studies of the karst features of this region. His works of particular importance which are considered presently, are *"Caves and Subterranean Hydrography in Eastern Serbia"* and *"Springs, Peat Bogs and Waterfalls*

Figure 2 Jovan Cvijic.

in eastern Serbia" published by the Serbian Royal Academy (Stevanovic, 1997). In the first mentioned article, Cvijic gives a classification of caves in eastern Serbia by hydrographic function (he explored a total of about 4000 m of channels in several caves). Detailed descriptions and plans of caves "from which periodically rivers and streams flow out" are of particular importance (Figure 3).

The first inventory of springs is given in the study "*Springs, Peat Bogs and Waterfalls in Serbia*". They are classified as:

1   "common springs emerging from caverns or fractures";
2   large springs and lakes, i.e. water-filled basins; and
3   intermittent springs from which "water intermittently flows and ceases to flow".

The group of "common springs" is divided into: (a) "large springs from caverns and fractures in limestones" and (b) "large springs at contacts of limestones and impermeable rocks".

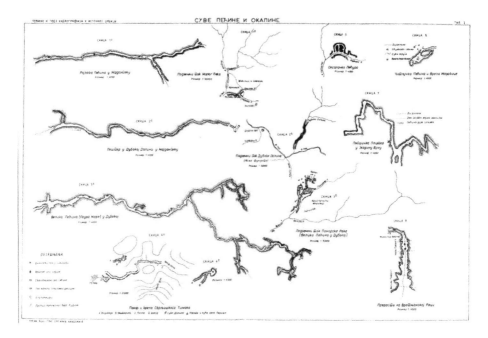

Figure 3 Cvijic's sketches of some caves in Eastern Serbia.

The inventory of 159 springs includes their names, localities, altitudes, rocks of emergence, and water temperatures but spring flows are mostly lacking. Cvijic gives a particularly detailed description of the Vauclusian type spring ("lake") with the examples of Modro Vrelo in Krupac, Jelovicko Vrelo, and Mlava spring – Zagubica Lake (illustrated by their plans and cross-sections). The description of intermittent springs in Zvizd and Homolje was followed by an explanation of their specific discharge mechanism which is perfectly clear and still scientifically acceptable. His observations of the behaviour of the Mlava spring in Zagubica subsequent to an earthquake are important mention (Figure 4).

Cvijic began his postgraduate studies of geography at the University of Vienna in 1889 tutored by Professor Albrecht Penk and finished in 1892 with the dissertation "Das Karstphaenomen". The dissertation was published in the following year by the Academy of Sciences in Vienna and aroused great interest among geoscientists all over the world. Cvijic is particularly merited for the acceptance of the concept of chemical corrosion as a dominant process in the morphogenesis of sinkholes, in contrast to the generally accepted theory of cave collapse as a major genetic factor. He also provided descriptive and genetic classifications of the caves, karst rivers and dolines. Praise came from France, Germany, Italy, and Austria-Hungary, and Cvijic himself mentioned in his autobiographical notes that this greatly encouraged and inspired him to carry out his investigations. Upon his return to Serbia, he was nominated Professor of Geography at the University of Belgrade. Later, as Rector, member and chairman of the Serbian Royal Academy he made an invaluable contribution to the development of education and

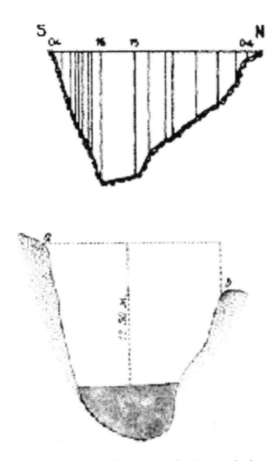

Figure 4 Cvijic's sketch of the Vauclusian Mlava spring after his own bathymetric measurements.

sciences in Serbia. His investigations extended to the whole of the Balkan Peninsula, the Mediterranean, and Europe, and his research covered not only the natural sciences, but also human geography.

In his later works and the numerous long excursions he made through the Dinaric holo-karst terrains, he paid special attention to the genetic analysis of the large karst poljes which characterised the whole region (Croatia, Bosnia & Herzegovina, Montenegro). Based on excellent geological knowledge, he considered bordering faults and epeirogenic movements, influenced also by specific paleoclimatic conditions, as major factors in the development of poljes (Figure 5). In "*Hydrographie souterraine et évolution morphologique du karst*" (1918) Cvijic identified four genetic types: 1. Polje–tectonic graben; 2. Polje–connected with a fault; 3. Polje – syncline; and 4. Erosional polje.

The most significant contribution made by Cvijic was to the theory of water circulation and the functioning of karst aquifers (Mijatovic, 1983, 1997). At the end of the 19th century, two contradictory theories existed: the first created by Katzer and

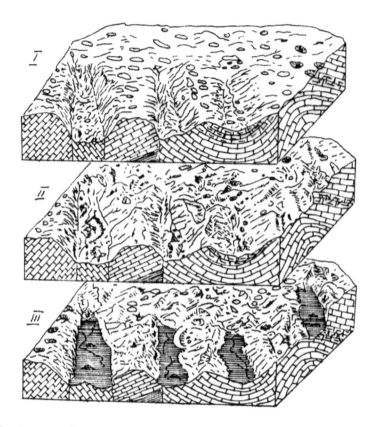

Figure 5 Development of karstic poljes (Cvijic, 1918) First phase – left: tectonic graben; mid: fault; right: syncline, second phase – deepening the bottom and forming initial valleys (uvalas); third phase – further deepening, the barriers between valleys have disappeared, poljes are formed.

Martel and the second by Grund and Penck. The first respects a traditional approach in which karst groundwater flows are isolated and drainage systems through karst-gerinne (channels) are separated in vertical sections. The second theory introduced the idea of two water layers. The deeper saturated part is stagnant, while the upper part of the aquifer is characterised by a common and continual groundwater table as well as by gravity drainage towards the nearest erosional base. Cvijic, with his considerable practical field experience, could not accept such a strict, almost dogmatic differentiation that both groups insisted on. He was the first to explain that arguments for both theories could be acceptable and tried to harmonise them through different case examples.

Cvijic (1918) also described his own theory of superposition of "hydrographic zones in karst" as a result of the influence of hydrologic and climatic factors. His coherent synthesis introduced three main zones (Figure 6) in specific dynamic coexistence which are connected and changeable following intensive groundwater level fluctuations throughout the hydrologic year. Vertical gravity circulation is dominant

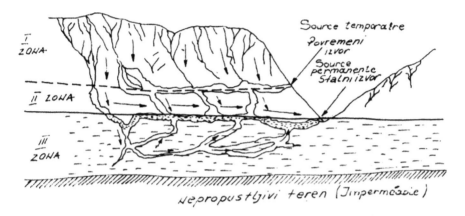

Figure 6   Schematic section of a karst aquifer with three super-positioned zones (acc. to Cvijic) First zone – Dry zone with percolation only; Second zone – Transition zone – presence of perco-lated water and groundwater flows, with variable level; Third zone – Permanently saturated zone (similar to Grund's grundwasser), often with ascending flows.

in the uppermost part, while deeper saturated zones are characterised by horizontal and/or ascending circulation. Cvijic also mentioned that permanent lowering of the saturated part of the carbonate rock sequence is a logical consequence of the dynamic evolution of the karst.

Most of Cvijic's hypotheses are still valid. His ideas have been confirmed by much of the more recent exploratory work such as, for example, the case of hydrostatic equilibrium for linear positioned submerged springs in coastal zones (Mijatovic, 1997). Cvijic's written works and exploration were fundamental to the development of modern karst geography and hydrogeology (or "subterranean hydrography" in his terms) in Serbia and the entire Balkan Peninsula. In some cases the descriptive character of his classification prevailed; in others, little attention was given to the effects of fluvial erosion on the karstification process. However, he created a very important and wide scientific base which led to further investigation of karstic phenomena (Ford, 2005; Zojer, 2005).

During his numerous excursions throughout the Balkan Peninsula and Southern Europe, Cvijic investigated not only such physical geographic features (Figure 7) as karstic or glacial landforms and relief, but also paid great attention to ethnographic and demographic issues. As a famous scientist and human geographer he also played a significant role during and after the First World War, when a new country, the Kingdom of Serbs, Croats and Slovenians (later becoming Yugoslavia) was formed from the kingdoms of Serbia and Montenegro and those parts of the Austrian – Hungarian empire populated by Slavic peoples. He contributed to the Versailles peace conference in determining Balkan and former Yugoslav state borders.

Cvijic received awards from the Royal Geographical Society of London, from the American Geographical Society (Figure 8) and many other academies, geographical

Figure 7   Photo of waterfall on Ljuberadja spring (Eastern Serbia) recorded by Cvijic (1896).

Figure 8   Jovan Cvijic's gold medal from the American Geographical Society – New York, 1924 "The recognition obtained for exceptional scientific results and published works in the field of physical geography of Balkan countries".

societies and institutions of countries including Czechoslovakia, Switzerland, Russia, Italy, Greece, Germany, Poland, Hungary, Romania and more. Few geoscientists worldwide have reaped such honours as Cvijic did. Serbia and Belgrade, where he spent most of his relatively short life (dying at 62) have gratefully acknowledged and honoured him by the establishment of the Museum Jovan Cvijic in 1965, by the construction of a large monument opposite to the entrance of the University headquarters and by naming one of the city's longest streets after him. The Geographical Institute

Figure 9 Suetolik Radanovic.

of Serbia, established in 1961 under the patronage of the Serbian Academy of Science, also bears the name of Jovan Cvijic.

"Groundwater" of Svetolik Radovanovic

Svetolik Radovanovic's excellent knowledge and technical background enabled him to take up any job with confidence (Figure 9). He was a state geologist in the Mining Department, a university professor, head of the Mining Department in the Ministry of National Economy, Dean of the Philosophical Faculty, Minister of National Economy, Academician, and the principal adviser to an important bank (Grubic, 1997).

He took the role of chair of the hydrogeological investigations of the thick alluvial deposits of the Sava River in Makis, which resulted in the construction of thirty drilled wells for the first centralized water supply system for Belgrade (Radovanovic, 1892). Soon afterwards, at the request of the directorate of the Serbian Literary Guild and with the aim of making certain scientific branches more popular, Radovanovic accepted the task of writing a book, which was to be a stepping stone when it came to the development of this scientific discipline in Serbia. Moreover, this is an important piece of scientific work as well as a monograph and textbook. Its dual function helped disseminate knowledge and increase interest in groundwater and thermal waters.

This book titled "*Ground Water, Aquifers, Springs, Wells, Thermal and Mineral Waters*" (1897) has 152 pages with 40 figures and is divided into 6 chapters. "There are few sciences of so many applications as geology. Some disciplines of this science, or more of them, have been jointly applied in agriculture and in architecture and in engineering and in mining and in technology ..." he wrote in the preface.

The first chapter describes the forms of water in the atmosphere, on the ground's surface, and in the Earth's crust and the importance of water; atmospheric water and the natural cyclic movement of water. The second and third chapters are concerned with the principles of fundamental hydrogeology. They contain explanations of the hydrogeological function of rocks; formation of aquifers; aquifer recharge and discharge; groundwater circulation; occurrence of groundwater in karst areas, as well as

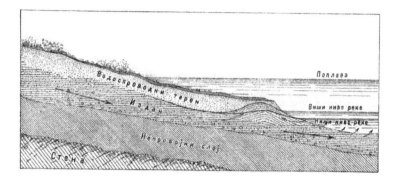

Figure 10 Radovanovic's cross-section showing the general conditions of an alluvial aquifer and hydraulic connections between surface and groundwater.

the hydraulic mechanisms of groundwater. This is followed by detailed presentations of confined i.e. artesian aquifers (with case examples from the world generally and from Serbia in particular). This part contains a descriptive definition of izdan, the equivalent term to the aquifer, which is colloquial and unique to the Serbian Language (similar to the French "nappe"). Radovanovic took the term from the native tongue and introduced it into scientific literature as the principal study object of hydrogeology.

Radovanovic was the first to present a purely hydrogeological section in domestic literature (Milivojevic, 1997). He also described groundwater elements, water table, aquifer thickness, elementary hydraulic mechanisms, groundwater regime, and gave definitions of springs and wells (as artificial "springs"). Considerable space in the book is dedicated to groundwater in alluvial deposits. The description of the relationship between groundwater and surface water followed by the riverbank-clogging problem is also included (Figure 10).

Radovanovic mentioned that the basis of a hydrogeological model is the geological model of the area in question. This is an issue currently under consideration because a computerised simulation model is successful in as much as the geological model is successful. The same conclusion also demonstrates his experience as an investigator: in hydrogeology in particular, there is no infallible method of prediction nor any promise that results will be positive, but a scientific approach can help minimize risk (Milivojevic, 1997).

The fourth chapter deals with the chemical properties of groundwater; derivation of their chemical composition; transformations of chemical composition during water flow and the geological and hydrogeological factors influencing these transformations; the quality of healthy water; groundwater pollution; and evaluation of the safety of groundwater. A sound knowledge of chemistry helped him to explain how the chemical composition of groundwater is transformed. The greatest value of the explanation lies in the detailed description of the dynamic process, which depends on the hydrogeological and geological properties of rocks. Radovanovic proposed the first national standards for drinking water; he listed the physical and chemical properties water should possess to be used for drinking, and which deviations are tolerable.

The remaining two chapters, chapters five and six, comprise half of the book and deal with geothermal resources. The principal classifications of mineral and medical waters; derivation of groups of mineral waters by chemical composition; the origin of spas and forms of mineral water occurrences, and geysers mechanism (based on best known geothermal regions of the world) are all fully described. Radovanovic's interpretation of the origin of thermal water may be described as revolutionary for that time. He explained its origin to be the same as that of fresh groundwater, thus meteoric and not with a juvenile nature as the principal origin. His classification of the origins and models of the occurrence of "mineral" springs is basically a classification of hydrothermal reservoirs with descriptions of their indications, which could be used in exploratory works (Milivojevic, 1997).

SERBIAN HYDROGEOLOGY IN THE TWENTIETH CENTURY

At the beginning of the 20th century mineral water continued to be a main subject of investigations involving hydrogeology. Forty spas were registered in the territory of Serbia at that time. Radojkovic's "*Mineral Waters of Serbia and Their Healing Properties*" was published in 1909, while the country's Law on Spas, Mineral and Warm Waters was declared in 1914.

Investigations and studies of mineral waters were intensified after the First World War. In the field of legislation, a number of laws were passed in the Kingdom of Serbs, Croats and Slovenes (including a new Regulation on Spas, Mineral and Warm Waters, 1920). Data collected earlier and new information were used in writing the first monograph "*Medicinal Waters and Health Resorts of the Kingdom of S.C.S.*" complete with a map (Leko *et al.*, 1922). The monograph describes 164 occurrences of mineral water and health centres. It gives the first classification of ten classes of mineral water, while mineral mud is classified into two groups (medicinal peat and medicinal mud). Though short, this was a period of flourishing balneal and health resorts in Serbia, as if to compensate for what was missed in the preceding period.

In 1936 the University of Belgrade and the Medical Faculty organized the XV Congres International de Hydrologie, de Climatologie et de Geologie Medicales. This was certainly a significant recognition of national scientific thought in this domain. In the same year Nenadovic wrote the important monograph "*Spas, Seaside and Health Resorts Places of Yugoslavia*". M. Protic and Mikincic prepared the Map of Ore Deposits, Coal Basins, and Mineral Springs in the Kingdom of Yugoslavia at a scale of 1:1 000 000; a worthy contribution to national hydrogeology.

In addition, low-mineralised water for drinking was obviously also investigated at that time, though a relatively small number of works on their large-scale investigation were published. Similar explorations were always local and for specific purposes: estimates of available quantities and analyses of quality and the conditions of water abstraction and distribution (Filipovic, 1992, 1997). Special attention was given to the exploration of deeper artesian aquifers in major basins (Leskovac, Krusevac, Zajecar, Smederevo, Paracin) and to water use for both domestic water supply and growing industry. A particularly important place in the development of hydrogeology between the two World Wars (and even more so after the Second World War) belongs to M. Lukovic. He published several important groundwater resource studies for selected areas of Yugoslavia (Loznica, Sabac, Obrenovac, Leskovac, Kragujevac, Prokuplje, Hvar, 1926, 1929). Mineral water resources was another field of hydrogeology studied by Lukovic. K. Petkovic, who closely collaborated with Lukovic, also made several notable contributions to the hydrogeology of Serbia at that time.

Some time after the Second World War, geological institutes were established in each of the six Republics of the Yugoslav Federation. Each had a Department of Hydrogeology and Engineering Geology, which began systematic and continuous hydrogeological investigations. The education of geologists also changed during this period. Hydrogeology was first introduced into the curricula of the existing schools for the education of engineers and technicians. In 1956 two major schools of higher level teaching in geology at Belgrade University were unified into the present Faculty of Mining and Geology. The initial Division of Engineering Geology and Hydrogeology was separated in 1971 and from that time the Department of Hydrogeology has existed as an organisational unit of the Faculty. Postgraduate studies in hydrogeological sciences were introduced in 1960. Since then, the Faculty has educated over 500 graduate hydrogeologists, and 50 doctors in this science (national and international). It has also become not only a leading educational and scientific centre for hydrogeology for the former Yugoslavia but also one of the largest in Southeastern Europe. Many research projects, maps, studies as well as text-books, dissertations and monographs have been produced and published by its professors and researchers. Professors Stepanovic and Milojevic played a crucial role in the development of the Department, followed later by Filipovic, Peric, Vukovic and several others who established a strong hydrogeological school and developed close collaboration with other similar scientific centres worldwide.

Stepanovic (1957) provided a new "sketch" of the hydrogeological provinces of the country, and wrote an important text-book "*Principles of general hydrogeology*" (1962). Milojevic wrote the first national "Hydrogeological terminology" in 1959 which included translation of the terms into English and Russian. His "*Hydrogeology*" published in 1967 (first edition) has been used for a long time as a basic text-book for students but also as a manual for professionals in hydrogeology and related disciplines. During the seventies Filipovic wrote several text-books related to specific disciplines of hydrogeology and together with others supported the completion of necessary educational references. In the 1950s and 1960s, to meet the needs in water supply, civil engineering, mining, forestry, agriculture, environment protection, and to prepare respective hydrogeological maps and studies at local and regional levels (for Water Master Plan and Development Regional Plans), intensive and extensive hydrogeological investigations were carried out in the country to make up as quickly as possible for what had been neglected for any reason in the preceding period.

Hydrogeologists, researchers and professionals have acquired affirmation in Yugoslavia and abroad through various forms of co-operation and in international competition. Food shortages due to the lack of water have stimulated Yugoslav companies and individuals (experts engaged through different national and UN organisations) to direct their business policies towards water development projects in arid and desert regions of Africa, Asia and South America. The consulting services and contracting works of Energoprojekt, Geozavod, "Jaroslav Cerni", Geosonda, Hidroprojekt,

Jugofund and other Serbian companies and institutes have been seen in Egypt, Tunisia, Libya, Algeria, Morocco, Jordan, Iran, Iraq, Cyprus, Peru and other countries.

The most significant role in hydrogeological research was taken by Geozavod-Belgrade, which was established as a state institute after the Second World War. Its researchers made important contributions to national hydrogeology not only through studies of selected areas, practical applications and provision of maps, but also in the theoretical domain. The work of Mijatovic, Komatina and other hydrogeologists has been well recognized on the international scene and their contributions have appeared in numerous publications and proceedings of symposia. A monograph concerning the Dinaric karst written by Mijatovic (1983) was been published by IAH and helped to spread knowledge and information about this specific and, for hydrogeologists, always attractive area.

During this time many specific hydrogeological disciplines became more developed. Important results were achieved in the definition of the hydraulic mechanisms of aquifers; mathematical modelling and its introduction as a standard method in hydrogeological research; groundwater balances; groundwater protection; assessment of geothermal potential, etc. Mineral water continued to be a subject of hydrogeological research, but to a much lesser extent than previously. Systematic exploration for deeper thermomineral waters in Vojvodina started in 1969 (Naftagas, Novi Sad) on the basis of the results from existing artesian wells and of geophysical and other data acquired in petroleum exploration. The main scope of the project was to evaluate the hydrothermal potential of the Vojvodina area and the applicability of this potential, primarily as thermal energy. Since 1969 seventy-two hydrothermal wells have been drilled, totaling 63,000 m (Milosavljevic *et al.*, 1997).

As karst has always been of central interest to hydrogeologists, its investigation has expanded to a larger scale due to numerous projects that included the construction of large and medium dams in the country and abroad. Several such dams were built with the support of Serbian companies and experts (Serbia, Herzegovina and Montenegro) and it was the first time that successful results were achieved in such porous media as karst. Technical applications for the control and regulation of karst aquifers through the construction of galleries, batteries of wells, and groundwater reservoirs (aquifer storage) represent an important contribution to international hydrogeological science.

The late seventies was also a period when a new generation of karst hydrogeologists led by Petar Milanovic grew up. Milanovic's "*Hydrogeology of karst and methods of investigation*" (first edition in 1979) soon became one of the important references dealing with problems of karst groundwater distribution and circulation. The book comprises wide practical experience collected during complex and detailed investigations of karst in Eastern Herzegovina where systems of cascade dams were built to control floods, generate hydro-power and provide water supply to the coastal area of neighbouring Croatia and Montenegro.

The main activities in hydrogeological science were coordinated in the past by the Hydrogeological Section of the Serbian Geological Society (founded in 1891), and the Committee for Hydrogeology and Engineering Geology at the National Union of Geological Engineers and Technicians. The latter has organised national hydrogeological symposia every three years since 1971. It should be noted also that IAH and its Karst Commission in particular is a place where the national hydrogeological school has always been well represented. This is also the reason why one of the first of the

IAH Congresses took place in Belgrade (1963), along with several other international conventions organised more recently.

Today, thanks partly to its well-known historical development, hydrogeology is the most attractive discipline offered within the science of geology at the University of Belgrade. More than 50% of its students regularly enroll in the Department of Hydrogeology, one of the eight existing specialisations of fundamental and applicative geology.

REFERENCES

Cvijic, J. (1893) Das Karstphaenomen. *Versuch einer morphologischen Monographie, Geograph. Abhandlungen Band*, V, Heft 3, 1–114, Wien.

Cvijic, J. (1895) Caves and Subterranean Hydrography in Eastern Serbia (in Serbian). *Glas of Serbian Royal Academy*, XLVI, 1–101, Belgrade.

Cvijic, J. (1896) Springs, Peat Bogs and Waterfalls in Eastern Serbia (in Serbian). *Glas of Serbian Royal Academy*, 1–122, Belgrade.

Cvijic, J. (1918) Hydrographie souterraine et evolution morphologique du Karst, Recueil. *Trav. Inst. geogr. alpine*, VI, fascicule 4, 1–40, Grenoble.

Filipovic, B. & Dimitrijevic N. (1990) Development of hydrogeology until World War II (in Serbian), Annales Geologiques de la Peninsule Balkanique, liv. LIII/1, Belgrade

Filipovic, B. (1992) Hydrogeology in Serbian Geological Society (in Serbian), Compte Rendu de Societe Serbe de Geologie, Le Livre Jubilee (1891–1991), Belgrade

Filipovic, B. (1997) Hystory of Yugoslav hydrogeology and level of hydrogeological investigation, In: Monograph "100 Years of hydrogeology in Yugoslavia", Spec ed. of FMG, pp. 31–46, Belgrade

Ford, D. (2005) Jovan Cvijic and the founding of karst geomorphology, In: Cvijic and karst/Cvijic et karst, Monograph) Spec. ed of Board of Karst and Speleology Serb. Acad. of Sci. (eds. Stevanovic Z. & Mijatovic B.) p. 305–321, Belgrade.

Grubic, A. (1997) Svetolik A. Radovanovic (1863–1928). Life and Work of Serbian Scientists (in Serbian), Biographies and Bibliographies, Serbian Academy of Science and Arts, II Department, b. 2, pp. 107–149, Belgrade.

Komatina, M. (1992) Hydrogeology in Serbia and worldwide during last hundred years (in Serbian), Compte Rendu de Societe Serbe de Geologie, Le Livre Jubilee (1891–1991), Belgrade.

Komatina, M. (1997) History and trends of development of hydrogeological exploration methods in Yugoslavia, In: Monograph "100 Years of hydrogeology in Yugoslavia", Spec ed. of FMG, pp. 47–55, Belgrade.

Lindenmayer, E. (1856) Opis mineralni i lekoviti voda (Description of mineral and medical waters). In: Opis mineralni voda i njino upotrebienie voobste a ponaosob lekoviti voda u Knjazestvu Srbiji dosada poznati, Pravitelj. knigopec. Knjaz. Srbskog, Belgrade.

Leko, T. M., Scerbakov, A. I., Joksimovic, M. Hr. (1922) Medicinal Waters and Health Resorts of the Kingdom of S.C.S. (in Serbian), Monograph, Belgrade.

Lozanic, S. (1874–1893) The analyses of Serbian mineral waters; Glasnici of Serbian Scientific Society, Belgrade.

Lukovic, M. (1926) Contribution to hydrogeology of Serbia (Loznica, Sabac, Obrenovac, Valjevo i Leskovac – in Serbian), *Vijesti of Geol. Survey-Zagreb*, I, Zagreb.

Lukovic, M. (1936) The geology of the thermal and mineral springs of Yugoslavia, *Proceedings of XV Congr. Int. Hydrol. Climatol. et Geol. Medical*, Belgrade.

Mijatovic, B. & Komatina, M. (1983) Geology of karst terrains and theory of karst in the works of Jovan Cvijic, Edition "Hydrogeology of Dinaric Karst", pp. 11–17, Belgrade

Mijatovic, B. (1984) Hydrogeology of the Dinaric Karst. IAH International Contributions to Hydrogeology, volume 4, Heise, Hannover, 254 p.

Mijatovic, B. (1989) Jovan Cvijic – precedent and founder of modern approach in karstology, Vesnik of Geozavod, Vol. 45, pp. 5–20, Belgrade.

Mijatovic, B. (1997) Ommage a l'oeuvre de Cvijic sur le karst, In: Monograph "100 Years of hydrogeology in Yugoslavia", Spec ed. of FMG, pp. 83–97, Belgrade.

Milivojevic, M. (1997) "Groundwater" – the first book and Serbian textbook of hydrogeology and geothermology by Svetolik Radovanovic, In: Monograph "100 Years of hydrogeology in Yugoslavia", Spec ed. of FMG, pp. 13–30, Belgrade.

Milojevic, S.P. (1952) Geological bibliography of Yugoslavia, I (from XIX cent. to 1944), Bibliographic Institute, Belgrade.

Milojevic-Kramzar, D. (1971) Geological bibliography of Yugoslavia, III (1959–1962); IV (1963–1967); V (1968–1970), Geozavod, Spec. ed., Belgrade.

Milojevic, N. (1959) Hydrogeological terminology with equivalent terms in Russian and English, Vodoprivreda, II, 6, Spec. ed, Belgrade.

Milojevic, N. (1967) Hydrogeology, Text-book, University of Belgrade, Belgrade.

Milojevic, N. & Kramzar, D. (1977) Hydrogeological bibliography of Yugoslavia, Spec ed. of Department of Hydrogeology FMG, Beograd.

Milosavljevic, S., Vasiljevic, M. & Vilovski, S. (1997) Hydrogeological explorations in Vojvodina (Explorations for drinking, industrial and thermomineral waters), In: Monograph "100 Years of hydrogeology in Yugoslavia", Spec ed. of FMG, Belgrade.

Nenadovic, L. (1936) Spas, Seaside and Health Resorts Places of Yugoslavia (in Serbian), Monograph, Belgrade.

Radovanovic, S. (1897) Podzemne vode (Groundwater. Aquifers, Springs, Wells, Thermes and Mineral Waters). Srpska knjizevna zadruga (Serbian Literary Guild), no. 42, pp. 1–152, Belgrade.

Radulovic, V. & Radulovic, M. (1997) Historic background of hydrogeological explorations of Montenegro, Monograph "100 Years of hydrogeology in Yugoslavia", Spec ed. of FMG, pp. 57–73, Belgrade.

Stepanovic, B. (1962) Principles of general hydrogeology (in Serbian), Geozavod, Spec. ed. vol. 11, Belgrade.

Stevanovic, Z. (1997) First studies of Eastern Serbian karst by Jovan Cvijic – basis of modern karst hydrogeology, In: Monograph "100 Years of hydrogeology in Yugoslavia", Spec ed. of FMG, pp. 99–114, Belgrade.

Stevanovic, Z. (1999) Jubilee and celebration "100 years of hydrogeology in Yugoslavia", Spec ed. of FMG, Institute of Hydrogeology, Belgrade.

Stevanovic, Z. (2000) Yugoslav hydrogeology – present state and perspective, Proceedings of "100 years of Hydrogeology in Romania", Spec Ed. of Romanian Academy of Science, Bucharest.

Stevanovic, Z. & Mijatovic, B. (2005) Cvijic and karst/Cvijic et karst, Monograph) Spec. ed of Board of Karst and Speleology Serb. Acad. of Sci. pp. 1–405, Belgrade.

Vilimonovic, J. (1997) Information on the history of hydrogeology and ecology in Serbia, In: Monograph "100 Years of hydrogeology in Yugoslavia", Vol. 2, Spec ed. of FMG, pp. 7–9, Belgrade.

Zojer, H. (2005) Jovan Cvijic, In: Cvijic and karst/Cvijic et karst, Monograph) Spec. ed of Board of Karst and Speleology Serb. Acad. of Sci. (eds. Stevanovic, Z. & Mijatovic, B.) pp. 322–325, Belgrade.

Zujovic, J. (1931) Springs and wells. Water supply in rural area (in Serbian), Serbian Royal Academy, Educational Library, b.5, Belgrade.

The history of South African hydrogeology: maybe short, but surely sweet

Marlese Nel
University of the Western Cape, South Africa

ABSTRACT

It was through Alex du Toit (Figure 1) geologist and philosopher, that groundwater science was initiated in South Africa. From 1903 already Du Toit was involved in irrigation water supply in the southern parts of the country. After that, John Enslin, also a geologist, made a profound impact with the use of geophysics, the quantification of resources and their sustainable management for two decades, onwards from the late 1940s. Enslin also investigated the exploitation of dolomitic aquifers in the western parts of the country. Another groundwater pioneer, J. R. Vegter, started his work in the 1950s. A geophysicist by training, he performed groundbreaking work in the application of modern techniques to characterise the hydrogeology of the country. He also, single-handedly, captured hydrogeology information on maps and in explanatory manuals. Before 1998 the national water laws hampered the sustainable uilisation and proper management of groundwater. During this time the need for water to meet national demands grew exponentially resulting in the growth of private groundwater consulting firms and unfortunately little co-ordination between different groundwater projects.

Figure 1 Alex du Toit.

INTRODUCTION

Water is the reason that South Africa is the country that it is today; not because of a shortage or a surplus, but because there is any at all. The maritime explorers of bygone days found Cape Town to be an undisputed stop on their journey; for both the splendour of the scenery, but mainly for the presence of sweet, freshwater springs on Table Mountain (Figure 2). After the settlers came from Europe, it was water that controlled the route of the travellers inland as they explored the country. In many cases it was not a river that formed the centre of a settlement, but towns would establish around a spring, the manifestation of the ever present groundwater.

As in many other parts of the world, the past 200 years have seen changes in our use of groundwater. Driving these changes in South Africa were the agriculture and mining industries in the 19th century and the increase in population in the 20th century. Prior to 1956, legislation governing groundwater use was almost non-existent and it was only after the promulgation of the 1998 Water Act that groundwater was addressed adequately in legislation. After going through exploratory phases of siting and drilling, followed by the quantification of resources and expanding into integrated management, the small group of hydrogeologists in South Africa is placing an emphasis on groundwater quality, geohydrological processes and sustainable use within an integrated environment.

Figure 2 Table Mountain.

For the purpose of this Chapter the history of South African hydrogeology has been divided into five eras to highlight specific activities of the different periods as the science developed over the past 30 years.

ERA 1: FINDING WATER USING LITTLE OR NO SCIENCE: 1820–1935

By the 1850s there were already some individuals claiming they could "see" or "find" underground water using various contentious methods. Some actually had an impressive strike-rate, given that no real alternative scientific methods were practiced at that time. Others however, built a reputation constantly swaying between failure and success. Later, Alex du Toit made no secret of his scepticism towards these "water finders" and ascribed their large public support to "plain, human superstition". For quite a few years this "feud" between the water-finders and the scientists – each with their avid supporters – continued. On some occasions, this led to fierce fights and even some interesting "boring contests".

In 1888 Thomas Bain was appointed as Irrigation and Geological Surveyor in the Department of Agriculture of the Cape of Good Hope. His first report, entitled *"Report on the prospects of water boring on Government Ground in Bushmanland"*, is probably the first hydrogeological report written in South Africa. Subsequently, the Inspector of Water Drills, H.P. Saunders, published an interesting booklet entitled *"Underground water supply of the Cape of Good Hope with special reference to the working of the diamond drill"*. (Saunders, 1897). The booklet was based on information gathered during four years of observation and practical experience with government drilling operations, and proved valuable to those in the drilling business.

The huge growth in the gold mining industry in the 1880s boosted the economy, but the toll on the water resources also increased. A terrible drought around 1895 made people consider innovative ways to provide settlements and the mining industry with water. These necessitated the enactment of various laws in Natal (Law 26, the Irrigation Law of 1886), Cape Colony (Act 32, the Irrigation Act of 1906) and the Transvaal (Act 27 of the Irrigation Act of 1908). In 1910 the country was declared a Union and water law was rationalized under one Act – the Irrigation and Conservation Waters Act 1912 (Union of South Africa, 1912). Specific provision was made for the regulation of dolomitic waters as the value of this water resource was recognised. With this new act, the government focused mainly on water for irrigation and did not anticipate the growing water demand from the mining industry.

Abstraction of water from the dolomites in the West Rand seemed a feasible option, although strong opposition was received from various individuals and organisations. In 1898, a huge underground water reservoir was located on the farm Zuurbekom by geologist David Draper. Pumps were installed and the Zuurbekom pumping station delivered up to 28 million l/day to the Rand from then on. Until the Vaal River scheme became operative in 1923, it was the Rand's main source of water supply.

The first drilling machine in South Africa was imported by the Cape Government in 1880. In 1895 the average depth of boreholes was 13.6 m (Venter, 1970). From then on things moved fast, and a drilling division was established in the Transvaal in 1904. The need for groundwater was growing rapidly, so much so that in 1924 there were already 115 active drilling machines in the country. These were manned by

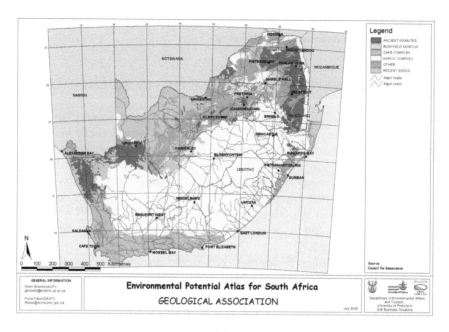

Figure 3 South African geology.

drillers, their technician/(s) and most of the time by a geologist. Regularly the drilling crews had to first explore the country, drill and get water and only then would other settlers follow. Real danger awaited these men and fighting off wild animals and deadly diseases was part of their expertise!

In 1903, Alex du Toit was appointed as a geologist within the Geological Commission of the Cape of Good Hope. This Cape Town native began to develop an extensive knowledge of the geology of southern Africa (Figure 3) by mapping large portions of the Karoo and its dolerite intrusions and publishing numerous papers on the subject. Even before the 1900s, researchers were intrigued by natural groundwater recharge. In 1906, du Toit made some valid comments on recharge after his study in the northern parts of the country. He stated that "Proportion of rainfall retained . . . clear that this depends on quite a number of factors, e.g. whether the rainfall is above or below the mean, on the porosity of the soil and the rock underlying, upon the depth down to composed material and thence to unfissured rock, upon the slope, upon the amount of vegetation it supports . . ." (du Toit, 1906). These ideas of du Toit are still supported today and a number of research projects on these different links with recharge have been carried out.

In 1920 de Toit joined the Union Irrigation Department as a water geologist and so became, probably, the first true hydrogeologist in the country. One of du Toit's papers, *"The geology of underground water supply with special reference to South Africa,"* was read in 1913 before the South African Society of Civil Engineers. In it he stated: "South Africa has proved so peculiar in matters scientific as judged from European standpoints, that it is no surprise to find the local problems of hydrogeology to be in many respects different from those dealt with in current literature." In another paper on groundwater, on work done in collaboration with the Union Irrigation Department, he reported on an analysis of data of some 10,000 wells that had been drilled up to that time (du Toit, 1928). H. F. Frommurze, du Toit's assistant at that stage, continued the statistical analysis work and further results were published (Frommurze, 1937).

Water legislation did exist, but was very vague and groundwater was not recognised as an exhaustible resource. In 1905 the Inter-Colonial Irrigation Commission compiled an interim report, in which they stated:

- That the owner of land has a *prima facie* right to the water under the surface of his ground, and that he can sink a well, or raise the water at his pleasure.
- If the water under the ground flows in a well-defined channel, he may take a reasonable use of that water; but he may not exhaust the supply as to deprive lower proprietors of a similar reasonable use. This will apply whether the water in such well-defined channels rises to the surface and feeds a river, or whether it continues its course underground.

ERA 2: GEOPHYSICAL BOREHOLE SITING: 1936–1955

In 1936 the Geological Survey started to use geophysical methods – magnetic and electrical resistivity – for the siting of boreholes. More and more geophysical work was carried out and, by the outbreak of World War II in 1939 the Geological Survey had a strong core of geologists trained in geophysical work. J. F. Enslin wrote his doctoral thesis and numerous papers on the application of geophysical methods in engineering and the interpretation of results, together with some focused work on the Karoo sedimentary rocks. As a result of the specific geological conditions in the country, a major challenge facing geologists was the problem of locating and tracing narrow vertical or inclined water-bearing structures such as fractures, fissures and faults beneath cover. This was resolved through the introduction of a galvanic electromagnetic technique developed primarily by Enslin (1952 and 1955) and Vegter (1962).

The first decade after World War II saw a steady increase in borehole siting and drilling. Geological logs and groundwater data of completed boreholes were compared with the expectations from the geological-geophysical investigations. This resulted in a better understanding of the occurrence of groundwater in different parts of the country. It further recognised faulty geophysical interpretation, particularly of electrical resistivity depth probing (Enslin, 1948 and 1963). Enslin (1961) published some results on the scientific selection of boring sites in secondary aquifers in South Africa.

During this time the first formal – though not comprehensive – investigations were undertaken on thermal waters. The need to dedicate more research to deep underground circulation was acknowledged even though such thermal springs occur mainly sporadically. Rindl (1916) was the first to classify known "medicinal" springs according to their chemistry. A comprehensive overview of temperature, dissolved solids and trace elements in thermal springs, artesian boreholes and mine waters was published (Kent, 1949).

Theron (1946) reported on lysimeter measurements of seepage through soil cover over a period of 15 years on a university experimental farm near Pretoria. Work by Van Eeden (1955) supported the results from Theron's work and concluded that recharge was 2% of mean annual rainfall in that geological setting. Van Eeden also investigated various factors that could govern groundwater recharge and estimated the country's consumption of groundwater, expressed in terms of mean annual rainfall.

With an increase in the use of groundwater, it was not long before receding water levels were reported. In 1953 a committee was appointed to investigate "groundwater supplies and related matters". Under du Toit this inter-departmental committee had a wide scope and examined:

- the different facets of the hydrological cycle
- the exploitation of surface water sources
- pollution, reclamation and re-use of water
- the exploitation of groundwater sources.

One of the most relevant recommendations from the committee was to strengthen the Geological Survey personnel and to establish an inter-departmental Co-ordinating Committee for Hydrological Research.

Prior to the Water Act of 1956, there was no legislation affecting the control and prevention of water pollution. However, groundwater quality had received some attention since the beginning of the 20th century and one of the first investigations on the geochemistry of groundwater ("*A geochemical survey of the underground water supplies of the Union of South Africa with particular reference to their utilisation in power production and industry*") was reported by Bond (1946).

The Council for Scientific and Industrial Research (CSIR) was constituted by an Act of Parliament in 1945 and is now one of the leading science and technology research, development and implementation organisations in Africa. The CSIR's main site is in Pretoria, while it is represented in other provinces of South Africa through regional offices. Various domains form the core focus of its scientific research, including natural resources and the environment. It provides specialist knowledge also in the ground-water sector and has continued to support innovative research as well as the training of students.

Around the 1950s the geology community realised that not only exploration, but also quantification of the groundwater resources were necessary. Enslin drew attention to the restrictive nature of South Africa's groundwater resources (Enslin, 1949) and went further to set the scene for future groundwater investigations in his presidential address before the Geological Society of South Africa. He also described procedures for conducting scientifically controlled yield tests and discussed the concept of safe yield and its determination (1961).

ERA 3: TWO DECADES OF WATERSHED ACTIVITIES – FROM QUANTIFICATION TO INSTITUTIONAL ESTABLISHMENT: 1956–1975

During this period mining activities – especially gold, iron ore and coal – began to have a severe impact on groundwater resources. Not only were large volumes of groundwater pumped out of the mine to reach the ore, but mining had a negative impact on water quality. After World War II South African cities experienced rapid growth and an increase in industrial development necessitated a new approach to water utilisation.

The Water Act No 54 of 1956 superceded the 1912 Act and a considerable number of amendments were made. State control over the abstraction of groundwater was introduced for the first time through the establishment of local control areas called Subterranean Government Water Control Areas. Over a 30-year period 16 of these areas were established throughout the country. This water management approach led to varying successes.

Some intensive geophysical investigations were carried out on mine dewatering, subsidence and sinkhole formation in the 1950s and 1960s. The experience gained and the information gathered has been of great value in later studies. Amongst the more important reports on the Far West Rand Enslin and Kriel (1967) discussed the assessment and possible future use of the dolomitic groundwater resources; Bezuidenhout and Enslin (1969) investigated surface subsidence and sinkholes in the dolomitic area; Kleywegt and Enslin (1973) looked at surface subsidence and sinkholes caused by lowering of the water table.

The value and the need for isotopic research was realised during this era. The Schonland Research Institute was founded as the Nuclear Physics Research Unit (NPRU) in 1958 under its first director J. P. F. Sellschop. It was part of the Physics Department at the University of the Witwatersrand in Johannesburg, until 1963 when it became a separate research entity funded directly from the University's research budget. Vogel and Bredenkamp (1969) added value to their findings in a study in the southern Kalahari using carbon-14 dating and isotope analysis. Balt Verhagen published several papers from his studies in the Kalahari, Botswana and in other southern African countries effectively showing how recharge studies can be substantiated by isotope analyses (Verhagen *et al.*, 1974, Mazor *et al.*, 1980).

In 1969 Government appointed a Committee of Enquiry into the groundwater situation in South Africa in the light of uncertainties about the status and availability of groundwater resources. The main recommendations of the Committee were that the following research should be initiated:

- the interaction between groundwater exploitation and land use practices
- a co-ordinated programme of long-term recording of groundwater levels
- the hydrogeological characterization of the country's geological formations

The Committee recommended that 1970 be declared a National Water Year to create general awareness of efficient water use and as a platform for discussions and to carry out proper assessments of the country's water resources. The most significant publication on groundwater in this period was by Enslin (1970) in which, using rainfall-recharge curves, he calculated the average yearly gross recharge for the country and concluded that approximately $2.5 \times 10^9 \, m^3$ groundwater per annum could be developed in the future. Another recommendation from the Committee that was implemented was the need for "effective coordination of water research conducted by the various organisations". The Water Research Commission (WRC) was established as an autonomous statutory body which would generate research funds from national water taxes. Since its establishment, the WRC has successfully funded many groundwater and groundwater-related research projects.

An attempt was made to get an overall view of groundwater levels of the country and the Geological Survey published a report that showed some examples of groundwater fluctuations (Kok, 1975).

FORMALISING GROUNDWATER TRAINING

By the 1960s geophysical techniques were being used extensively to locate gold reefs for the mining industry. The need for training was recognised and the University of Pretoria established geophysics as part of its post-graduate degree in Geology. By that time, J. R. Vegter had been a practicing geologist for just over 10 years and was approached to lecture the subject part-time. This he did, as well as taking on the first formal teaching – as a post-graduate optional subject – on groundwater in South Africa. The only other institution offering groundwater training at that time was Rhodes University in Grahamstown under the leadership of Andrew Stone. This was initially only as an undergraduate course, presented by the Department of Geography.

With the rippling effects of the 1970 National Water Year activities still visible, it was an event two years later that really brought groundwater science to public attention. With the reassurance that the last time it had contained water was in 1888, people in the town of Daniëlskuil started building houses on a dry pan in the mid-1900s. However, during the months of January and February 1974 most parts of the country had exceptionally good rains. In the naturally drier northern parts such high rainfall events tend to have catastrophic environmental impacts. About 5 months after the heavy rains, in the dark of night, people were woken by slowly rising water in their homes. The flooding of the pan area left a handful of people dead and many houses destroyed. This event puzzled not only the inhabitants of the town but also the municipal engineers and managers. The inhabitants took their concerns to higher authority and demanded that provision be made for the training of groundwater scientists. Fanie Botha, then the Minister of Water Affairs, made R7 million available for this purpose. With support from the recently established WRC, Botha approached all the universities to find an academic institution that would take up this training and research challenge, with little initial success.

However, Ben Botha, who headed the Department of Geology at the University of the Orange Free State, Bloemfontein, recognised the opportunity and decided to take up the challenge. Lacking a suitable person to lead this groundwater section, he approached his student, Frank Hodgson, who had recently submitted his doctoral thesis. With little hesitation, Hodgson took the job and left almost immediately for training in Tucson, Arizona. When he returned to South Africa in 1974, the sub-department of Geohydrology was established within the Geology Department. Hodgson and Botha approached the Water Research Commission for funding, and secured R1.25 million for seven years for groundwater research and training. The research grew steadily at the Institute for Groundwater Studies (IGS) at the university, whilst the training was carried out within the Geology Department, both these functions being managed by Hodgson. In 1985, when Botha retired, Hodgson combined both functions under the umbrella of the IGS and it is still managed in that manner.

Under the guidance of Hodgson, joined by Gerrit van Tonder at IGS, a good understanding of fractured rock aquifers has been developed over the years. Unique software has been developed by the Institute and aquifer testing techniques for fractured rocks have been optimized. Many of the students graduating focused on some aspect of the fractured rock environment. Hodgson also did ground-breaking work on Acid/Base Accounting in the mining environment. Hodgson and Krantz (1998) found that open-cast and shallow underground coal mining in the Witbank area had the greatest impact of all mining activities on groundwater quality.

J. F. Botha joined IGS in 1979 as research professor. With his strong mathematical background he brought a new dimension into hydrogeology – numerical modelling. He developed around 10 models for different areas in South Africa and Namibia, including the model that led to the licensing of the South African Radioactive Waste Disposal Site at Vaalputs. But it was his work on the Karoo aquifers which gained particular attention amongst his peers. Traditionally, it had been accepted that the main water-bearing fractures in the Karoo aquifers were vertical. However, core drilling on the University testing site and more drilling and testing on other sites in the central parts of the country, provided Botha with the data he needed and, despite skeptics, it was soon accepted that the main water-bearing fractures in the Karoo aquifers were actually the horizontal structures.

In search of uranium the Atomic Energy Corporation (AEC) dealt indirectly with the occurrence of groundwater in the main Karoo Basin and in the Kalahari region. Geohydrological considerations were of major importance when the AEC embarked on a search for a site for a National Radioactive Waste Disposal Facility in 1979. Levin (1988) described the in-depth hydrogeological investigations that were required to verify the integrity of the Vaalputs site in the Namaqualand District for disposal of low level radioactive waste. Other hydrogeological spin-offs from AEC investigations included the use of radon emanations as an indicator of fracture zones (Levin, 2000) and the capability of modelling the unsaturated zone (Van Blerk, 1994).

Whilst the volume describing the surface water resources of South Africa has seen a number of revisions, no similar work was available for groundwater. Early attempts at quantifying the groundwater resources of South Africa, (Enslin, 1970 and Vegter, 1995a) were largely educated guesses and not based on algorithms. The figures for sustainable groundwater yield derived by these pioneers of hydrogeology in the country were 2.5×10^9 m^3/a and 5.4×10^9 m^3/a respectively The first attempt to provide a synoptic and visual representation of the groundwater resources of South Africa was that of Vegter (1995b). The borehole prospects, saturated indices and recharge maps that he produced, in particular, provided a valuable indication of regional scale availability of groundwater. This work was built-on by Baron, Seward and Seymour (1998) with production of the Groundwater Harvest Potential Map. Harvest Potential was defined by the authors as "the maximum volume of groundwater that may be abstracted per annum without depleting the aquifer".

Artificial recharge was practiced as early as 1950/60 – even though sometimes inadvertently – and de Villiers (1972) discussed South African case studies and the feasibility of the process. A success story in the implementation of artificial recharge was at Atlantis on the West Coast (Figure 4). The town of Atlantis has been fully supplied by groundwater since 1975 for all their municipal water needs. Artificial recharge amounts to nearly 2 million m^3 per year and this is not only used to prevent salt water intrusion, but also to ensure sustainability.

Figure 4 Atlantis scheme.

The application of mathematical models abstracted from international publications, e.g. Theis (1935), Cooper-Jacob (1946) and Kruseman and De Ridder (1991) resulted in headaches for South African hydrogeologists. The hard-rock conditions found in much of the country are not compatible with the assumptions required for closed-form analytical solutions. Kirchner and Van Tonder (1995) developed practical methods for the analysis of pumping test data of South African fractured rock aquifers. Several groundwater systems have been modeled and examples are described by Muller and Botha (1987); van Tonder (1993) and Janse van Rensburg (1994). Mass transport models are widely used in the evaluation of groundwater pollution. Many of the applications are linked to the mining industry who is a consumer of large volumes groundwater in the country.

ERA 4: A SHIFT TO ASSESSMENT AND EXPANDING ACTIVITIES: 1976–1997

The Groundwater Division of the Geological Survey was transferred to the Department of Water Affairs and Forestry (DWAF) in January 1977 and was merged with the Division of Geohydrology. From 1970 an overall growth of hydrogeological and engineering consulting firms took place. To keep track with what was happening on the international scene, a small group of groundwater experts from various institutions went on a study tour to Israel, the Netherlands, the United Kingdom and the United States in 1980. Some valuable recommendations and ideas arising from the visit influenced local research.

Dave Bredenkamp, who obtained his doctorate in 1978 from IGS, is an expert on processes – especially recharge – in the dolomitic aquifers of South Africa. He has combined his interest in isotopes with his recharge research and published joint papers with the isotope experts. As a result, a prominent report has been published under his guidance on the quantitative estimation of groundwater recharge and aquifer storativity (Bredenkamp et al., 1995). This examined the wide range of methods that had been employed in South Africa. Once again, this work evoked comments and criticism. Vegter raised specific issues of concern, which included the formalised concept of effective or net recharge, problems arising from dual porosity conditions and the considerable degree of uncertainty that is attached to different methods of recharge estimation.

Another prominent activity taking place during this period was a programme of hydrogeological mapping that started in 1993. In 1995 Vegter published a set of national maps together with an explanatory document for the Water Research Commission (WRC) and DWAF (Vegter, 1995). DWAF also embarked on another mapping programme, a 1:500 000 hydrogeological map series, covering the entire country. Today this series of 23 maps and accompanying brochures are all available from the DWAF.

In 1984 the IGS was awarded a project by the WRC to develop a national groundwater database for the country. The project was completed in 1984 (Cogho et al., 1989) and the database implemented on the then existing mainframe computer of the Department of Agriculture. The database evolved and is now operated from within the Department of Water Affairs and Forestry.

Some major water quality investigations executed in the 1990's include:

- Continuous ambient groundwater quality on a national scale, funded by the WRC.
- Classification of South African aquifer systems to support a national groundwater quality management strategy (Parsons, 1995).

- Impact of agricultural practices on South Africa's groundwater resources (Conrad *et al.*, 1999).
- Nitrate content of groundwater (Tredoux, 1993).

After the election of a democratic government in 1994, one of the policy papers from that year stipulated that each person must have a minimum of 25 l of water per day, within walking distance of 200 m or less from their household. Surveys indicated that over 12 million people distributed over some 11 000 rural villages did not have this basic water supply. The Directorate of Geohydrology in the Department of Water Affairs and Forestry played a prominent role in meeting this daunting challenge. Eberhard Braune, then the Head of the Directorate made the following statement: "... at the local level, river and spring flows are decreasing during the dry season ... problems of surface water pollution. All this makes groundwater, despite its low yield nature, the most feasible option to meet the massive backlog in domestic water needs everywhere in the country. For the Eastern Cape Province planners now agree that groundwater has the potential to serve over 80% of the 5700 communities there." A 3-year Norwegian-assisted programme, with the DWAF as lead agent, got underway in 2000 to tackle this challenge. Numerous groundwater consulting firms, local government authorities and national departments worked together and to date have made significant progress – recognizing the inevitable problems and setbacks. Practical and user-friendly operational manuals have been produced as outputs of the programme and are used to combat past groundwater system failures in rural areas.

Some prominent groundwater-related organisations were established during this period. They are:

1   The Groundwater Division of the Geological Society of South Africa – 1978.
2   The Borehole Water Association of Southern Africa – 1979.
3   A South African Chapter of the International Association of Hydrogeologists – 1997.
4   Drilling Contractors Association of Southern Africa – 1998.

All these organisations are still active today and provide diverse support functions for the groundwater environment.

ERA 5: INTEGRATED MANAGEMENT: 1998 UNTIL THE PRESENT DAY

The new government of 1994 was faced with a situation in which the majority of South Africans, through past laws, were excluded from owning land and also denied direct access to water for productive use. Changes to the water law were inevitable. The National Water Act of 1998 does away with previous concepts and establishes principles that have far-reaching impacts. The more important ones related to groundwater include:

- Interdependence of its elements establishing the unity of the water cycle.
- All water is a resource common to all. Groundwater is no longer considered a private property.

- Water required for meeting basic human needs and the needs of the environment is identified as "the reserve" and must enjoy priority of use by right.

With the guidance of the "state of the art" Water Law of 1998, the focus of hydrogeological research changed somewhat towards the end of the 20th century. This follows the international trend to a growing awareness and concern that has developed about the environment over the past two decades. Unfortunately a natural resource, like water, cannot be used without altering itself and the linked environment. This presents big challenges for hydrologists, geohydrologists and ecologists. Quantification of the groundwater contribution to an aquatic or ecological environment or to surface water bodies is now one of the main challenges groundwater scientists have to resolve. Other research includes understanding groundwater/surface water ecosystem links, the protection of drinking water and the capacity to implement these and other matters.

DISCUSSION

Learning to understand, even only partly, the hydrogeological environment in South Africa has proved to be an interesting and sometimes surprising journey; in the early days of drilling and exploring the country, truly a dangerous journey. Then, over a period of about two centuries, this country with its complex fractured rock hydrogeology saw some extraordinary geologists, geophysicists, chemists and geohydrologists making an impact on groundwater science. Especially from the 1970s onwards, after formal groundwater training was secured and national structures established to provide research funding and institutional support, hydrogeology really developed rapidly as a science. Today, a small yet strong group of scientists are focusing on groundwater quality, geohydrological processes and sustainable use within an integrated environment.

In the words of J.R. Vegter, arguably the most experienced hydrogeologist in the country, "... in the past and even still today, far too little attention is given to the evaluation of groundwater information obtained from research". He emphasises the need for "observations, observations!", because, at the end of the day, that is where the proof for all theories and assumptions lies. Frank Hodgson also echoes this approach, insisting constantly to "show me your data!". Groundwater information (quality and quantity) is captured on a national database, but with a short observational period of 30 years, groundwater systems in South Africa are only now starting to be understood. Some longer records are available, such as the water level data from the Wondergat sinkhole, dating back 100 years though this is a real exception.

Challenges will always be there, and preferably so, otherwise no growth or improvement can be achieved. Currently hydrogeology is somewhat of an orphan in terms of national institutional structures. However, it is suggested that the passion and knowledge of the scientists themselves will ensure that groundwater will never end up as an "undercover organisation" but will soon take a strong, more united front into the environmental science sphere.

The majority of currently practicing hydrogeologists are concentrated in the private sector. Water resource management functions are slowly being decentralised to a catchment level and the inputs from these groundwater scientists are becoming more and more crucial. Within this group of scientists are definitely some potential rivals to Charles Theis and Henry Darcy; and who knows, when this chapter is rewritten a century hence, the history of South African hydrogeology will be just as sweet but not as short.

ACKNOWLEDGEMENTS

Many thanks to everybody who responded with an e-mail or even agreed to an interview. A special thanks to Mr. Vegter for his time and providing the baseline information for the structure of this chapter.

REFERENCES

Baron, J., Seward, P. & Seymour, A. (1998) *The Groundwater Harvest Potential Map of the Republic of South Africa.* Pretoria, Directorate Geohydrology, Department of Water Affairs and Forestry. Technical report GH 3917.

Bezuidenhout, C.A. & Enslin, J.F. (1969) Surface subsidence and sinkholes in the dolomitic area of the Far West Rand, Transvaal, Republic of South Africa. In: Tison, L.J. (ed.) *Land Subsidence, V.2. IAHS/AISH Publication no. 89.* pp. 482–495.

Bond, C.W. (1946) A geochemical survey of the underground water supplies of the Union of South Africa. *Memoirs Geological Survey South Africa,* 41.

Bredenkamp, D.B., Botha, L.J., van Tonder, G.J. & Janse van Rensburg, H. (1995) *Manual on Quantitative Estimation of Groundwater Recharge and Aquifer Storativity.* Pretoria, Report Water Research Commission, TT73/95.

Cogho, V.E., Kirchner, J. & Morris, J.W. (1989) *A National Groundwater Database for South Africa: Development of the database.* Pretoria, Report Water Research Commission 150/1/89.

Conrad, J.C., Colvin, C., Sililo, D., Gorgens, A., Weaver, J. & Reinhardt, C. (1999) *An Assessment of the Impact of Agricultural Practices on the Quality Of Groundwater Resources in South Africa.* Pretoria, Report Water Research Commission 641/1/99.

Cooper, H.H. & Jacob, C.E. (1946) A generalized graphical method for evaluating formation constants and summarizing well field history. *Transactions of the American Geophysical Union,* 27, 526–534.

De Villiers, J. (1972) Aanvulling van grondwater deur middel van gronddamme, oos van Alldays, Distrik Soutpansberg. *Annals of the Goelogical Survey,* 9, 139–141.

Du Toit, A.L. (1906) Underground water in south-eastern Bechuanaland. *Transactions of the South African Philosophical Society,* 16, 251–262.

Du Toit, A.L. (1928) Borehole Water Supplies in the Union of South Africa. *Minutes and Proceedings of the South African Society of Civil Engineers,* 31, 31–42.

Enslin, J.F. (1948) Lateral effects on electrical depth probe curves. *Transactions of the Geological Society of South Africa,* 51, 249–270.

Enslin, J.F. (1949) Die beperkte ondergrondse watervoorraad van die Unie. *Suid-Afrikaanse Akademiese Tydskrif vir Wetenskap en Kuns,* Oktober 1949. pp. 143–164.

Enslin, J.F. (1952) Waterare en 'n nuwe tegniek om dit op te spoor vir die aanwys van boorplekke. *Suid-Afrikaanse Akademiese tydskrif vir Wetenskap en Kuns,* October 1952.

Enslin, J.F. (1955) A new electromagnetic field technique. *Geophysics,* 20, 318–334.

Enslin, J.F. (1961) Secondary aquifers in South Africa and the scientific selection of boring sites in them. *Proceedings Inter-Africa Conference on Hydrology, Nairobi, Publication 66, Section IV.* pp. 379–389.

Enslin, J.F. (1963) The hydrogeological interpretation of electrical resistivity depth probe surveys. *Annals of the Geological Survey South Africa,* 2, 169–180.

Enslin, J.F. & Kriel, J.P. (1967) The assessment and possible future use of the dolomitic ground-water resources of the Far West Rand, Transvaal, South Africa. *Conference Proceedings, Water for Peace, Washington, USA, May 1967.* p. 908.

Enslin, J.P. (1970) Die grondwaterpotensiaal van Suid-Afrika. *Convention: Water for the Future, November 1970.*

Frommurze, H.F. (1937) The water-bearing properties of the more important geological formations in the Union of South Africa. *Memoirs of the Geological Survey South Africa,* 34.

Hodgson, F.D.I. & Krantz, R.M. (1998) *Groundwater quality deterioration in the Olifants River Catchment above Loskop Dam with specialized investigations in the Witbank Dam sub-catchment.* Pretoria, Report Water Research Commission 291/1/98.

Janse van Rensburg, H. (1994) Management of the Grootfontein aquifer in Western Transvaal. *Report Water Research Commission Contract: The development of risk analysis and groundwater management techniques for Southern African aquifers,* 3.

Kent, L.E. (1949) The thermal waters of the Union of South Africa and South West Africa. *Transactions of the Geological Society of South Africa,* 52, 231–264.

Kirchner, J. & van Tonder, G.J. (1995) Proposed guidelines for the prosecution, evaluation and interpretation of pumping tests in fractured-rock formations. *Water SA,* 21 (3). Pretoria, South Africa, Water Research Commission.

Kleywegt, R.J. & Enslin, J.F. (1973) The application of the gravity method to the problem of ground settlement and sinkhole formation in dolomite on the Far West Rand, South Africa. In: Wolters, R. (ed.) *Symposium of the International Association of Engineering Geology, Hannover, Germany.* Paris, IAEG, Paris. pp. 1–15.

Kok, T.S. (1975) Examples of groundwater level fluctuations in the Republic of South Africa. *Memoirs of the Geological Survey South Africa,* 1, 1–33.

Kruseman, G.P. & de Ridder, N.A. (1991) *Analysis and Evaluation of Pumping Test Data,* 2nd ed. ILRI publication 47.

Levin, M. (1988) *A Geohydrological Appraisal of the Vaalputs Radioactive Waste Disposal Facility in Namaqualand, South Africa.* Ph.D. thesis. Bloemfontein, University of the Orange Free State.

Levin, M. (2000) The radon emanation technique as a tool in groundwater exploration. *Borehole Water Journal,* 46, 22–26.

Mazor, E., Bielsky, M., Verhagen, B.Th., Sellschop, J.P.F., Hutton, L. & Jones, M.T. (1980) Chemical composition of groundwaters in the vast Kalahari flatland. *Journal of Hydrology,* 48 (1–2), 147–165.

Muller, J.L. & Botha, J.F. (1987) *A Preliminary Investigation of Modelling the Atlantis Aquifer.* Pretoria, Report Water Research Commission 113/187.

Parsons, R. (1995) *A South African Aquifer System Management Classification.* Pretoria, Report Water Research Commission KV77/95.

Rindl, M.M. (1916) *The medicinal springs of South Africa. South African Journal of Science,* 13, 528–551.

Saunders, H.P. (1897) *Underground Water Supply of the Colony of the Cape of Good Hope with Special Reference to the Working of the Diamond Drill.* Report of the Inspector of Water Drills [G30-98].

Theis, C.V. (1935) The relation between the lowering of the piezometric surface and the rate and duration of discharge of a well using groundwater storage, American Geophysical Union Transactions, 16, 519–524.

Theron, J.J. (1946) *Lisimeterproewe.* University of Pretoria, Agricultural Research Institute Series 16. Science bulletin 283.

Tredoux, R. (1993) *A preliminary investigation of the nitrate content of groundwater and the limitation of the nitrate input.* Pretoria, Report Water Research Commission 368/1/93.

Van Blerk, J.J. (1994) *Moisture Content Flow Analysis at the National Radioactive Waste Disposal site.* Pretoria, Atomic Energy Commission. Report GEA-1100.

Van Eeden, O.R. (1955) Die verbruik en aanvulling van ondergrondse water in die Unie van Suid Afrika. *Suid-Afrikaanse Akademiese Tydskrif vir Wetenskap en Kuns*, Nuwe Reeks deel 15.

Van Tonder, G.J. (1993) *Modelling Groundwater Flow and Recharge in the Sishen compartment*. Internal report. Bloemfontein, Institute of Groundwater Studies.

Vegter, J.R. (1962) Locating zones of weathering and fracturing by the electromagnetic technique using a long earthed cable. *Annals of the Geological Survey of South Africa*, 1, 219–226.

Vegter, J.R. (1995a) *Groundwater Harvest Potential Map of South Africa and Explanation Report*. Pretoria, South Africa, Department of Water Affairs.

Vegter, J.R. (1995b) *An Explanation of a Set of National Groundwater Maps*. Pretoria, Report Water Research Commission TT74/95.

Vegter, J.R. (2000) *Groundwater Development in South Africa and an Introduction to the Hydrogeology of Groundwater Regions*. Pretoria, Report Water Research Commission. TT134/00.

Venter, F.A. (1970). *Water*. Afrikaanse Pers Boekhandel.

Verhagen, B.Th, Mazor, E. & Sellschop, J.P.P. (1974) Radiocarbon and tritium evidence for direct rain recharge to ground waters in the northern Kalahari. *Nature*, 24, 1643.

Vogel, J.C. & Bredenkamp, D.B. (1969) *A Study of Subterranean Water in the Southern Kalahari with the Aid of Carbon-14 Dating and Isotope Analysis*. Unpublished report. Pretoria, Council for Scientific and Industrial Research.

The history of hydrogeology in Spain

Emilio Custodio

Department of Geotechnical Engineering and Foundation International
Centre for Groundwater Hydrology, at the Technical University of Catalonia,
Barcelona, Spain

ABSTRACT

In Spain, interest towards groundwater dates from ancient times, mostly to solve problems in the dry areas for agricultural irrigation and town supply and also for mining activities. However, until more recent times, its scientific development may be considered as relatively poor, although some interesting groundwater developments should be mentioned. This chapter comments on the hydrogeological background, scientific development and technological achievements, with most emphasis on this last aspect. The 20th century is only considered in an abridged form, especially the second half, when groundwater research and development grew exponentially in spite of scarce funds, lack of interest by public organisations, and a clear bias towards surface water in the government arena. The chapter also introduces the historical evolution of groundwater legislation, management and teaching.

INTRODUCTION

Spain is geographically and hydrologically a varied country which extends over a large part of the Iberian Peninsula and two archipelagos, the Balearic islands in the Mediterranean sea and the Canary islands in the Central Atlantic Ocean close to the African coast. The geology is also varied and includes large carbonate rock areas and recent sedimentary formations. The Canaries are fully volcanic.

Although the north and northwest of the Iberian Peninsula is humid, the Mediterranean and central parts comprise semi-arid areas with short and discontinuous rivers and creeks which are nevertheless capable of producing large floods. In the islands the highlands receive significant rainfall, although the coastal areas are semiarid or even arid in the Canaries. A recent review of water in Spain can be found in Garrido and Llamas (2010).

These factors, the fact that a large proportion of population lives close to the sea, good conditions for intensive agricultural production and an important tourist industry determine people's attitudes towards water and water supplies, which vary conspicuously from one area to another, and consequently help to define the history of hydrology and hydrogeology in Spain.

BACKGROUND AND EARLY HISTORY

Water works and groundwater use have probably existed since the fifth century B.C., for example by the Tarthesians in the south (mentioned in the Bible) and the different

Iberian groups in the Mediterranean and southern areas of the Iberian Peninsula, and perhaps by the Celts in the northwest and the Celtibers in the centre. Water was needed for town supply, agriculture and mining. Unfortunately there are neither records nor recognisable water works from this earlier period. Iberian inscriptions, still to be fully deciphered, seem to refer mostly to trade agreements and not to the way of living.

During Roman times, from the second century B.C. until early in the fifth century A.D., Iberia (Hispania) was an important Roman Province and shared in the progress of water technology of the Imperium (Manzano, 1995). The most important figure from that time in terms of contributions to hydrological thinking is Seneca, although he spent most of his life in Rome (Biswas, 1970).

The coming of the Visigoths, a barbaric people from the north, ended Roman times and established a feudal system that was not propitious to important water developments. During this epoch the outstanding Saint Isidore of Seville included in his writings some hydrological concepts, following classical Greek thought.

The invasion of the Arabs in the 8th century A.D., accompanied by berber people and troops recruited in North Africa, occupied most of the Iberian Peninsula and the Balearic islands. Afterwards, little by little they were displaced by the northern Christian kingdoms, and the Arabic period finally ended with the capture of the last refuge late in the 15th century. The Arabs improved and established agricultural irrigation and water supply systems for the large towns, such as Córdoba, a large city and an important cultural and social centre of that time. Important places with irrigated gardens were developed in many places, such as Granada, Jaen and Medina Azahara (near Córdoba). However it seems that no especial scientific developments were introduced and no outstanding documents with hydrological concepts remain. However, during the occasional peaceful periods, translations into Arabic of the outstanding old Greek writings were performed, both in the muslim and christian kingdoms. They were later rewritten in Latin at translation schools, such as those of Toledo, Córdoba and Tortosa, in which Muslim, Christian and Jewish experts worked together. Many of these translations were carefully continued in the Christian monasteries, where they were preserved for western civilisation. They include some texts that contain thoughts on hydrology, mostly following the ideas of the Old Greeks.

After the extinction of the last islamic refuge, the different Christian kingdoms in the Iberian peninsula and the Balearic islands were finally reunited late in the 15th century, and including Portugal in the 16th century which seceded later on. Thus, the modern times started, although with a century long transition time.

The first human settlements in the Canary Islands were established a few centuries B.C. by people coming from Berberia in northwestern Africa. They probably did not develop specific water works and instead relied on springs and the few associated, small water courses. The first settlement of people from the Iberian Peninsula started early in the 15th century. When the islands were finally incorporated into the Crown of Castille late in the 15th century, the population grew fast and irrigated agriculture was introduced. However, groundwater works especially adapted for local conditions were not developed until the second half of the 19th century (Gavala and Goded, 1930; Díaz Rodríguez, 1998; Custodio and Cabrera, 2002).

The expansion of Spain and Portugal in what is now the Ibero–American countries was accompanied by the exportation of water technology to supply towns and for irrigated agriculture, often applied there with more emphasis and technology than in the mainland, and incorporating early developments by the Amerindians. Especial developments were introduced for mining (Garcia Tapia, 1989). The religious orders (eg the Jesuits and Franciscans) introduced irrigation systems to upgrade the amerindian way of life.

THE MODERN AGE – FROM THE 15TH CENTURY

Most water and groundwater developments during the Modern Age and up to the relatively scarce present refer to technological developments for water abstraction. Scientific developments were scarce and Spanish science derived mostly from developments in other European countries or remained hidden in poorly diffused writtings. From being a relatively open country at its beginning, Spain slowly drove towards isolation (but for some exceptions) and this situation has changed only in the last decades. The consequences of this isolation is that there has been a long way to catch up with developments abroad and a low rate of investment in science and technology. The latter is still the prevailing national situation in spite of it being repeatedly identified as a serious gap. This is at odds with the often scarce water availability and salinity problems in Spain relative to other western European countries. The large surface water developments (Spain is the country with the largest density of large dams) have been carried out by engineers who did not translated their knowledge into theories and writings for widespread readership. Only a few of them produced technical papers, but they were written in castillian or catalan, as internal reports or included in local publications. An often quoted and sadly celebrated saying of the early 20th century was: "let others invent for us", and to some extent is still alive in some ambients.

However, descriptive records of rainfall events and floods were available in parishes and monasteries since earlier times, with data from exceptional situations being available from the 12th century. Unfortunately some of them were destroyed during the French invasion of the Napoleon army in early the 18th century, in Catalonia early in the 20th century in a popular upheaval, and later just before the Spanish Civil War of 1936–1939 by revolutionaries and anarchists. That which has been saved is now the subject of ongoing studies, and yield interesting results.

Very few writings on the history of hydrology and hydrogeology in Spain are available, most of them referring to specific subjects. The most detailed work dealing with groundwater is probably that of Puche (1996) which is extensively used in the present paper. Other papers on the history of hydrogeology were produced by Martinez Gil (1990; 1991). Other information can be found in the book of the 150 years of the Geological Survey of Spain (Martín Municio, 2000; López Geta, 2000). The author has also relied on internal documents prepared by F. Ribera, E. Batista and A. Galofré, for the Barcelona International Course on Groundwater Hydrology. In what follows old writings are only mentioned, and the full reference is given only for some of them.

HISTORY OF HYDROGEOLOGICAL SCIENCE IN SPAIN

There have been no major developments in water science in Spain through history, except perhaps the Al Burundi explanations of evaporation, in about the year 1000

(Puche, 1996). Old Greek principles, mostly those of Aristotle more or less modified, lasted until the nineteenth century, when new ideas from Europe were little by little permeating the minds of Spanish professors and teachers. The main interest was in technology to solve water problems. Military and civil engineers responsible for water works and other developments were neither trained in scientific curiosity and experimentation nor motivated to spread their knowledge, with some exceptions.

Divining for groundwater, probably introduced by the Arabs, was a common practice up until recent times, and was advocated even by trained people and continued to be practiced in rural areas from time to time. The castillian (Spanish) name of water diviners, "zahorí", comes from Arabic. Despite the common usage, this "wisdom" was challenged as early as the first quarter of the 18th century by Jerónimo Feijoo. References to water divining "technology" was still included in some of the books of the first half of the 20th century, as such of those of Darder (1932) and Murcia (1953), which are otherwise quite interesting and with up to date hydrogeological concepts.

An interesting early book introducing some concepts of groundwater science is that of Ardemans (1724). In 22 chapters it deals with the origin of springs and the quality of water, and goes further to comment on water prospecting, water mine construction and water transport (Davis, 1973). Another book, by Aznar de Polanco, appeared in 1727 and commented on the origins of the water quality of the water mines of Madrid.

In the 17th century authors still followed classical Greek ideas about the existence of caverns, although Torres Villaroel introduced new ideas in line with those of Palissy, the French authority of the time. The writers of the 18th century were mostly interested in groundwater technology, although Francisco Javier Gamboa theorized about the origin of springs, and in 1833 José Mariano Vallejo explained the hydrological cycle and said that runoff was about 30 percent of rainfall, a figure close to current ones. Silvino Thos y Codina, in a short paper published in the third quarter of the 19th century, explained evaporation and infiltration. Additional details of these works can be found in Puche (1996).

HISTORY OF HYDROGEOLOGICAL TECHNOLOGY IN SPAIN

Groundwater withdrawal by means of galleries (water mines) has been quite well developed in the semiarid and arid areas of continental and insular Spain since ancient times. Some are from Roman times and earlier, although there are no clear documents about them, except for that devoted to mine drainage, as in the huge gold mines of Las Medulas, in north–central Spain (Juncà, 1993). Wells were excavated along the Roman ways, water galleries were constructed to supply some towns, such as Tarraco (now Tarragona), up to 3 km long, and baths using thermal spring water were developed. Mechanical bucket–chains ("norias", "sènies", "sinies") were introduced to abstract water from especially constructed large diameter wells. Some wells had bottom galleries. The drainage of some wetlands for agriculture commenced. Drainage of mines, a main interest of the Romans, was carried out by means of water galleries, excavations, bucketchain wheels, Archimedes screws and double–effect Ctesibio pumps, powered by animals or slaves. During Arabic times the construction of water mines was active, although it is not clear if this follows from what was found from earlier times (Greco–roman technology from Roman and Visigoth times) or resulted from the importation of technology from the Middle East, or was just a local development. It seems that the water mine to supply Medina Azahara was the only large work of that period.

The name Madrid seems to derive from the Arabic Mayrat, later Mayrit (Oliver Asin, 1959), applied to drainage galleries for water supply, although some think this name is just an arabisation of "Matrix Aquae", the Roman designation. These "viajes de agua" (water ways) of Madrid were the main water supply sources to the city up to the middle the 19th century (Mallada, 1926; Llamas, 1983; Lopez–Camacho et al., 1986) (see Figures 1 & 2).

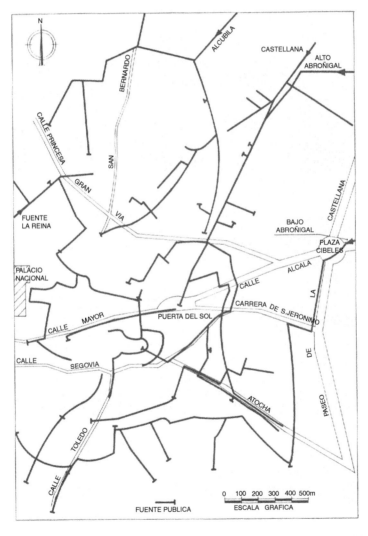

Figure 1 Layout of the "Viajes de agua" (underground water ways or water galleries) in Madrid (Spain), starting in the 8th or 9th centuries and enlarged until the 19th century. Modified from Custodio and Llamas, 1976.

Figure 2 Fountain constructed in 1780 as a new public discharge of an older "viaje de agua" (underground water way, or water gallery). The notice says that currently it cannot be used for drinking purposes since the water is contaminated (Courtesy of M.R. Llamas).

Drainage galleries (water mines) were common in the Mediterranean area and in Mallorca island (Viladrich, 2004) and in other areas of Iberia, many of them iniciated by its muslim settlers (Barceló *et al.*, 1986; 1998). The construction technology was similar to that of the "qanats" and "foggaras" of Persia, Oman and Yemen. A net of water mines draining the surrounding elevations supplied public fountains in Barcelona of which an inventory was made in 1650 by Socias, and private dwellings (Custodio, 1976); some of these date from the Middle Ages or earlier, such as the Rec Comtal (Conillera, 1991, although there is who claim that they may be initiated in Roman times. Water mines existed in many other locations of the Iberian Peninsula (Hermosilla, 2006; 2008; Hermosilla *et al.*, 2004), as around Tarragona and Reus, and are documented in areas around Barcelona since the 14th century (Ferret, 2002); they were used for town supply and to power mills to produce flour. Many of the existing old springs are really the outflows of old water mines whose memory has been lost (Figure 3).

Long water galleries have been developed in the highlands of the Canary Islands since late the nineteenth century, and especially in Tenerife (Custodio and Cabrera, 2002). A total of about 3000 km have been drilled, and are still maintained and extended to drain the volcanic formations. Typical mine technology has been applied, with special measures to stabilize sections of swelling unconsolidated rock, but lacking

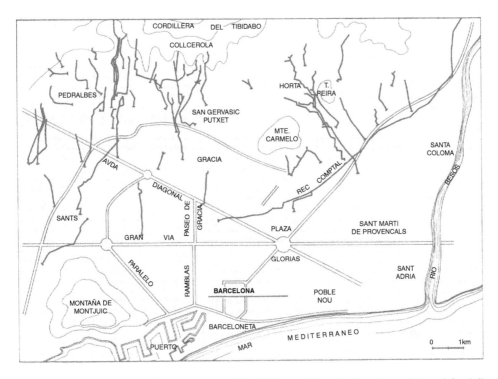

Figure 3 Several centuries old water mines (galleries) around in the Barcelona plain and foothills (Catalonia, Spain). Some of them ended in fountains for public water supply. The Rec Comptal is a centuries-old transfer canal from a water mine draining the Besós river water through its alluvium (from Custodio and Llamas, 1976).

ventilation shafts due to the steep slopes. Ventilation needs mechanical means, and this delayed the construction of deeply–penetrating ones until technology allowed for this.

Wells for irrigation were known near Barcelona from the 10th century. The first wooden agricultural bucket–chain wheels (Figure 4) were introduced in the 16th century (Ferret, 2002), although this device was described in Spain as early as the year 1200 by the muslim Ibm al Razzar al Gazari. Developments can be found in many areas, especially in dry mountaineous zones (Avila, 1993) and irrigated flat lands.

Apart from traditional large diameter, shallow wells in flat and coastal areas, deep wells were not developed until late in the nineteenth century, due to the difficulty of pumping water from them. Some old wells were operated by mechanical bucket–chains moved by animal force. Quite sophisticated models were developed in the lower part of the Llobregat Valley, and were also installed in other parts of the Mediterranean area (Montaner, 1982).

Wind power was also used to abstract water by means of wind turbines and pumps. Most of them have now disappeared, although many of them are preserved as a monument. Some of them are still in use around Palma de Mallorca. In order to obtain high flows, large diameter, deep wells were excavated to hold the cumbersome pumping

Figure 4 Old bucket-chain well with ceramic pots. Rifá area, Camp de Tarragona (Catalonia, Spain). From EC.

machinery inside, which was moved by steam engines above at ground level (Figure 5). Most are from late in the nineteenth century (1881 in Cornellà, Barcelona, after Ferret, 2006), and remnants are preserved around Barcelona. Well construction was local, using common practice rather than science, but the machinery was mostly imported.

Artesian wells were in fashion after the good results obtained in Paris in mid the 19th century became widely known. Numerous studies were carried out. Such wells enjoyed great success in the area of Prat de Llobregat, near Barcelona, starting late in the 19th century, and this was rapidly imitated (Roigé i Badia, 1895). The wells were used for town supply, agricultural irrigation and industrial uses. In the report of Santa Maria and Marin (1910) more than 200 flowing wells are mentioned.

The construction of artesian wells was attempted in Madrid, which was suffering water supply and water quality problems in the mid nineteenth century. These wells failed as the hydrogeological model used was incorrect due to an undue extrapolation of the Paris artesian basin structure to that of the Madrid basin. Madrid's aquifer is really an aquitard which needs advanced technology and very deep wells to develop groundwater. This was not accomplished until the 1980's, when Madrid again paid attention to groundwater, now as a complementary supply source in droughts (López–Camacho and Cabrera, 1993; Sánchez *et al.*, 2005; López–Camacho and Iglesias, 2000). The failure to develop artesian wells in Madrid, where the very centralised government resided, produced a general lack of interest in groundwater in the technical

Figure 5 View of the building to contain the Central Besós (Barcelona, Catalonia, Spain) large diameter well and the 19th century steam-powered marchinery (partly dismantled) to move the
submerged vertical piston pumps. Courtesy of SGAB.

departments of the Spanish government, and also in the engineering schools, at that
time most of them in Madrid, and consequently the dismissal of groundwater as a
source of supply. The notable developments in other parts of the country, especially
those mentioned in Barcelona and other Mediterranean areas, were just ignored. This
caused, at least in part, the lagging behind of Spanish hydrogeology until recent times,
as interpreted by Llamas (1983; 1994).

In the Canaries, and especially on Gran Canaria Island, deep, large diameter wells
(shafts) have been drilled since the late nineteenth century using mining technology
(Figure 6). An early description is that of Ascanio y León (1926). Since volcanic rocks
are highly heterogeneous and often of low permeability, in order to increase the yield,
galleries were excavated at the well bottom. These were later substituted by horizontal
drilled drains ("catas"), which had to be repeated as groundwater levels progressively
decreased (Custodio and Cabrera, 2002). The shaft depth was limited by the then
availability dewatering capacity during the construction works. Most of these shaft
wells, numbering about 3000, are still in operation, although machine drilled deep
wells are now replacing the old ones, both as completely new ones or by deepening
those existing. Pumping machinery for the shaft wells, initially imported, started with
mechanical bucket–chain wheels, then followed steam engines, internal combustion
oil motors and finally electric motors (Fernández González, 1974).

An important treatise containing details of well and gallery construction in Spain
is that already mentioned of Ardemans (1724). In the early 17th century Jerónimo

Figure 6 The first deep, large diameter well in Gran Canaria (Canary Islands, Spain), near Telde, constructed in the late 19th century. Inside there is a heliocoidal path that allowed water abstraction by means of dunkies, before pumping machinery was available. Courtesy of the CIA-GC and MCC.

de Ayanz tried to build a steam machine to pump water from wells and from mines (García Tapia, 1989), but it was rarely used until new engines developed in England were imported a century later. The exploitation of mercury at Almaden in central Spain was very important to the Crown as the mercury was used to obtain gold in the mines in America. This fostered drilling improvements by using augers and percussion tools and dewatering systems.

In 1829 Cristobal Bordiu published papers on how to obtain groundwater and a translation and improvement of the book of the Frenchman Garnier (1829) which was a reference book at that time. In 1833 José Mariano Vallejo published a treatise on movement and use of water in which he proposed the areas in Spain favourable for the drilling of artesian wells. Starting in 1835, at the initiative of the Spanish Government who wanted to improve water supply by means of groundwater, the engineer Joaquin Ezquerra del Bayo was commissioned to survey several Spanish provinces.

During the 19th century geology was relatively well developed while other branches of science lagged behind the level existing in other European countries, and this helped the development of hydrogeology. Mining and the need to develop groundwater boosted interest in hydrogeology. Chapters on groundwater were included in

1841 in the geological treatises of Francisco de Luxan and in 1848 by Julian López de Novella.

In 1849 Queen Isabel II approved the creation of the Commission for the Geological Map of Spain, which later became the Geological Survey of Spain (Instituto Geológico y Minero de España, IGME). The memoirs accompanying the geological maps regularly included a section on hydrogeology. This was accompanied by the publication of papers in Spanish journals on the results of studies fostered by the Government and surveys of the possibility of drilling artesian wells. A large part of the work was carried out by mining engineers. For more details see Puche (1996). However, the technical staff of the Public Administration, mainly the Civil Engineers in charge of water works and water supply, lost interest in groundwater due to the failure of early wells constructed for Madrid in 1830 and 1834, and in Barcelona in 1834, but especially due, as commented before, to the failure to obtain artesian wells in 1856 in Madrid, which was suffering water shortages and water quality problems. As a consequence, importation of surface water from other basins by a long canal was the solution adopted for Madrid and from the success of this transfer began the official disregard for groundwater. In fact, in the last third of the 19th century groundwater development for public supply and factories was mostly fostered by foreign companies, and some private promoters.

An outstanding figure of that time (mid to late the 19thcentury) was the physician Juan Vilanova y Piera (see Martínez Gil, 1994), a Valentian who was later elected to the chair of geology at the Central University of Madrid, and coined the name underground hydrography to describe what is now called hydrogeology. Underground hydrography was used for some time (Calvo, 1908). It seems that the name hydrogeology was first used in 1873. His writings in the mid 19th century are interesting documents that deal with groundwater, and some conclusions referring to groundwater resources were advanced for this time, and currently they still look modern in many aspects. His knowledge was collected together in a book of outstanding quality (Vilanova y Piera, 1880). Another interesting figure is that of Horacio Bentabol y Ureta, who produced an interesting paper on well drilling describing new methods (Bentabol y Ureta, 1880). Later he published an outstanding book on groundwater in Spain and Portugal (Bentabol y Ureta, 1900), in which some principles of quantitative hydrogeology were introduced. Cortázar (1896) produced a compilation of the existing territorial knowledge.

The mining engineer Lucas Mallada produced a book (Mallada, 1890) under the title "Los males de la patria" (misfortunes of the mother country), a personal reflection on Spain's economy drawing on his extensive knowledge of the country, in which he vehemently advocated the use of groundwater to enhance agricultural production to reactivate the economy and create jobs, and moreover urged the Government to sponsor this development. Once more this was largely ignored.

Quantitative hydrogeology was applied by the engineer Moragas (1896) to the discharge of a water gallery draining the Besós river alluvium near Barcelona to solve a court dispute. But this is an isolated example. Most other works up to this time were largely descriptive.

As commonly happens in arid and semiarid lands, little attention had been paid to groundwater quality compared with quantity aspects. Studies and surveys were often limited to reporting compliance with existing water quality norms and rules for

Figure 7 Cover and page of the Alfonso Limón Montero (1697) book on "Crystaline mirror of waters in Spain". From historical books in IGME library, Madrid, IGME (facsimil reproduction).

public water supply. These norms were neither strict nor detailed until late in the 19th century. The exceptions were water quality considerations for thermal and mineral waters, which were mostly the dominion of physicians, and for bathing establishments and water bottling factories.

Thermal and mineral springs were known and used since ancient times. The Romans developed bathing stations and the Arabs used them extensively. There is a continuous tradition of use, and their healing properties have been often mentioned, suh as that of Sacedón (Salam–Bir), by AGMER been Abdalá in 1054, as in the translation of Pizzi Franceschi, done in 1759. A first treatise on such springs is that of Limón Montero (1697), which was still descriptive but introduced ideas about the origin of springs along classical lines (Figure 7). This was followed in 1757 by Gayan y Santoyo (Figure 8), and by the book of Gómez de Bedoya (1764) which had an inventory of 214 springs, wells, rivers and lagoons (Figure 9). In 1850 Rubio published a paper on the temperatures of springs, and later the same author published a book on mineral springs in Spain (Rubio, 1853). The important thermal and mineral springs of Caldes de Malavella (Girona) were described and studied in 1882 by Luis Mariano Vidal and the results published in the Boletín de la Comisión del Mapa Geológico.

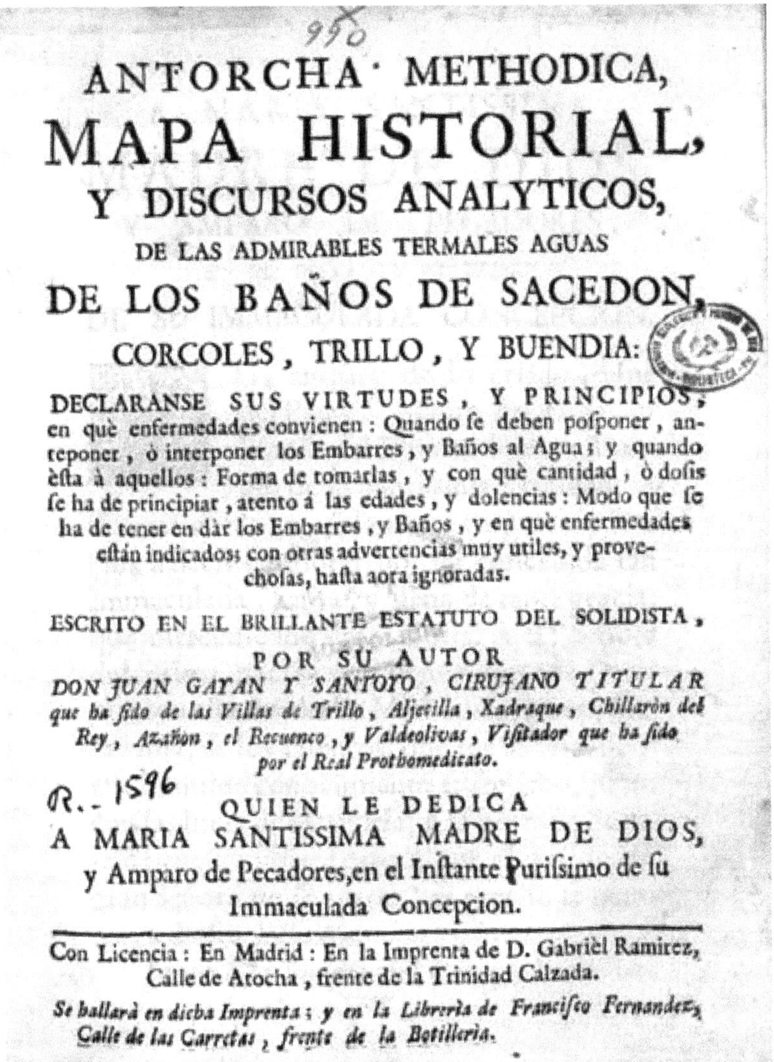

9 5 0

ANTORCHA · METHODICA,

MAPA HISTORIAL,

Y DISCURSOS ANALYTICOS,

DE LAS ADMIRABLES TERMALES AGUAS

DE LOS BAÑOS DE SACEDON,

CORCOLES, TRILLO, Y BUENDIA:

DECLARANSE SUS VIRTUDES, Y PRINCIPIOS,
en què enfermedades convienen : Quando fe deben pofponer, an-
teponer, ò interponer los Embarres, y Baños al Agua; y quando
èfta à aquellos : Forma de tomarlas, y con què cantidad, ò dofis
fe ha de principiar, atento á las edades, y dolencias : Modo que fe
ha de tener en dàr los Embarres, y Baños, y en què enfermedades
eftàn indicados; con otras advertencias muy utiles, y prove-
chofas, hafta aora ignoradas.

ESCRITO EN EL BRILLANTE ESTATUTO DEL SOLIDISTA,

POR SU AUTOR

DON JUAN GAYAN Y SANTOYO, CIRUJANO TITULAR
que ha fido de las Villas de Trillo, Aljecilla, Xadraque, Chillarón del
Rey, Azañon, el Recuenco, y Valdeolivas, Vifitador que ha fido
por el Real Prothomedicato.

R.- 1596

QUIEN LE DEDICA

A MARIA SANTISSIMA MADRE DE DIOS,

y Amparo de Pecadores, en el Inftante Purifsimo de fu
Immaculada Concepcion.

Con Licencia : En Madrid : En la Imprenta de D. Gabrièl Ramirez,
Calle de Atocha, frente de la Trinidad Calzada.

Se hallarà en dicha Imprenta; y en la Libreria de Francifco Fernandez,
Calle de las Carretas, frente de la Botilleria.

Figure 8 Cover of the Juan Gayan y Santoyo (1757) book on "Historical map and analytical comments of the wonderful thermal waters of Sacedón, Córcoles, Trillo and Buendia baths". From historical books in IGME library, Madrid.

An interesting story is that of the Fuente Santa (holy spring) in Fuencaliente, La Palma island in the Canaries. This was a well known thermal spring which was famous for healing serious skin and venereal illnesses since the time of incorporation to the Crown of Castille late in the 15th century. In the 17th century it was covered by about 50 m of lavas and scoria from the eruption of the San Antonio volcano. The island lost an important source of income and this fostered different failed attempts to unearth

Figure 9 Cover of the Pedro Gómez de Bedoya y Paredes (1764) book on the "Universal history of mineral springs in Spain". From historical books in IGME library, Madrid.

it. Drilling technology was not advanced enough to penetrate through the alternating highly unconsolidated, unstable blocky volcanics and very hard lavas. The spring was finally relocated in 2006 by means of a gallery (Soler Liceras, 2007).

Interest in thermal and mineral waters from the medical point of view has continued with ups and downs until the present. Madrid's Complutense University has a hydro–therapeutic service at the hospital of the Faculty of Medicine. In 1912 a Chair of Medical Hydrogeology was created there (San Martín, 1994).

Groundwater recharge has been a concern since early times in the arid areas. Old, small undertakings existed in alluvial valleys to enhance recharge by techniques of unknown origin, perhaps brought by the Arabs, or engineered spontaneously by local people. Remnants of these works, comprising mainly a transverse dike to retain water and artificially inundate the land during the occasional creek floods, are found throughout the Mediterranean area and perhaps in the Monegros dry region between Catalonia and Aragón (Figure 10). These works were quite well developed in the Alpujarras (Granada), locally called "careos", probably originating from Arabic or perhaps from late Roman times (Pulido–Bosch and Ben Sbia, 1995). Similar works named "gavias y nateros" are found in the Canaries, especially in the arid Fuerteventura island where they are still in operation for agricultural purposes. The idea was probably imported by the first settlers from the Iberian Peninsula.

A special technique for conserving water from the scarce rains in Lanzarote and Fuerteventura islands is to cover the soil with a 10 to 30 cm layer of eolian sand or recent (black) volcanic lapilli ("picón") (Figure 11). This allows a portion of the infiltrating

Figure 10   Creek terrassing to favour occasional storm runoff infiltration to increase soil humidity and recharge, in the arid Monegros area (Aragón, Spain). Unknown age. From EC.

Figure 11   Cultivation of crops (grapes, figs) in the arid environment of La Geria (volcanic Lanzarote island, Canary Islands, Spain) by using a cover of young, black lapilli ("picón") to facilitate rain infiltration and delaying evaporation. The method is several centuries old. From EC.

rain water to be stored in the soil below, delaying its evaporation. This mimics what happens in Nature in some eolian sand covered areas and is a traditional way of greatly enhancing crops by restricting soil evaporation that resembles the current technology of mulching.

SUMMARY OF DEVELOPMENTS DURING THE 20TH CENTURY

There were no special groundwater developments during the first half of the 20th century. The Geological Survey of Spain continued to produce geological maps with a chapter on groundwater in the accompanying memoir, and some interesting reports. That of Santa María y Marin (1910) on the great success of flowing wells in the Llobregat delta in the southwestern area of Barcelona is especially interesting. It reviewed about 200 such wells and documented the close to natural situation of this small but important aquifer which is now a keystone of groundwater research, development studies and management. Artesian wells and groundwater were also the subject of the papers and books of Mesa y Ramos (1909; 1929) and Fernández Navarro (1914; 1922).

Things began to change in the 1940s and 1950s. The National Institute of Colonization was created in 1939 to foster agricultural development and it drilled a large quantity of wells with its modern park of machinery, sponsored studies and introduced improved irrigation methods. Benitez (1956) published what can be considered a first book in Castilian (Spanish) on pumping tests and quantitative hydrogeology. As a consequence of the Barcelona International Course of Groundwater Hydrology, an extensive, 2 volume book was produced (Custodio and Llamas, 1976) which has been re-edited several times and also translated into Italian. It has become the reference book for groundwater in Spain and Ibero-America.

Shortage of water supply in dry periods for Barcelona, which used mostly groundwater, was improved by the Barcelona's Water Supply Company by means of artificial recharge (Figure 12). This was achieved by carefully scraping the Llobregat river bed (since the late 1940s), by recharging and recovering canal water from an upstream water gallery in the Besós river by means of a special well to control clogging (constructed in the 1950s and operating up to the 1990s) and from the late 1960s by recharging treated river water into the Llobregat river aquifer by means of wells to store the water for latter recovery. This was an early development of the ASR (aquifer storage recharge) concept. Currently reclaimed water, with ultrafiltration and reverse osmosis treatment, is injected in a line of wells to halt seawater intrusion. Several basin recharge plants are in operation, using river and reclaimed water.

The Public Works Geological Service was created in the 1940s to carry out geotechnical studies related to public water works, but excellent hydrogeological studies of many areas were also carried out all over the Spanish territory, with interesting early characterizations in the Canaries (Macau, 1957). The Eastern Pyrenees Water Authority in Barcelona created in early the 1960s the first groundwater group in such organizations and they developed extensive studies jointly with the Public Works Geological Service, in their own territory and also in other areas of Spain.

In the 1960s, two major international groundwater development projects were initiated, one in the Guadalquivir Delta-Doñana area and the other in the Canaries.

Figure 12 Recharge well of the 1950s with a back flow to help in back flushing daily cleaning. Central
Besós (SGAB). Barcelona (Catalonia, Spain). From EC.

The Geological Survey of Spain launched a series of regional hydrogeological studies
to cover a large part of the Spanish territory and the Public Works Geological Service
started successful aquifer studies as a first, effective survey (Llamas, 1968). Numerous engineering and consulting firms created hydrogeological teams to collaborate in
many of these projects. Key figures from that time were M.R. Llamas, A. Sahuquillo,
J. Porras and A. Dominguez. R. Fernández Rubio fostered studies of groundwater in
mining activities.

The development of groundwater, mostly for agriculture, has been exponential in
many areas, leading to intensive exploitation and even groundwater mining in some
places (Llamas, 2005; Llamas and Custodio, 2003). The benefits and drawbacks of
this development have not been correctly understood nor addressed by water authorities, but it has fostered the growth of hydrogeology and helped to incorporate into
water legislation norms to manage existing groundwater overexploitation, groundwater quality and seawater intrusion, and to take into account, little by little, the
essential role of groundwater stakeholders (Galofré, 1991; Codina, 2004). Part of current research is in the field of groundwater management and related social sciences,
with a worldwide focus, lead by Dr. M.R. Llamas, from the Madrid's Complutense
University, jointly with the Marcelino Botin Foundation.

In the 1970s and later, groundwater science flourished in several university departments, led by young hydrogeologists. The most active departments in hydrogeology
are now in the Technical Universities of Catalonia, Valencia and Cartagena, the

Complutense University of Madrid, and the Universities of A Coruña and Almería. Most research is experimental in the field and in numerical flow and mass transport model development. Laboratory developments have been scarce. There is currently participation and leading of scientific and technical projects, both national and international, and numerous meetings are convened. Spain has one of the largest groups of the International Association of Hydrogeologists.

HISTORY OF GROUNDWATER LEGISLATION AND ADMINISTRATION

Up to the mid 19th century water law in Spain derived mostly from Roman rights. No special consideration existed towards groundwater. After the fall of the Roman Empire the feudal system brought by the Goths (barbars) imposed the concept of gifts ("regalias") that lasted through the Medieval Age. Generally all waters were the property of the nobles and Church organizations and their use was granted by means of written permissions. General regulations were very scant. The most relevant are the Rules ("Partidas") of Alfonso X, "the Wiseman", King of Castile in the 13th century and in the "usatges" (common law) of Catalonia. The Partidas explicitly mention that the property of water is from top to bottom.

During Arabic times, water and groundwater development by local communities or individuals created rights for them, especially in the Mediterranean area. Later, governors moved to a feudal system, restricting popular initiatives to the benefit of high class persons, thus ending development initiatives by ordinary people.

The property of water rights signified social prestige (Alegria Suescun, 2006). How to get, transfer and sell them was the subject of detailed rules. This situation was later modified in practice by the common use of untapped resources, with legal actions against neighbours to protect what was assumed to be their property. This was abolished in 1836 by the Crown but continued to exist in some traditional areas. Jordà (1875) commented on the acquisition of groundwater rights in the Tarragona area. A traditional court, resolving issues on the spot once a week at one of the gates of the cathedral of Valencia, has administered justice uninterruptedly for at least 700 years, but deals mostly with the use of surface water for irrigation (Figure 13). This is an example of civil institutions with a long tradition of effective ruling that have survived.

The first Water Act of 1866 was a bold document which declared all water to be a public domain. This proposal was too advanced for the time, and finally the Water Act of 1879, while maintaining most surface water in the public domain, declared most groundwaters to be the private property of who abstracted them. This was included in the Civil Code, thus allowing inscription in the Property Register. The Water Act of 1879 contains some rules about groundwater withdrawal and establishes the minimum distance between two wells, galleries or shafts. Water from mines was also the property of the mine owner, as specified in the 1841 regulation from the Crown, and this was reflected in the Mining Act of 1868. Local regulations acquired importance.

Starting in 1926, the Hydrographic Confederations (authorities for public water management and administration) were created, dividing Spain into ten river basins and two archipelagos. Thus, water management was carried out inside hydrographical basins, which was a clear advance at that time. The Hydrographic Confederations were

Figure 13 Ordinarily weekly meeting of the Water Court ("Tribunal de les Aigües") in one of the gates of Valencia's cathedral. This is a 14 centuries-old form of administering justice on local water issues. Courtesy of A. Sahuquillo.

conceived as entities to establish joint management by the Government and the users, who were mostly farmers irrigating their lands. However, they were soon dominated by the civil engineers of the administration and as a consequence interest was almost exclusively in surface water, disregarding groundwater since it was a private affair. This greatly delayed the establishment of groundwater management. Problems created between groundwater owners, or with surface water users, were solved by agreements between them or in the civil courts. An improvement was introduced in 1956 by the creation of Water Authorities to deal with public water administration, but in 1976 they were integrated as a division within the Hydrographic Confederations. This was a step back, since both development and control were the responsibility of the same organization, that also started to be politically controlled.

Difficulties in applying the Water Act of 1879 soon began to appear, and different attempts to change it were proposed. However, the Act solved many problems in spite of the scientific and technological developments produced during the 20th century. The clearly insufficient consideration of water quality and of groundwater, the accumulation of decrees and orders to fill gaps, and the intensive development of groundwater in some areas led to the passing of the Water Act of 1985.

The Water Act of 1985 declared all water to be a public domain, to be administered within hydrographical basins. To avoid expensive compensation to the owners of groundwater rights, they were asked to voluntarily give up their rights in exchange for administrative protection (a poorly defined concept) or to continue with their rights, forever, after registering them in a public catalog. In theory no improvements in the groundwater capture works can be made in the latter case, even to maintain the yield, but in many cases this is ignored or is incorporated into local legislation, as in The Canaries. There, earlier legislation which was too restrictive produced a fall of the local government. Since most groundwater users opted for the second possibility of the Water Act, or just ignored the regulation since they were protected under the Civil Code by the Property Register. The result has been poor management and a series of conflicts which still continue, although things seem to be improving little by little. However, in spite of past achievements, current governmental weakeness in water administration and too much political involvement is to some extent a step back.

The Water Act of 1985 has suffered from different reforms intended to adapt it to the establishment of Autonomous Regions, whose limits do not coincide with hydrographical boundaries, and later to incorporate the requirements of the European Water Framework Directive of year 2000. The process is still going on and is awakening attitudes in the public administration towards groundwater.

Hydrogeologists working for the public water administration have been traditionally very scarce, their role being carried out by well trained but often groundwater–ignorant civil engineers. However, things are now changing for the better, at least in some Water Authorities from the State and the Regional Governments, and even in large Municipalities. In fact the Water Agency of Catalonia has now a large technical staff of hydrogeologists.

The creation in 1976 of the Lower Llobregat River and Delta User's Association was a milestone in stakeholder participation that has definitively helped groundwater management in this complex area near Barcelona (Codina, 2004; Galofré, 2000). This association was created under the 1879 Water Act. Other such associations, recognized as public entities by the 1985 Water Act, have been created but the development process in slow and difficult due to a lack of understanding of the problems, the reluctance of stakeholders to give up part of their rights for the common benefit, and the "red tape" of public administration which misjudges groundwater and still fears losing its traditional quasi–dictatorial power and privileges. However, a country–wide private association exist to coordinate efforts and share experience. In the Canaries there are old institutions and tradition of public involvement in groundwater (Hernández Ramos, 1954), and imperfect but working private water markets.

DEVELOPMENT OF HYDROGEOLOGICAL EDUCATION IN SPAIN

No structured teaching of groundwater existed in Spain until recently. During most of the 16th to early 20th century, groundwater scientists and engineers traveled to Europe, mostly to France and Germany, to learn about current developments. Official teaching at universities and engineering schools was, at most, a few lectures and conferences. However, in 1880 a school of drillers for artesian wells was created in Alcoi (Alacant). The Public Works Geological Service had a school for borehole drillers in the 1940s which lasted until the 1980s, initially for geotechnical studies and later extended to hydrogeological exploration.

The first chair of hydrogeology was created in 1921 at the Mining School of Madrid, and Pablo Fabregat y Coello was the first professor. The Mining Schools included some lectures on groundwater (Pendás, 1994). It is necessary to wait until 1969 for the first hydrogeology chair in a classical university, in the Faculty of Geology at Madrid's Complutense University, and the first professor was M. Ramón Llamas. After that, many university faculties (mostly in geology) and engineering schools (mostly in mining and civil engineering) have offered hydrogeology courses. Currently there are at least twenty different active groups, although only a few have acquired the critical size to effectively combine teaching and research. The universities have played a key role in groundwater development (Villarroya, 1994) and have influenced water administration and regulations.

A keystone in groundwater teaching and also in research was the call by UNESCO at the start of the 1965–1975 Hydrological Decade, to improve hydrological teaching worldwide. In Spain, two courses on groundwater were created in 1966, starting in 1967, one in Madrid (at the Complutense University), which lasted until year 2000, and another in Barcelona, at what is now the Technical University of Catalonia (Batista, 1991; Custodio, 1994), which is still very active and offers both attended and on line teaching, in the latter case by reporting to reference centres in different countries. Also a Master in Groundwater Hydrology is offered. These courses have trained a large number of professionals (geologists, engineers and others), especially in Spain, Ibero–America, Portugal and Italy.

ACKNOWLEDGEMENTS

The author is highly indebted to Dr. Andrés Sahuquillo, Eng. Josep Ferret and Dr. M. Ramón Llamas for reviewing the manuscript and making suggestions, and also to Dr. Fidel Ribera for providing personal documents. Also thanks to the late geologist Andreu Galofré, who tried to start a historical review some years ago.

REFERENCES

Alegria Suescun, D. (2006) El agua en la Edad Media: presencia, poder, explotación y trabajo [Water in the Middle Age: presence, power, exploitation and work]. *Tecnología del Agua*, 278, 102–109.

Ardemans, Th. (1724) Fluencia de la tierra y curso subterráneo de las aguas [Flow from the land and groundwater path]. *Francisco del Hierro. Madrid*, 1–278.

Ascanio y León, R. (1926) Gran Canaria y sus aguas subterráneas [Gran Canaria and its groundwater]. In: Sucesores de M. Curbelo & La Laguna de Tenerife. pp. 1–131.

Avila, J. (1993) En busca del agua, norias del Maestrazgo [Looking for water, buckett–wheels in the Maestrazgo]. *Rev. MOPT, Madrid, June 93*. pp. 78–80.

Barceló, M., Carbonero, M.A., Martí, R. & Rosselló Bordoy, G. (1986) Les aigües cercades [The search for water]. *Els qanatts de l'Illa de Mallorca*. Institut d'Estudis Balearics. Palma de Mallorca. pp. 37–46.

Barceló, M., Kirchner, H., Martí, R. & Torres, J.M. (1998) The design of irrigation systems in Al–Andalus. Universitat Autònoma de Barcelona, Bellaterra (Barcelona).

Batista, E. (1991) Enseñanza y formación en hidrología subterránea: situación general [Groundwater teaching and training: general situation]. *Hidrogeología, Estado Actual y Prospectiva*. Barcelona, CIHS/CIMNE. pp. 331–336.

Benítez, A. (1956) Captación de aguas subterráneas [Groundwater wells]. *Dossat*. Madrid. pp. 1–157. Reedited 1963.

Bentabol y Ureta, H. (1880) Investigaciones subterráneas y alumbramientos de aguas por medio de sondeos [Groundwater investigations and water abstraction by means of bores]. *Revista Minera, Madrid*, XXXI, 209–242.

Bentabol y Ureta, H. (1900) Las aguas de España y Portugal [Water in Spain and Portugal]. *Bol. Com. Mapa Geológico de España*, XXV. Madrid, 1, 3–47.

Biswas, A.K. (1970) *History of Hydrology*. Amsterdam, North-Holland. pp. 1–336.

Calvo, L. (1908) Hidrografía subterránea: conocimiento sobre los terrenos para la investigación de manantiales [Underground hydrography: land knowledge for spring investigation]. Madrid. pp. 1–289.

Codina, J. (2004) Las aguas subterráneas: una visión social. El caso de la Comunidad del Delta del Llobregat [Groundwater: a social vision. The case of the Llobregat Delta Community]. *Rev. Real Acad. Cienc. Exact. Fis. Nat. (Esp). Madrid*, 98 (2), 323–329.

Conillera, P. (1991) *L'aigua de Montcada [Water of Montcada]. L'Abastament Municipal d'Aigua a Barcelona: Mil anys d'història*. Barcelona, Institut d'Ecologia Urbana de Barcelona.

Cortázar, D. (1896) Noticias referentes a estudios hidrogeológicos [News about hydrogeological studies]. *Bol. Com. Mapa Geológico, Madrid*, XXVII, 1–190.

Custodio, E. (1976) Galerias de agua, zanjas de drenaje y pocos excavados [Water galleries, drainage trenches and excavated wells]. Hidrología Subterránea. In: Custodio, E. & Llamas, M.R. (eds.) *Ediciones Omega, Barcelona*. Vol. II. pp. 1791–1808.

Custodio, E. (1994) Tendencias en la docencia e investigación hidrogeológica en España [Trends in hydrogeological teaching and research in Spain]. *Bol. Real Acad. Cienc. Exact. Fis. Nat. (Esp). Madrid*, LXXXVIII (4), 127–148.

Custodio, E. & Llamas, M.R. (eds.) (1976) Hidrología subterranean [Groundwater hydrology]. *Ediciones Omega, Barcelona*, 2 Vols. pp. 1–2350. Revised 1983; reeditions.

Custodio, E. & Cabrera, M.C. (2002) ¿Cómo convivir con la escasez de agua? El caso de Canarias [How to live with water scarcity. The case of the Canaries]. *Bol. Geológico y Minero, Madrid*, 113 (3), 243–258.

Custodio, E., Llamas, M.R. & Sahuquillo, A. (2011) La investigación hidrogeológica española en el contexto mundial [Spanish hydrogeological research in the world context]. 100 Años de Hidrogeología en España. IGME. Madrid (in press).

Darder, B. (1932) Investigación de aguas subterráneas para usos agrícolas [Groundwater investigation for agricultural uses]. *Salvat, Barcelona*, 1–360.

Davis, S.N. (1973) Teodoro Ardemans pioneer water supply engineer of Spain. *Water Resources Bullettin*, 9 (5), 1028–1034.

Díaz Rodríguez, J.M. (1988) Molinos de agua en Gran Canaria [Water mills in Gran Canaria]. *La Caja de Canarias. Las Palmas de Gran Canaria*, 1–649.

Fernández González, E. (1974) Un poco de historia; curiosidades sobre las captaciones de agua en Gran Canaria [A piece of history; curiosities on water winning works in Gran Canaria]. *Simposio Intern. Hidrología Terrenos Volcánicos. CEDEX. Madrid, II.* pp. 1151–1168 (edited 1985).

Fernández Navarro, L. (1922) Aguas subterráneas, régimen, investigación y aprovechamiento [Groundwater, regime, investigation and exploitation]. *Espasa Calpe. Madrid*, 1–203.

Ferret, J. (2002) Els antics aprofitaments d'aigua del Delta del Llobregat, 1600–1900 [The old water works of the Llobregat Delta, 1600–1900]. *Com. Usuaris Aigües de la Vall Baixa i Delta del Riu Llobregat. Prat de Llobregat (Barcelona)*.

Ferret, J. (2006) El primitius pous artesians del delta del riu Llobregat, 1893–1912 [The first artesian wells of the Llobregat river delta, 1893–1912]. *Com. Usuaris Aigües de la Vall Baixa i Delta del Riu Llobregat. Prat del Llobregat (Barcelona)*, 1–89.

Galofré, A. (1991) Las comunidades de usuarios de aguas subterráneas: experiencias en la gestión y control de los recursos hidráulicos en Cataluña [Groundwater users' communities: experiences in water resources management and control in Catalonia]. *Hidrogeología, Estado Actual y Prospectiva.* Barcelona, CIHS/CIMNE, pp. 337–357.

Galofré, A. (2000) Las comunidades de usuarios de aguas subterráneas en la Catalunya del 2000: desarrollo histórico y realidad actual [Groundwater users' communities in Catalonia in the 2000s: historical development and current situation]. La Economía del Agua Subterránea y su Gestión Colectiva. In: Hernández Mora, N. & Llamas, M.R. (eds.) Publ. Fundación Marcelino Botín / Ediciones Mundi Prensa, Madrid.

García Tapia, N. (1989) Inventores españoles del Siglo de Oro [Spanish inventers of the Gold Century]. Investigación y Ciencia, Barcelona. September 1989.

Garnier, F. (1829) Traité sur les puits artesiens et sur les diferentes espèces de terrains [Treaty on artesian wells and on the diverse kinds of terrains]. Spanish translation by C. Bordiu. Tratado de las fuentes ascendentes y de las varias especies de terrenos en que pueden buscarse aguas subterráneas [Treaty on the ascending springs and the diverse kinds of terrains in which to look for groundwater]. Impr. León Amarita. Madrid. 1–192.

Garrido, A. & Llamas, M.R. (2010) *Water Policy in Spain*. CRC Press/Balkema. pp. 1–234.

Gavala, J. & Goded, E. (1928) Aprovechamientos de aguas en las Islas Canarias [Water winning in the Canary Islands]. *Bol. Instituto Geológico y Minero de España, Madrid*, 52, 1–104.

Gómez de Bedoya, P. (1764) Historia universal de las fuentes minerales de España, sitios en que se hallan, principios de que constan, análisis y virtudes de las aguas, modo de administrarlos y de ocurrir a los accidentes que suelen nacer de su abuso, . . . [Universal history of the mineral springs of Spain, place where they are found, principles on them, water analysis and virtues, how to manage them and to solve accidents that may occur when abused, . . .]. Imprenta de Ignacio Aguayo, Santiago de Compostela. (published 1764 and 1765).

Hermosilla, J. (2006) Las galerías drenantes del Sureste de la Península Ibérica [Draining galleries of Southeastern Iberian Peninsula]. Ministerio de Medio Ambiente. Madrid. pp. 1–227.

Hermosilla, J. (2008) Las galerías drenantes de España: análisis y selección de qanats [Draining galleries of Spain: analysis and selection of qanats]. Ministerio de Medio Ambiente, Rural y Marino. Madrid . pp. 1–269.

Hermosilla, J., Iranzo, E., Pérez Cueva, A., Antequera, M., Fernández, J.A. & Aguilar, P. (2004) Las galerías drenantes de la Provincia de Almería : análisis y clasificación tipológica [Draining galleries of the Almeria Province: analysis and typological classification]. *Cuadernos de Geografía. Valencia*, 125–154.

Hernández Ramos, J. (1954) Las heredades de agua en Gran Canaria [Water heritages in Gran Canaria]. *Impr. Sáez, Buen Suceso. Madrid*, 1–106.

Jordà, R. (1875) Reseña histórica sobre el modo de adquirir la propiedad de las aguas subterráneas en la comarca del Campo de Tarragona, desde los tiempos antiguos hasta nuestros días [Historical note on the form to get groundwater property in the country of Campo de Tarragona from old times until present]. *Revista Minera, Madrid*, XXVI, 109–123.

Juncà, J.A. (1993) L'enginyeria subterrània a l'època romana [Underground engineering in Roman times]. *Espais, Barcelona*. Part 1: 1993(37): 53–56; Part 2: 1994(38): 52–56.

Limón Montero, A. (1697) Espejo cristalino de las aguas de España, hermoseado y guarnecido con el marco de la variedad de fuentes y baños, cuyas virtudes, excelencias, y propiedades se examinan, disputan y acomodan a la salud, provecho, y conveniencias de la vida humana [Cristaline mirror of waters in Spain, made pritty and accompanied by the framework of springs and baths variety, whose virtudes, excellencies, and properties are examined, concurred and accomodated to health, profit and conveniencies of human life]. Impr. Francisco García Fernández. Alcalá de Henares: 1–432. Facsimil publ. by the Instituto Geológico y Minero de España, Madrid, in 1979.

Llamas, M.R. (1968) Los estudios regionales de recursos hídricos totales [Regional studies of regional total water resources]. *Bol. Inform. Ministerio de Obras Públicas. Madrid*, 123, 17–23.

Llamas, M.R. (1983) Las aguas subterráneas de Madrid y la política hidráulica española [Groundwater in Madrid and water policy in Spain]. *Estudios Territoriales. Madrid*, 10, 113–130.

Llamas, M.R. (1994) La influencia del reducido o impropio uso de las aguas subterráneas de Madrid en la política del agua española [Influence of the small and improper use of groundwater in Madrid in water policy in Spain]. *Bol. Real Acad. Cienc. Exc., Fis. y Nat. Madrid*, LXXXVIII (4), 91–125.

Llamas, M.R. (2005) La revolución silenciosa de uso del agua subterránea y los conflictos hídricos en España [Silent revolution of groundwater use and water conflicts in Spain]. Libro Homenaje al Profesor D. Rafael Fernández Rubio. Instituto Geológico y Minero de España. Madrid. pp. 79–86.

Llamas, M.R. & Custodio, E. (2003) *Intensive Use of Groundwater: Challenges and Opportunities*. Dordrecht, Balkema. pp. 1–478.

López–Camacho, B., Bascones, M. & Bustamante, J. (1986) El antiguo abastecimiento a la Villa y Corte: los viajes de agua [The old supply to Madrid: the water mines]. *Bol. Inform. y Estudios 46*. Servicio Geológico de Obras Públicas. Madrid.

López–Camacho, B. & Cabrera, E. (1993) Investigación de aguas subterráneas en el acuífero detrítico de Madrid [Groundwater investigation in the Madrid's detritic aquifer]. *Tecnoambiente, Madrid*, October. pp. 61–65.

López–Camacho, B. & Iglesias, J.A. (2000) Las aguas subterráneas en los abastecimientos, un decenio de experiencias del Canal de Isabel II [Groundwater in supply, a ten years of experience in the Canal de Isabel II]. *Revista de Obras Públicas, Madrid*, 3403, 16, 41–56.

López Geta, J.A. (2000) Contribuciones del Instituto al conocimiento y protección de las aguas subterráneas en España [Contributions of the Institute to the knowledge and protection of groundwater in Spain]. Estudio e Investigación en las Ciencias de la Tierra. Instituto Geológico y Minero de España, Madrid. pp. 199–233.

Macau, F. (1957) Estudio hidrológico de Gran Canaria [Hydrological study of Gran Canaria]. *Anuario de Estudios Atlánticos. Madrid–Las Palmas*, 3, 9–46.

Mallada, L. (1890) Los males de la patria [The evils of motherland]. Impr. José Esteban. Madrid. Reedited 1990, Fundación Banco Exterior, Madrid.

Mallada, L. (1926) Aguas y pozos en los barrios bajos de Madrid [Water and wells in Madrid's low quarters]. *Bol. Com. Mapa Geol. España*, XXVIII. Madrid.

Manzano, R. (1995) El agua en la antigüedad púnica y romana [Water in Punic and Roman old times]. El Hombre y el Agua en la Geografía y la Historia de España. Ed. FCC, Madrid. pp. 37–63.

Martín Municio, A. (2000) Sesquicentenerio del Instituto Geominero de España [150 years of the Geo–mining Institute of Spain]. Estudio e Investigación en las Ciencias de la Tierra. Instituto Geológico y Minero de España, Madrid. pp. 37–48.

Martínez Gil, J. (1990) Historia de la hidrogeología española [History of hydrogeology in Spain]. Historia de la Ciencia. *Real Acad. Cienc. Exac., Fís. y Nat., Madrid*, 197–238.

Martínez Gil, J. (1991) Historia de la hidrogeología española [History of hydrogeology in Spain]. Hidrogeología, Estado Actual y Prospectiva (Ed. Anguita, Aparicio, Candela & Zurbano). Curso Internacional de Hidrología Subterránea. Barcelona, CIMNE/UPC. 391–418.

Martínez Gil, J. (1994) Don Juan Vilanova y Piera: su persona y su obra hidrogeológica [Don Juan Vilanova y Piera: the person and his hydrogeological work]. *Rev. Real. Acad. Cienc. Exac., Fís. y Nat. (Esp). Madrid*, LXXXVIII (1), 19–46.

Mesa Ramos, J. (1909) *Pozos artesianos [Artesian wells]*. Madrid, Estud. Tip. de V. Tordesillas, 5 ed. pp. 1–238.

Mesa Ramos, J. (1929) *Pozos artesianos y pozos de petróleo [Artesian wells and oil wells]*. Madrid, Librería de Roma.

Montaner, M.E. (1982) *Norias, aceñas, artes y ceñiles en las vegas murcianas del Segura y Campo de Cartagena [Water winning devices in the Murcian meadows of the Segura and Campo de Cartagena]*. Murcia, Editorial Regional de Murcia.

Moragas, G. (1896) Estudio general del régimen de las aguas contenidas en terrenos permeables que ejercen los alumbramientos por galerías y pozos y especial del régimen de la corriente subterránea en el delta acuífero del Besós (Barcelona) [General study of the regime of water in permeable terreins due to abstraction by means of galleries and wells, and especial of the groundwater flow in the Besós delta aquifer (Barcelona)]. *Anales Rev. Obras Públicas. Madrid*, 1–133.

Murcia, A. (1953) Aguas subterráneas: prospección y alumbramiento para riegos [Groundwater: prospecting and winning for irrigation]. Ed. Ministrio de Agricultura. Manuales Técnicos 51 no. 18. Madrid. pp. 1–360.

Oliver Asin, J. (1959) *Historia del nombre de Madrid [History of the name of Madrid]*. Madrid, Consejo Superior de Investigaciones Científicas.

Pendás, F. (1994) La enseñanza de la hidrogeología en las escuelas de minas de España [Teaching of hydrogeology in mining schools in Spain]. *Rev. Real Acad. Cienc. Exac. Fis. y Nat. (Esp). Madrid*, LXXXVIII (1), 47–72.

Puche, O. (1996) Historia de la hidrogeología y de los sondeos de agua en España y en el Mundo, desde sus orígenes hasta finales del Siglo XIX [History of hydrogeology and water boreholes in Spain and the World, from the origns until the end of the 19th Century]. *Bol. Geol. Minero, Madrid*, 107 (2), 180–200.

Pulido-Bosch, A. & Ben Sbih, Y. (1995) Centuries of artificial recharge on the southern edge of the Sierra Nevada (Granada, Spain) *Environmental Geology*, 26, 57–63.

Roigè i Badia, R. (1895) Los pous artesians de Prat de Llobregat [Artesian wells of Prat de Llobregat]. La Pagesia, Institut Agrícola Català de Sant Isidre. Barcelona.

Rubio, P.M. (1853) Tratado completo de las fuentes minerales de España [Complete treatise of mineral springs in Spain]. *Est. de la P.R. de Ribera, Madrid*. pp. 1–740.

San Martín, J. (1994) La hidrología médica en España [Medical hydrology in Spain]. *Rev. Real Acad. Cienc. Exac., Fis. y Nat. (Esp). Madrid*, LXXXVIII, 85–90.

Sánchez, E., Iglesias, J.A., Cabrera, E., Muñoz, A. & López-Camacho, B. (2005) Los campos de pozos del Canal de Isabel II: construcción, equipamiento y resultados [Wellsfields of the Canal de Isabel II: construction, equipment and results]. Libro Homenaje al Profesor D. Rafael Fernández Rubio. Instituto Geológico y Minero de España, Madrid. pp. 369–385.

Santa María, L. & Marín, A. (1910) Estudios hidrológicos en la cuenca del río Llobregat [Hydrological studies in the Llobregat river basin]. *Bol. Com. Mapa Geológico de España, Madrid*, 31–52.

Soler Liceras, C. (2007) La historia de la Fuente Santa [The history of Fuente Santa]. Ed. Turquesa. Santa Cruz de Tenerife. pp. 1–431.

Viladrich, M. (2004) Tecnologia hidràulica agrícola i preindustrial de la societat andalusina a l'àrea dels Països Catalans [Agricultural and the pre–industrial water technology of the Andalusian Society in the Catalan Countries area]. In: Venet, J. & Parés, R. (eds.) La Ciència en la Història dels Països Catalans. Barcelona, Universitat de València/Institut d'Estudis Catalans. Vol. I, 575–595.

Vilanova y Piera, J. (1880) Teoría y práctica de pozos artesianos y arte de alumbrar aguas [Theory and practice of artesian wells and the art of extracting water]. *Impr. M. Tello, Madrid*, 1–593.

Villarroya, F.J. (1994) Historia reciente de la hidrogeología en España: el papel de la Universidad [Recent history of hydrogeology in Spain: role of the University]. *Rev. Real Acad. Cienc. Exac., Fís. y Nat. (Spain), Madrid*, 88 (1), 149–159.

Hydrogeology in Sweden

Chester Svensson

Department of Geology, Chalmers University of Technology, Gothenburg, Sweden

INTRODUCTION

The Swedish word *grundvatten* (groundwater) was used for the first time in a book by Urban Hiärne (1702, p. 52). In fact, Swedish hydrogeology was more or less started by him in 1679, when he reported on the first mineral spring water in Sweden (Hiärne, 1679). At that time, there were many spas in Central Europe and the Swedes eagerly searched for a good spring within the country. Occasionally, a farmer found a peculiar spring, at Medevi in the county of Östergötland, whence the land owner, Councillor Gustaf Soop, sent two bottles of water to the chemists in Stockholm. Urban Hiärne had to take care of them, analyze the water and then go to Medevi. He had found many indices for a promising mineral water. During the inspections in Medevi his thoughts were confirmed and very rapidly the owner arranged for the establishment of a spa. This spa became a popular resort for prominent people in Sweden, including royalty. This was the start of the field of groundwater chemistry, which since then has been one of the main themes of Swedish hydrogeology. Other aspects of hydrogeology started much later, such as groundwater prospecting and mapping, artificial recharge, and studies of frost in soil.

URBAN HIÄRNE (1641–1724)

Hiärne was born in Ingermanland, in Finland, close to St. Petersburg (at that time Finland was under Swedish government). He was the son of a priest from a Swedish family. His father died when Urban was 12 years old Three years later, a war started and he went to Sweden and searched for help from his father's relatives and friends. He managed to survive and to become a student in medicine in Uppsala. He was full of enterprise and had several wealthy friends.

After he obtained his medical degree, he became assistant to the Swedish Colonel in Riga, Colonel Stahl, who travelled widely, giving Hiärne the opportunity to visit Dortmund, Aachen and Spa during four months in 1667. Later, Hiärne became a physician at the Swedish embassy in Poland and friend of a priest on duty there. This priest persuaded Hiärne to follow him to England 1669. Hiärne obtained a letter of introduction, which opened the doors to "Doctor Willis, Doctor Charleton, Doctor Celadon" and several others. He became a member of the Royal Society in November 1669 and visited the meetings every Wednesday. The following year, in spring, Governor-general

Count Claes Tott ordered him to go to Paris for some months. These months lasted for almost three years and Hiärne had a very pleasant and instructive time there. He was eager to learn French but the Swedish social intercourse was too intensive so he had to make a journey. During this he came to Angers and at the university there was persuaded to make his dissertation in medicine, in 1670. Back in Paris he studied chemistry eagerly, taught by Cristopher Glaser at the Jardin des Plantes. He learned how to make chemical experiments and took part in scientific conferences (Hiärne, 1916; Olsson, 1971).

Back in Stockholm, Hiärne became a practising doctor and chemist. During his visits to Central Europe, he became knowledgeable about mineral waters and he was thinking a lot about water and its characteristics. Though he was fond of Paracelsius, he had an open mind and tried to form his own opinions. He worked out different analytical methods which he used at several springs, but he never got the results he wanted – every water lacked some ingredient. At that time the bottles of mineral water arrived from Medevi. He analyzed the water, was satisfied, and reported to the owner Gustaf Soop. Councillor Soop compelled him to go to Medevi and on July 25 1678 he made field analyzes there. He found *acidum universale, sulphur martis,* some iron vitriol, and a little alum. He declared the water to be a subtle and complete mineral water (Hiärne, 1679; 1916).

Medevi became very popular and others tried to find their own waters. Two young doctors were very eager and announced loudly that they had found their own mineral waters close to Stockholm. Hiärne knew that their accounts were false, but had no evidence to contradict them. Again he went to Central Europe and visited many different spas and mines. He used his equipment and analyzed waters everywhere. He also met many scientists and learned even more about chemistry and waters. At home again he wrote a book – *Den lilla wattuprofvaren* ("The small water tester"; Hiärne, 1683) – in which he listed all of the springs around Stockholm and in some other places, with all their characteristics. The two young doctors were exposed and dishonoured due to the new published facts.

In the same year, Hiärne was appointed head of a newly founded governmental *Laboratorium chemicum* in Stockholm and in 1684 he also became royal physician-in-ordinary. He directed the laboratory as long as his strength allowed him to, until 1720. He made a lot of good work for the king, the army (pharmacology) and the mining industry. In 1694 he wrote an inquiry to be answered by anybody. The inquiry contained questions about Swedish natural conditions of different specified and grouped types. He promised a very high award for a hot spring – but nobody got that award as the Swedish ground does not contain any volcanic area. The groundwater temperature is close to the year mean air temperature. He obtained many answers and started to compile them "using many sleepless nights". He managed to publish two volumes, in 1702 and 1706, containing his thoughts about different topics and exemplified with the answers. The first 60 pages discussed water – surface water and groundwater – from different aspects. He used the Swedish term for groundwater (grundvatten) for the first time herein. He also presented his view of the earth – a figure being a mixture of the ideas presented by Paracelsus, van Helmont, Becher and Kirchner (Frängsmyr, 1969) (Figure 1).

As most people at that time, Hiärne believed in the four elements earth, fire, air and water and that the transition water-earth was a reality. He worked a lot in laboratories

Figure 1 The Earth by Hiärne 1702. A = hot, thick, fatty and muddy water; B = roof of the central water chamber A; C = A-water flowing from the sea; D = porous and spongy earth; E = subterranean hollows filled with water and fine sand; F = *linea trivialis*. Above this line the vapours are free; G = solid earth, minerals and metals; H = the mountains; I = the sea; II = Pacific ocean; 12 = Arctic sea; 13 = Antarctic sea; K = volcanoes (Hiärne, 1702, p. 50).

and aimed to publish his results in chemistry in an acta-series, but he and most of his late works are still concepts in the archives. Only one issue was published, in 1712.

THE FIRST PROFESSORSHIPS IN CHEMISTRY

During the following decades there were some people handling water in different ways. In 1737–46, Nils Wallerius (1706–1764) made many experiments on evaporation. He found that evaporation was proportional to the area of exposed water surface, not to the volume of water (Wallerius, 1746, p. 13). However, he did not follow up on his research in water science – he obtained a professorship in logic and later in theology (SBL).

His younger brother Johan Gotschalk Wallerius (1709–1785) wrote a book on liquids called "*Hydrologia*", in Swedish and later in German and French (Wallerius, 1748). This book is written in the spirit of the German philosopher Christian Wolff and is a classification of all liquids, named with Latin nomenclature. He defines water as "a composite body constantly decreasing". The scientists knew that water consisted of small round or oblong mucous particles with something in between. The water could go into other things and it could evaporate, and in nature, water was a constantly diminishing (due to land upheaval, not verified yet). He used many different literature sources in this book: Aristotle, Pliny the Elder, the Bible and more than 60 books and papers published between 1547 and 1747. There is also a manual for water analyses in the book. At that time there seem to have been standardised reagents: Wallerius used distilled water, nitric acid, lead acetate, sal-ammoniac, gall apples, leaves of oak, tea, egg white, galmeja, vitriol, tournesol, auripigment, silver, brighty iron, urine, and several other substances.

Wallerius obtained the first Swedish professorship in chemistry in 1751 in Uppsala. He built up a laboratory and had numerous students defending many dissertations on water, soil and chemistry (most of them in Latin). The dissertations were written by Wallerius (the rule at that time) and many of them contain a defence for the transformation water-earth. He wrote a famous book on agricultural chemistry (Agriculturae Fundamenta chemica/Chemical Foundations of Agriculture, 1761 and 1778), but in all scientific writings he always referred to his religious faith. The chemists' methods where rather rough and they had difficulties when they vaporized water. They were left with small amounts of powder even from totally clear waters and Wallerius never abandoned his belief in the water-earth-transformation.

Wallerius had to leave the professorship in1764 due to health problems and was followed by Torbern Bergman (1735–1784). He was more open-minded than his predecessor and there were several disputes about several chemical matters between Wallerius and him. Bergman became a popular teacher and a famous scientist – he had as many international contacts as Linnaeus or even more. Bergman had some problems with his health too, and used a lot of mineral waters, mostly bottled water imported from Germany. He thought that it must be possible to make artificial waters as good as those imported and much cheaper. He experimented a lot and realised how to put carbon dioxide into the water and thus invented artificial bottled mineral water. Bergman was a close friend of Carl Wilhelm von Scheele, a pharmaceutical chemist, who in 1772 discovered oxygen, in parallel with Priestly, and later also chlorine. Bergman and von Scheele acted as a couple of tracker dogs in the world of chemistry – von Scheele first and Bergman one step after (Lindroth, 1967, SBL).

In 1779, one of the best known Swedish scientists was born, Jöns Jacob Berzelius (1779–1848). He studied medicine and incidentally in 1799 had the opportunity to try to analyse waters from Medevi. The equipment and methods used by Bergman, who had carried out some analyses on the Medevi water in 1778 were out of date, so Berzelius had to use his initiative and judgement. He presented his results together with Bergmans', without special remarks (Berzelius, 1800 in Trofast, 1992, SBL).

When Berzelius published the first volume of his Swedish chemistry textbook *Lärobok i chemien* in 1808, he was of the following opinion regarding the existence of chemical elements in groundwater: "Along its path, the water in the veins, in the fissures in the rock and in the depths of the Earth, dissolves numerous elements. The water then appears at the surface, influenced by these elements. Such elements take the form of silica soils and vaious salts and acids, along with extracts from the organic soil in the outermost layer of the earth' crust. How the water is supplied with all these, has yet to be elucidated, since quite a number of springs contain so much of certain elements, that the areas surrounding the veins could not possibly supply the water to them for more than a couple of months. ... These waters probably flow down into rock, the inner mass of which has been gradually disintegrated and dissolved by the water, whereupon the water consequently starts a chemical process, thus providing it with these elements; sometimes in such quantities that it would scarcely possible even in the world of art to reproduce it. The content of dissolved elements in the waters thus will vary considerably because of the different conditions and over long periods of time, perhaps several centuries, these springs need to change their content, to the extentsuch dissolvable elements have been removed, withvarying amounts still remaining to be dissolved." (Berzelius, 1808, pp. 217–218). Berzelius knew that chemical processes in the ground are slow, and could not sufficiently explain the high contents of several elements in the spring waters.

The methods for chemical analyses were successively improved during the following decades. Wimmerstedt (1866), who also analyzed the Medevi waters, obtained the same results as Berzelius, but with a little more precision.

WATER FOR THE CITIES, 1860s

In the middle of the nineteenth century the cities in Sweden were still rather small, but growing. In most places, local wells were used within the urban areas. In Stockholm several rather deep wide wells in the esker Brunkebergsåsen were maintained by the authorities since at least A.D. 1500. The cities needed water for fire-fighting and better drinking water. Epidemics due to poor quality water and bad hygiene were problems in Sweden as well as in the countries in central Europe. During 1866–67 N. P. Hamberg made an examination of 27 community wells in Stockholm on behalf of the public health committee. He found that the water from all the wells contained such large amounts of calcareous salts that they must be called hard waters. He also found a lot of other elements and organic matter. In order to grade the wells, he listed them according to their content of organic and mineral matters (by weight). The best water, from the well at Österlånggatan in the old city, contained 3.64 organic and 66.81 mineral parts per 100 000. The worst water, from awell at Pilgatan in the Södermalm district, contained 74.73 and 364.47 parts per 100 000 respectively. His conclusions were that all but one water contained too much organic matter, half of them were too hard, and all contained too much nitric and sulphuric acids (Hamberg, 1868). People used the bare surface in the city for anything and the dung from all horses had to be laid somewhere too. Though this was the situation in our largest city, it was probably representative for most densely populated areas at that time.

Some cities had organised water pipes from springs or lakes/rivers at some distance. In Malmö they used a water pipe from a small river, from 1582. Göteborg engaged a Scottish engineer, Andrew Blackwood, to build a water pipe from a spring at Kallebäck, 5 km away, to two central water taps in 1786. On the slopes of the sedimentary mountain of Billingen, some springs were used for a pipe to Skövde around 1850. Also some other cities started using springs for their first waterworks: Uppsala (1875), Askersund (1876), Örebro (1886) and Falköping (1889) (Hansen, 1903).

THE FIRST WATERWORKS SUPPLIED WITH GROUNDWATER

The springs in Sweden are normally rather small, and subsequently very few could be used for more densely populated areas. The first city to make thorough investigations for its water supply was Malmö. Malmö had tried to enlarge its surface water facilities two times, but the water was still not sufficient. The borough finance department started in 1882 to try to find a solution. Five years later, investigations of groundwater conditions started near the city, supervised by Colonel J. Gabriel Richert, in service in Göteborg. From 1890 the investigations were supervised by the civil engineer Adolf Thiem, assisted by Johan Gustaf Richert (son of J. Gabriel and he called himself J. Gust. in all papers), also employed in Göteborg.

The initial investigations were made by the geologist Jöns Jönsson in 1888. He collected data about more than 100 deep wells on the farms east of Malmö. The wells were mostly drilled through till and boulder clay down to the limestone. Between the till and the boulder clay was normally a layer of sand. All wells gave considerable quantities of water from the sand layer and some water from the limestone surface. The normal depth was 30 m; the deepest 90 m and the shallowest 12 m. The wells were artesian and a new well often had an influence on wells in a certain direction and within a certain distance Jönsson, 1889.

This compilation of data was followed by three well borings, 38–88 m deep, and all three were artesian. In 1890 "Germany's most prominent specialist within the groundwater branch, Baurat Adolf Thiem from Leipzig", was sent for as an expert engineer. This was prepared for by Professor Otto Torell, in Lund, during a personal visit to Leipzig in 1888 (Torell, 1889). Thiem inspected the site and had the opinion: "that the flow direction of the groundwater seems to follow the hollow in the limestone from Romele Klint to Lomma, that the investigation area ought to be chosen close to the discharge of the flow, e.g. in close vicinity and north of the city, the borings ought to start at Bulltofta and continue northwards" (Richert, 1893, p. 4). During the project, Thiem gave more instructions and wells were mainly drilled at two locations, not far from the city: 4 km E and 8 km NE of Malmö. The first was in an area with flat limestone under the till. The latter with tills over a thick layer of sand. From the results and the investigations of Jönsson, the city's chief engineer proposed that wells should be bored further east at Torreberga, in an area with the same geology as those 8 km NE and containing the same groundwater stream. The limestone surface in this area contains a valley, in SE-NW direction, about 5 km wide and at least 50 m deep, as shown in the cross section in Figure 2. This valley, invisible at ground level, is a graben called the Alnarp Valley. While drilling through the 30–50 m of preglacial fluvial sand, pieces of amber were found – thus people often call it the Amber River. The groundwater was rather salty, containing 26 up to 236 mg Cl/l (Ramberg, 1912). In Figure 3 is a profile with the two well areas and alternative performance. This pioneering groundwater investigation project was finished in 1901 using 13 wells at Torreberga (Richert, 1898, 1911). The result profile after 70 years of extraction from the aquifer is shown in Figure 4.

The aquifer in the Alnarp Valley and its watershed is used for many municipalities. This is probably the most examined and utilized aquifer in Sweden. In 1968 there were 16 more municipality small well fields here, two factory wells and some farmers using 7 Mm^3/year (Vattenbyggnadsbyrån, 1969). Occasionally, the utilisation is lowering due to chemistry problems (Figure 5).

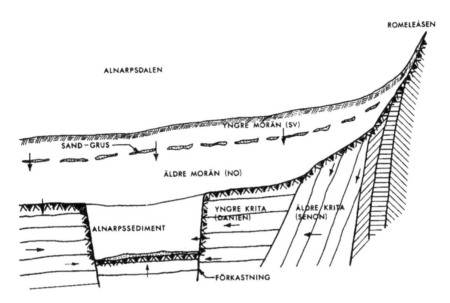

Figure 2 Cross section of theAlnarp valley, 15 km E. Malmö (Sweden), water source for the waterworks of Malmö. 10–20 m of till and boulder clay on river sand from Tertiary. Rock surface of limestone. (Mohrén in Vattenbyggnadsbyrån 1969) Yngre morän =Younger till; Sand-grus = Sand-gravel; Äldre morän = Older till; Förkastning = Fault.

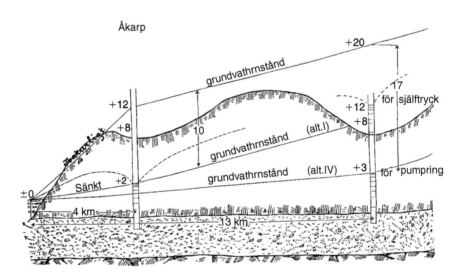

Figure 3 A profile outline along theAlnarpValley (SE-NW), Sweden, with two well areas and alternative performance – withdrawal atVinninge (=Torreberga=Grevie), 15 km from Malmö (right), or at the end, Åkarp 8 km from Malmö. grundvattenstånd = ground water level; naturligt = natural, sänkt = lowered; för själftryck = when used at natural pressure; för pumpning = during pumping (Richert, 1898) Cf. Figure 5.

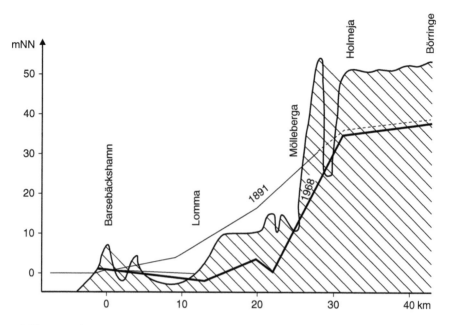

Figure 4 The groundwater levels in the Alnarp Valley (Sweden) after 70 years of withdrawal (Leander, 1971).

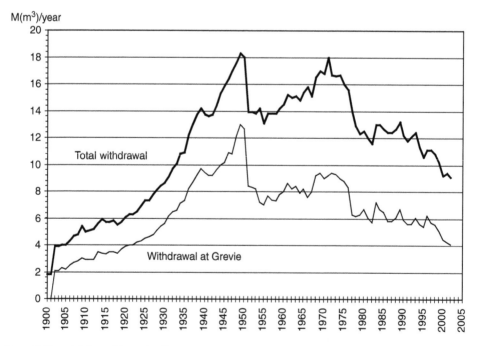

Figure 5 The withdrawal from the Alnarp groundwater flow in SW Skåne, southern Sweden. At Grevie are the wells supplying city of Malmö. Several smaller municipalities, factories and farmers have their wells within the watershed too. At Grevie the withdrawal is lowering due to some difficulties in mixing this water with the water from the main waterworks, situated in another watershed with different geology (Source: SWECO Viak AB, Malmö).

J. GUST. RICHERT IN MALMÖ AND GÖTEBORG

In Malmö, J. Gust. Richert fulfilled the work of Adolf Thiem. Richert was examined as civil engineer from the Royal Institute of Technology in Stockholm 1884. The year after, he got his second employment at the waterworks in Göteborg. His father, Colonel J. Gabriel Richert, was the director and gave his son instructive and responsible mandates. The main problems were water supply and wastewater. Göteborg took its water from two lakes and partly from a spring. The growing population needed more water than the discharge from the lakes. J. Gust had the opportunity to make a study tour to Germany and Austria in 1886, where he studied water supply and wastewater facilities in several cities. He made a second study tour to Germany in 1889. At some time during the tours he must have become acquainted with Thiem, whom he later calls "my friend and teacher". The travel book after the second tour was written as a textbook on groundwater occurrence and withdrawal (Richert, 1891).

In Göteborg the situation was going from bad to worse – the city needed more water. Richert's office suggested that the water supply would be better with an intake of water from the river Göta Älv, a little north of the city, and then filtered and purified. The problems were the clay particles in the river water and the risk of incoming sea water during gale occasions. A committee was set up, which called for a German specialist, Baurat B. Salbach from Dresden. Salbach came to Göteborg and was presented with the plans. They looked fine, but during the visit Salbach had looked at the geological environments of Göteborg and noted gravelly hills on the valley slopes, especially at the valley mouths. Probably they looked like remnants of old river beds, rather common in Germany, and at many places used for groundwater withdrawal. Why not try to make some wells in the valley upstream of the planned water intake? He worked out a plan for the borings and went home again. The city handled this quickly and by the next spring these inspection wells were completed. They gave unexpected success: the borings "showed, that groundwater flow was found, but the nature of the materials in the ground were unfavourable for making wells in, and the composition of waters extremely variable" (Richert, 1891).

However, one of the wells was promising and Salbach suggested that a well should be bored close to Lärjeholm, but on the other side of the river. Meantime, Richert had made some inspection wells at Lärjeholm. Water samples were taken and analysed by the city's chemical officer. He found considerable amounts of ammonium and 410 mg Cl/liter and the contents seemed to be increase with time. Ammonium and salt were at that time considered as signs of urine and bacteria. Richert reported Oct. 11th 1890 to the City Council: "Technically, the nice idea of Salbach has all possibilities for success, but unfortunately the analyses of the characters of the groundwater quality have not been as favourable as the borings." The project seemed to fail.

Four years later, Richert wrote a new memorandum to the City Council about groundwater investigations. "During the last years …many investigations have been carried out and the results can be said to be surprising. Under old graveyards and cities, which during centuries had been filled with impurities, sterile waters are found at just a few meters depth, and an almost universal rule has been set up, that bacteria do not go deeper than 4 to 5 meters" (Richert, 1895).

In a speech in 1897 Richert said that at Lärjeholm, they had several very good artesian wells, with a piezometric level several meters above the surface of the river, and they were powerfully flowing constantly. There was an interesting observation made before the well borings, people had found groundwater while digging in a sand pit in the hill of glacial sand at Lärjeholm (one of those hills seen by Salbach). The water table was constantly at the same level as that of the artesian wells, but when the wells were open for owerflow the water table in the quarry sunk. That means that there must be a direct connection between the sand pit and the sand layer under the clay at the riverside (Richert, 1897).

In a report in 1895 to the City Council, Richert explained the favourable situation at Lärjeholm: The big sand pit is one of the points where groundwater has its natural water inflow. The amount of infiltrated water is depending partly on the sand area and partly on the amount of precipitation. The net precipitation available for infiltration is, at the highest, 500 mm/year. This is not more than the sand is able to let through during a day or less. The yearly capacity can be multiplied by a hundred if water is let in artificially. If water from the river is poured in the sand pit, the pit will be transformed to an excellent filter. The filtration bed in an artificial filter is only 1–2 m, passed through during a day, but here the water has to go through a sand layer several hundreds meters long. It will take at least a month for a water particle to move from the sand pit to the well. During this slow filtration, the water will get rid of bacteria and all organic matter – and the temperature will be modified and the water will in most aspects be as a groundwater. Probably the water will not be able to take up so much salt and ammonia as the groundwater in situ contains. This is very like the situation for the filtration plant in Chemnitz, made by Adolf Thiem – and his proposal for Strahlsund – but in much better conditions.

In order to be allowed to try this, he said that the planned filtration of the river water in a filtration plant would cost a lot of money every year. If the Lärjeholm sand pit were used as a natural filter, the savings will be higher than the costs for the new proposed well field.

He made a full scale experiment during two months in 1896 and found that a $65\,m^2$ basin in the sand pit could take care of $1300\,m^3$ per day continuously. After this he could set out the wells on a line 200 m from the sand pit (Richert, 1897). The waterworks weres complete in 1901 and used up to the 1960s (Figure 6).

Richert learned a lot during his study tours in Europe, and even more during the complicated work with the waterworks in Malmö and Göteborg. His reputation soon gave him more to do. Though he had his position in Göteborg he got commissions from several cities. This lead him to start a firm on his own in 1897 and he moved to Stockholm. In 1902 Richert and his engineer employees together founded Vattenbyggnadsbyrån AB (VBB), the first consulting firm in civil engineering in Sweden. Richert obtained a professorship at the Royal Institute of Technology in 1903 which he gave up 1909 as the firm took all his time – there are many water power stations planned by Richert. He was executive at VBB up to 1925 when he retired. Richert was also elected city councillor in Stockholm and member of parliament.

Richert was very good at writing, and published a lot of articles, reports and a couple of books about groundwater and artificial replenishment. His report from the second study tour and a book about groundwater in Sweden became widely used (Richert, 1891, 1911). He managed to have articles published in several languages – Swedish, English, German, French and also Russian (one) (Richert, 1900, 1902).

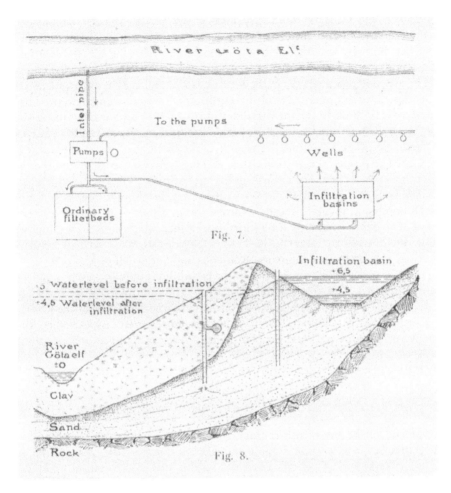

Figure 6 Plan and section of the waterworks at Lärjeholm in Göteborg, called Alelyckan, Sweden. The water is pumped from the river, and poured into the two infiltration basins in the sand pit (Richert, 1900).

DRILLING EQUIPMENT

The well boring equipment in earlier times was more or less man-driven and heavy.

A blacksmith started on his own a drilling company in 1866 in Skåne Hult 1991. In 1880 he was boring a first well using his invented tools. His undertakings grew successively and following generations took over. A hundred years later, the firm was the most capable in drilling wells in sedimentary rock and deep earth layers, the Malmbergs i Yngsjö AB. Many fairly deep and wide diameter wells were installed by this firm for municipalities and farmers in areas with sedimentary rocks, especially in Skåne, and in areas with deep glaciofluvial sediments.

In 1886 a firm was founded by two mining engineers, Per Anton Cælius (1854–1905) and Gustaf Abraham Granström (1851–1941), called Svenska Diamantbergborrningsaktiebolaget. Their customers were principally from the mining industry, but new facilities gave new opportunities. During the 1880s Adolf Erik Nordenskiöld, the famous Arctic researcher, proposed that the new boring equipment should be used for making wells in the hard bedrock. During a visit to Spitsbergen, he had the idea that the upper parts of the Earths crust are constantly exposed to great temperature variations during the year. For this reason the rock mass has to expand and shrink. The rock must crack and a horizontal displacement fissure ought to be found at some depth, e.g. 30 m. He initiated a meeting of the Geological Society of Stockholm in January 1891 to discuss the possibilities of making wells on the island in order to give the aircraftsbetter service. The discussion led to the first well in 1891 – which gave sea water. The next well, drilled in 1894, succeeded and in 1896 Nordenskiöld reported about 28 boreholes, giving 27 wells (Nordenskiöld, 1896). The idea of a horizontal fissure at 30 m depth lived only for a short time: joints and fissures do not follow such a simple rule. During the years 1894–1955, many thousands of wells were drilled in Sweden and abroad by this company (Rothelius and Sund, 1955).

In the 1960s, scientists were aware of the possibilities of making many wells in the crystalline bedrock. The growing microelectronic capabilities made several engineers think about well logging. The first usable instrument was developed at the Royal Institute of Technology, Stockholm. Depth, diameter and resistivity measured continually were new variables to take into account when analysing groundwater occurrence in the bedrock (Houtkamp and Jacks, 1972).

FROM IRON TO RADIUM

Up to the end of the nineteenth century, the interest in mineral waters was focused on the content of iron, carbon dioxide and sulphur compounds. At that time, Marie Curie and others worked on radioactivity and this was also found in many waters. Some results were published in Nature in 1903 and 1904 by some English researchers. This caused Professor Hjalmar Sjögren to let his student Naima Sahlbom to study this phenomenon. During 1905 Sahlbom analyzed water from, in the beginning, iron springs, and had very varying results even from springs close together. Next summer she examined the water from many springs in the south and the middle of Sweden, to a large extent sponsored by the spas. The English results, indicating that the gas was more radioactive than the water itself, were confirmed, especially in the water from an 83 m deep well in sandstones from the Lower Jurassic and Upper Trias in Hälsingborg. Sahlbom used methods and a new instrument from Germany ("Engler's fontaktoskop"). The first results, based on waters from 59 springs and wells, showed that there was no relation between the radioactive emanation and the chemical content or the well depth. But there was some relation between radioactivity and geology. The highest radioactive emanation were from waters from wells in granite (no data from other types of crystalline bedrock) and from the sandstone below Hälsingborg. After some laboratory experiments, she drew the conclusion that the emanation was accumulated in the water during its flow through radioactive bedrock and it was most related to radium. Surface waters from lakes had no radioactive emanation (Sjögren and Sahlbom, 1907). Some years later, Sahlbom reported that the radioactivity was highest in acid rocks (granite, syenite, pegmatite), i.e. those which can contain accessory radium minerals. Waters from metamorphosed sedimentary rocks and mafic rocks had low emanations (Sahlbom, 1915).

Naima Sahlbom (1871–1957) studied geology in Uppsala and was recommended to Professor Alexander Classen in Aachen, were she studied radioactive methods. Further, she studied in Basel and Neuchâtel and obtained her PhD in Neuchâtel in 1910. Sahlbom became a skilful analytical chemist and started an analytic laboratory of her own in 1914. She made advanced analyses on waters, minerals, and rocks, especially on radioactivity, on behalf of many scientists during many years. Besides this, she became famous because of her great engagement in the Swedish part of Women's International League for Peace and Freedom (SBL).

PROPERTIES OF CLAYS

The agricultural aspects of soils were treated in 1760–1785 by Wallerius, but then little was done until a hundred years later. In 1877, Albert Atterberg (1846–1916) became manager of a State-owned chemical and seed control office established in Kalmar. He analysed the relationship between the nutritional content of crops and the chemistry of the soil. He gradually became more aware of the need to study the soil particles not only chemically but also physically and more in detail. He started with sands and published in 1903 his results on permeability and capillarity. By systematic elutriation, he partitioned the sands into fractions, in his now well known 2-system (Atterberg 1903). Thereafter, he attended to the clays, to try to organise a system for them also. The elutriation was impractical and time consuming so he tried to find a faster and simpler method for the description of the physical qualities of soils. He thought that a method used by Bischof 1904 was the only one to be used. He modified the method and the water content limits for clays. He summed up that the correct limits for consistency are the plastic limit and the liquid limit, the difference between these water contents is the best value for the rate of consistency. He also described the shrinkage limit, a measure of the possible volume change in a wetted soil (Atterberg, 1911). We are still using these parameters today.

FROST IN SOIL AND FROST HEAVING

Simon Johansson (1881–1944), for some time assistant to Atterberg, studied water movement in the laboratory and in the field, becoming a skilful scientist. His most essential discovery was early, in his freezing experiments on silt – the water flows to the frost front and accumulates within the frozen zone – reported in his dissertation (Johansson, 1914). In Swedish literature, Urban Hiärne had questions about frost in his inquiry and said "In fields where boulders have been picked out and the ground cleared, more boulders will be generated" (Hiärne, 1706). The next frost related text is by Ephraim Otto Runeberg (1765). He experimented with freezing soil and found that silt could take up almost four times as much water in the form of ice as the weight of the dry silt. He was astonished.

Figure 7 Professor Gunnar Beskow (1901–1991), Geologist at Swedish Geological Survey during the frost research project 1927–39, Departmental manager at Swedish Institute of Roads 1939–49, Professor in Geology at Chalmers University of Technology, Göteborg, Sweden 1949–1967.

In 1885–86 Captain Knut Lindmark used a ground freezing in order to be able to dig a tunnel for pedestrians trough the glaciofluvial esker Brunkebergsåsen in Stockholm. He got the idea from the sheep-transporting liners from Australia to England (Schütz, 1961). The tunnel is still in use.

During the first decades of the twentieth century, roads and railroads needed to be improved because the cars and the trains became bigger, stronger and faster – but the roads and railroads were sometimes in bad, or locally in very bad, condition due to frost activity. During winter, frost-heaving made the roads uneven and stones seemed to grow, coming up from beneath. The most severe problem, however, was the softening of the road surface during spring, and small "volcanoes" of supersatured silt being pumped up from lower layers. The road engineers had found that two methods gave better results – drainage and isolating with layers of sand or brushwood.

The discussion between engineers and scientists forced a meeting to be held in Oct. 1925 in Luleå, with delegates from State offices and others (Nordendahl, 1926). After this conference, the Swedish Geological Survey and the Swedish Institute of Roads decided to undertake jointly a research project, carried through at the Survey. Simon Johansson became leader of the project and Gunnar Beskow (Figure 7) was employed in 1927 (Johansson, 1927).

Capillarity in soils was essential for understanding the impacts of frost. The main hypothesis in the project was that originating from Johansson, that water is transported up to the frost's lower limit though capillary flow. Beskow studied this using precise particle size fractions named by Atterberg (Beskow, 1930). He made capillary measurement apparatus (at the same time but independent to Engelhardt) which he used to verify a formula for calculating capillary water rise. During the experiments, he found that the permeability must be directly proportional to the square of the particle size (the works by Hazen seem not to have been known). He also tested an idea, saying that most water was transported from below as air moisture. In the first stage, Beskow also thought that this was an essential part of the transported water volume, but later he abandoned this idea, convinced by laboratory tests by Taber 1929 and his own tests. This process occurred, but with much lower amounts of water than that from capillary flow. One of his first diagrams shows a silty clay with a lot of thin ice bands, making the silt contain more than 100% water (Figure 8) (Beskow, 1929). This figure was compiled after field observations and moisture analyses. In order to simulate this in the laboratory, Beskow used a refrigerator of special design. By this equipment, Beskow could freeze the soil during controlled conditions – temperature, freezing velocity, water flow etc. The frost thickness in soil is dependent on several factors: the soil's capillarity and hydraulic conductivity, depth to groundwater level, pressure (vertical load). The most frost active soils are (medium) silts with a high groundwater level.

In spring, the frost thaws at the top and at the bottom and the ice layers melt. If the water content is high, the surface becomes very soft as the water cannot flow downwards through the frost. If there is a dry or hard surface layer, the saturated soil will have a high pore water pressure and becomes a liquid when loaded with, e.g., traffic. It is easily pressed up through cracks in the surface layer giving mud volcanoes. This still occurs on old gravelled roads in northern areas where the till has high a silt content. To get rid of this problem, the road has to be rebuilt with better drainage and isolating layers and/or layers of coarse sand and gravel to break the capillary transport.

Beskow's work was accurate and exhaustive and the final report in 1935 (Beskow 1935) became, in its English version (1947), a bestseller. The report was retyped and reprinted, as one of the most essential frost texts together with Tabers, in the USA in 1991. After the frost project, Beskow became departmental manager at the Swedish Institute of Roads 1939–49, and obtained the professorship in Geology at Chalmers University of Technology, Göteborg, 1949. Beskow also was a poet and published several collections of poems.

Later, Rune Gandahl invented simple frost limit measuring equipment. With this you can very precisely get the lower and the upper frost lines (Gandahl and Bergau, 1957).

ARTIFICIAL RECHARGE

Richert's first infiltration project was only one in a long series. The next was in Örebro, where the water consumption in 1911 had doubled in the previous five years. The engineers at Vattenbyggnadsbyrån AB had at first assumed that it would be possible to use infiltration basins close to the river. However, the local circumstances – fine sand

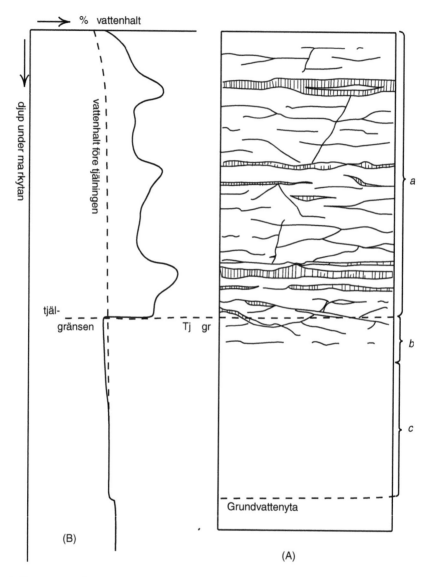

Figure 8 Frost in a surface layer of clay: (A) Ice layers in a frozen medium clay (cracks are normal close to the surface). (B) Water content distribution curve. vattenhalt = water content; djup under markytan = water depth below surface; vattenhalt före tjälningen (dashed line) = water content before freezing; tjälgränsen = frost line; grundvattenyta = ground water table; a = frozen part, thin and thicker ice bands (vertical lines); b = dried up zone below the frost line; c = vadoze zone. The continuous line denotes the water content after freezing, >100% at levels with thick bands of ice (Beskow, 1929).

and some risk of pollution – made the water manager try different solutions to the water supply problem. One was artificial infiltration in an esker. After four years of different experiments, Teodor Glosemeijer, Richert's successor at the waterworks in Göteborg, was consulted. He did not hesitate to advise Örebro to enlarge the waterworks this way, which he called the "Frankfurt System", as it was tried for three years in Frankfurt am Main, Germany, with very good results (Hållén, 1920). The method was published by Scheelhaase 1911. Unlike the situation in Göteborg, the infiltration basins were localized at a high level, several meters above the groundwater table. Thus the water had to percolate through a thick aerated zone, where the organic matter was oxidised. This version was fulfilled and became such a success that almost every replenishment plant was designed like this thereafter. The engineers at Vattenbyggnadsbyrån AB and other consultants used this concept in many municipalities in Sweden. In 1952, among 74 municipalities, 13 had artificial replenishment plants in operation and 13 had plants in different stages of construction, Table 1 (Jansa, 1952).

"During the hydrological investigations preceding the design of an artificial recharge plant, infiltration experiments will often be possible by re-circulation of groundwater pumped out of wells and infiltrated in a temporary basin. The observed changes of the groundwater level during such a test will often be of great value when making hydrological computations" (Jansa, 1952). "It has been generally stated from experience of the Swedish artificial groundwater recharge plants that the method of artificial infiltration is a most efficient method of removing impurities. The ground has also an outstanding capability of removing and destroying the abundant micro-organisms always to be found in the raw waters applied. Provided that the artificial infiltration and recovery of the infiltrated waters is carried out in the right way, the raw water can be refined into a genuine groundwater" (Jansa, 1954).

Successively the infiltration method has become a normal solution at many places and in different ways. It is, so to say, used as a optimising system for almost all groundwater plants – for surface water purification, for enlargement of the natural groundwater resources, for removal of iron and manganese (after aeration) and so on (Agerstrand, 1968, Winqvist 1968). In 1975, groundwater withdrawal in Sweden by municipal waterworks was 250 Mm^3 plus the artificial recharge at 70 water-works 200 Mm^3, together 47% of the total water production (Svenska vatten- och avloppsverksföreningen, 1976).

GEOLOGICAL SURVEY OF SWEDEN, SGU

The Survey was founded in 1858 and its commission was to discover and map the geology of the country. In the beginning, the field geologists only had to map rocks and soils and document whatever they found. Groundwater conditions were noted occasionally. Some geologists were interested in groundwater, such as Otto Torell in Lund, Hermann Hedström at Gotland, and Simon Johansson, during the first decades of the twentieth century. The first groundwater text in a description for a geological map was by Fredrik Svenonius, about an area north of Lake Vänern. He found that water from one third of the springs and half of the wells contained considerable amounts of iron: 3–5 mg Fe/l. He pointed out the problem of clogging in pipes due to the high content of iron. He also found that the content of salts was often higher in water from wells in the bedrock than in spring water (Sandegren, Högbom and Svenonius, 1922).

State geologist Gunnar Ekström (1891–1961) worked a lot with soils and groundwater. His classification of Swedish agricultural soils is a thorough description and grouping of all types of earth materials (Ekström, 1927). He also worked out a list of groundwater terminology in Swedish (Ekström, 1938). Most of the terms are still in use (TNC 86).

"The Geological Survey of Sweden in 1953 founded a groundwater department with an archive to compensate a certain lack of general and concise data on groundwater conditions in Sweden. The purposes of the groundwater department are 1) to collect data on ground-water, e.g. levels, yield, geology, permeability and storage coefficients of the underground, chemical composition i. e. chloride contents of subsurface water at various depths, 2) to study the collected data and to publish the results of such studies, and 3) to supply data on groundwater conditions for persons interested" (Tullström, 1960).

Tullström emphasises the need for more knowledge on the condition of groundwater in the bedrock, such as structure and mineralogical composition, porosity and tectonic history, weathering and metamorphosis, and hydrothermal fissure fillings. He also points out the lack of detailed information and understanding of groundwater conditions in superficial materials, so important for the groundwater supply in Sweden. The intention of the work at the department is also to compile the information into maps showing both the geology and different groundwater related conditions and parameters.

THE NATIONAL WELL DATA ARCHIVE

During the 1940s, accessible data from wells in Skåne and from Cambrosiluran rocks in the county of Östergötland were collected by state geologists interested in groundwater. The first state geologist employed as a hydrogeologist was Helge Tullström in 1953. He had to go further with the well data collection, cooperate with the well boring companies and work as a consultant. In 1958, the well archive contained data on 19,500 wells from all over the country. In 1966, the archive was organised formally. During the initial period, data were collected manually by geologists visiting the well boring companies regularly in the field. This was expensive and time consuming, so SGU asked for a law, which was passed in 1975. Since then, all well borings should be reported to the archive (Grundvatten, 2000).

THE NATIONAL GROUNDWATER OBSERVATION NETWORK

A national groundwater observation network was established 1967 within the framework of the International Hydrological Decade (IHD) at SGU with funds from the Swedish Natural Science Research Council. At the beginning of 1974, the network contained 57 sub areas containing a total of 538 stations for monitoring groundwater level, 58 stations for groundwater chemistry and 36 stations for frost penetration depth. 57% of the wells were situated in till, being the most common earth superficial material. The use of glaciofluvial deposits for groundwater withdrawal explains the high percentage of wells in those deposits: 32%. Only 24 wells in the bedrock were included in the network – half of them in limestone and half of them in hard crystalline rocks. Less than 7% were situated in deposits of other types. Several areas included was "abandoned" ground water investigations. The sub areas close to Stockholm and

Göteborg were established during the STEGA project as reference areas for these cities. The longest measurement series starts 1952, in the Abisko area in the far north. Normal initial time is around 1965. The standard measurements were manual and made twice a month: only a few recorders were used (Nordberg and Persson, 1974). It soon became apparent that groundwater level variation patterns could be divided into four regimes, typical of four parts of the country. One simple regime, with one maximum and one minimum per year in the south and in the north – but just the opposite! In the south, the maximum level was normally during the autumn, caused by thunderstorms in late summer and heavy rains during autumn, followed by a rather dry spring giving minimum levels during early summer. In the northern regime, the snow melting in spring gives high levels in summer and the frost, functioning as a dry period giving low levels during the winter. The two other regimes are more complex, having two maxima and two minima a year, the levels in the southernmost affected by rains and in the northern most affected by the winter dryness.

HYDROGEOLOGICAL MAPPING

The geologic maps in Sweden published up to 1965 were combined maps for unconsolidated and consolidated rock. The only groundwater related objects taken into consideration were springs. In the texts following each map in series Aa (1:50 000), almost 12 000 springs are listed during 1862–1974 on 201 map areas over the southern half of Sweden (De Geer, 1979). Ten issues of agrigeological maps in series Ad (1:20 000) were edited 1947–1966. The descriptions contains some pages on wells and springs.

SGU started regular mapping of hydrogeological properties in 1969 for maps at the scale of 1:50 000. The first maps in this series, series Ag, were printed in 1971, covering the area around Örebro (Möller *et al.*, 1971) and the next year the first map for a part of southern Skåne (Gustafsson, 1972). For the first five maps in the series, the Quaternary maps were used, with hydrogeological information, such as the groundwater divide, groundwater levels and flow directions, point information on large wells and their capacity, Fe, Cl and Mn-contents, as an overprint. Numbers Ag 6–15 have a legend based on hydrogeological conditions: extensive and productive aquifers, median capacity of known wells, and the geology in light colours in the background. Thus the last ones are more like the maps in the following Ah-series.

Mapping at the scale of 1:50 000 is expensive and time consuming. After some consideration and marketing research the next maps edited were series Ah, maps at a scale of 1:250 000 covering whole counties using the UNESCO legend. The first was finished in 1981 (Pousette *et al.*, 1981) and now there are maps in this series, printed or in press, covering all counties.

WATER IN THE VADOZE ZONE

The treatment of the hydrogeology of the vadoze zone is found in the literature published by agricultural and forestry scientists. Drainage of different kinds of soils in the fields and the forests is necessary to take care of, together with the understanding of conditions for the flow and storage of nutrients. Atterberg's classification of grain size in 1903 was a first step. Some of the state geologists took part at the beginning (Johansson, 1916; Ekström and Flodkvist, 1926; Flodkvist, 1931). In 1952 Sigvard Andersson, at Department of Soil Sciences, Swedish University of Agricultural Sciences in Uppsala, started a series of papers on the physical properties of soils, presented in *"Grundförbättring: Journal of Agricultural Land Improvement" (1947–1976)*. The papers are theoretical and experimental texts on pores, pore water and the water-grain interacting forces based on many years of experiments on soils, e.g. Andersson & Wiklert 1972.

REGIONAL HYDROGEOLOGY

Studies of regional hydrogeology in Sweden are related to water supply problems in different areas. Meier and Sund (1952) presented a list of the water supply deposits/formations of Sweden. The till is the most frequent unconsolidated material (75% of land surface in Sweden), but one well does not give enough water for more than a couple of households or a small farm. Next are sandstones of different ages and fissure zones in the bedrock.

The crystalline bedrock was considered by Ingemar Larsson (1913–2002), in Lund, later professor at the Royal Institute in Stockholm, who used tectonic concepts from Professor B. Sander, Innsbruck, to model the bedrock in order to optimise the success of well boring (Figure 9). Successively, he and his colleagues laid the scientific base for the great bedrock research project, i.e. the OECD/NEA International Stripa Project, which started in 1980.

In Skåne, two regions have been the aim for thorough investigations, 1) the SW part close to the cities of Malmö, Lund, Landskrona and Hälsingborg – all on sedimentary rocks with rather complicated structure, and 2) the Kristianstad Plain – a basin with Quaternary sediments on Cretaceous limestones and uncemented sands. The southwestern part of Skåne has been evaluated more or less continuously since Jönsson started the investigations for Malmö waterworks (e.g. Leander 1971).

The 1955 the Cooperating Committee of Hydrological research of the Kristianstad Plain started the investigations on the 700 sq. km basin due to a rising awareness of pollution risks (von Feilitzen, 1954; Larsson, 1962; Nilsson, 1966).

The Swedish islands in the Baltic, Gotland and Öland, consists of Cambrosilurian rocks and have thin superficial layers – boulder clay and some glaciofluvial sediments. As mentioned above, Richert and Hedström worked for Visby at Gotland in the beginning of the twentieth century. In the 1950s, the Geological Survey had to do more work on the hydrogeology of Gotland (Tullström, 1955; Statens Naturvårdsverk and Viak AB, 1969).

The glaciofluvial deposits are used for municipal water supply and thus the consultants have investigated them more in detail than the Geological Survey usually does. There are many consultants' reports listed in the descriptions to the maps in the *series Ah* published by SGU. The most common superficial deposit, the till (75% of land area), has been less well treated. First more thorough studies by Gert Knutsson (Knutsson, 1966, 1971).

The interest in the chemistry of groundwater blossomed again in the 1960s. Erik Eriksson (soil water chemistry), Gunnar Jacks (soil, groundwater in bedrock wells) and

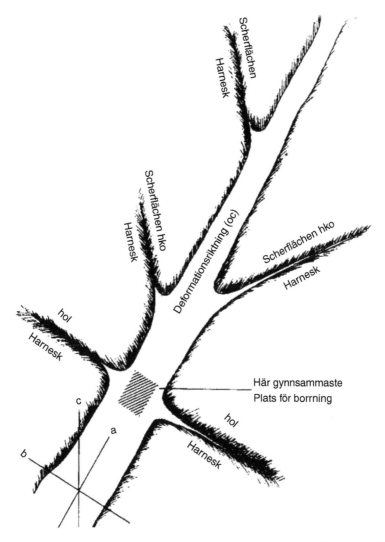

Figure 9 "Idealised cleavage system belonging to a postcrystalline ruptural deformation." One of t first bedrock tectonics by Ingemar Larsson; based on the ideas of B. Sander, Innsbruck. Här gynnsammaste platsen för borrning = most suitable boring localisation for a well (Larsson, 1959).

Leif F. Carlsson (regional geochemistry) presented interesting contributions (Eriksson and Khunakasem, 1968; Jacks, 1972; Carlsson, 1974).

URBAN HYDROGEOLOGY

Several cities in Sweden are located in places where the ground contains clay layers of different types and thickness. During expansion and renewal of roads, pipes and cables, the cities had problems concerning settlement and groundwater levels (Hellgren, 1959). During the 1960s the problems became troublesome in Stockholm and Göteborg. In 1968, a research group was established in Stockholm with the aim to make a thorough study of groundwater problems in Stockholm and other cities. The group named itself after their second names: Sund, Tyrén, Eriksson, Gustafsson, Arnborg, i.e. STEGA. They obtained a considerable amount of money from the Swedish Building Research Council, engaged several more engineers and made many investigations of groundwater problems. In Stockholm and Göteborg, with thick soft clays, more or less severe damage could be found almost everywhere due to settlement. In the records of groundwater levels during different types of work in the ground, deep in the bedrock or close to the surface, it was clear that the confined aquifers were very dependent of the amount of infiltration in their unconfined areas. Just a small amount of water tapped in a tunnel under the city could cause groundwater lowering, as the coefficient of storativity in overlying unconsolidated material is very low, Lindskoug and Nilsson 1974.

In the 1970s, a research group on urban hydrology was formed at Chalmers University of Technology financed by the Swedish Building Research Council. This was a collaboration between four departments at the university, lasted for 15 years and became a forum for many PhD students.

EDUCATION IN HYDROGEOLOGY

At the Royal Institute of Technology (founded 1827) in Stockholm, the professors in hydraulics had to teach the students how to handle groundwater. The real opening of hydrogeological science at the Institute was done by Yngve Gustafsson (1912–2000) (Figure 10) who obtained the professorship at Department of Land Improvement and Drainage in 1955. The department became the most prominent hydrogeology research centre in Sweden in 1960th and is still so.

At Chalmers University of Technology (founded 1829) in Göteborg, Gunnar Beskow continued his frost research. In 1960 Gert Knutsson was employed and started up ground-water research – a starting-point for the future activities in hydrogeology at Chalmers. His first student and collaborator was Leif F. Carlson, who studied ground water flow in glacial deposits and groundwater chemistry. Knutsson moved in 1969 to the Swedish Institute of Road and Traffic and later to the Geological Survey. When Gustafsson retired, Gert Knutsson obtained his chair in Stockholm in 1980.

At the universities, the teachers made mimeographed texts for their students but nothing was systematically edited for public use. Handbooks for different professions were edited – for civil engineers the handbook series BYGG, in several volumes and editions – but with rather short chapters on groundwater and frost in soil. The first real textbook in Swedish on hydrogeology was written in 1973 by Gert Knutsson and Carl-Olof Morfeldt and has been reedited later (Knutsson and Morfeldt, 1973). The vadoze zone is thoroughly treated in another textbook by Troedsson and Nykvist (1973). Theoretical treatises of the hydrologic cycle are compiled in a third book by Lindh and Falkenmark (1973).

Figure 10   Professor Yngve Gustafsson (1912–2000), Dr. of agriculture and Head of the Department of Land Improvement and Drainage, Royal Institute of Technology, Stockholm 1955–1977. Photo: Maria Evertsson.

AQUIFER ANALYSES

The most advanced analysis for a hydrogeologist is the evaluation of an aquifer. Richert learned a lot from Adolf Thiem and contributed to the articles in the scientific journals. At least one of them, the *"Journal für Gasbeleuchtung und Wasserversorgung"* ought to have been in his office, because the volumes in Chalmers University Library have labels with his office's name and several articles have pencilled notes, especially those by A. Thiem. In Richerts book, published in 1891 are several formulae deduced and he refers to Darcy, Dupuit and the German engineers Salbach, Smreker and A. Thiem. In the textbook of 1911 Richert refers to Darcy, A. Thiem and G. Thiem (Richert, 1911).

A small textbook on waterworks by Westerberg (1916) contains some text on aquifer analysis. Up to the 1960s there are just a few articles in Swedish journals about aquifer analyses and they are more or less reviews, or trying to use methods published in foreign journals. During the 1950s, isotope techniques were developed and adopted in hydrology. By using these techniques groundwater flow velocities are possible to calculate using time series. Several scientists have used these facilities (Knutsson *et al.*, 1963; Knutsson, 1967; Persson, 1974; Nordberg and Modig, 1974).

In 1974, the state geologists Per Engquist and Tommy Olsson presented a paper on a method to evaluate dug wells in till. Glacial till is the most common unconsolidated material in Sweden and thus essential to water supply in the countryside. The method is an evaluation of the relation between residual levels after stopping pumping, and a late part of the recovery curve. A simplified evaluation using a plot and a type curve is also presented. The method was used in hydrogeological mapping in order to obtain data on well capacities (Engquist and Olsson, 1974).

Another type of aquifers in unconsolidated materials are to be found in the glacial eskers – long ridges of sand and gravel, often more or less buried in fine sediments. These aquifers are often large enough to be used for waterworks. Methods for the evaluation of the hydraulic properties of esker aquifers (Figure 11) were presented by Gunnar Gustafson (Gustafson, 1974, 1978), who later received the professorship in Geology at Chalmers University of Technology (Beskow's chair).

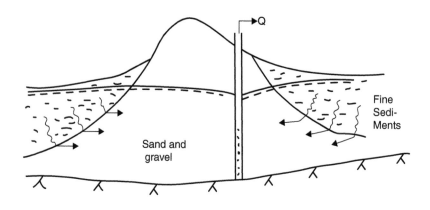

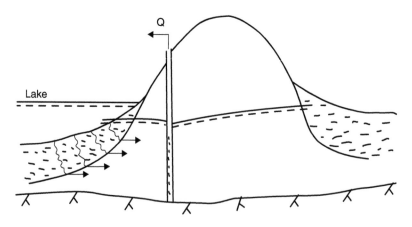

Figure 11 Groundwater in an esker, a long ridge of glaciofluvial gravel and sand, formed during the melting of the land ice. Fine sediment on both sides, sometimes a lake. Water leakage though the fine sediments. The groundwater normally flows along within this esker type as in a ditch (after Gustafson 1974).

REFERENCES

KVA = Kungl. Vetenskaps-Akademien (Royal Academy of Sciences), Stockholm.

SGU = Sveriges Geologiska Undersökning (Geological Survey of Sweden).

Agerstrand, T. (1968) Ground water draft from earth layers. In: Eriksson, E., Gustafsson, Y. & Nilsson, K. (eds.) *Ground Water Problems*. Oxford, Pergamon Press. pp. 181–195.

Althin, T. (1947) Vattenbyggnadsbyrån 1897–1947. Stockholm. 182 p.

Andersson, S. & Wiklert, P. (1972) Markfysikaliska undersökningar i odlad jord. XXIII. Om de vattenhållande egenskaperna hos svenska jordarter. [Physical examinations of agricultural soils – On the water retention characteristics of Swedish soils] *Grundförbättring*, 25, 53–143, Uppsala.

Atterberg, A. (1903) Studier i jordanalysen. [Studies in soil analyses] Kongl. *Landtbruks-akademiens handlingar och tidskrift*, 42, 183–253.

Atterberg A (1911) Lerornas förhållande till vatten, deras plasticitetsgränser och plasticitets-grader. [The clays' relations to water, their plasticity limits and plasticity degrees] Kungl. *Lantbruks-akademiens handlingar och tidskrift*, 50, 132–158.

Berzelius, J.J. (1800) Nova analysis aquarum Medeviensium. [New analyses of the Medevi waters] Diss. 15 p.

Berzelius, J.J. (1808) Lärobok i kemien. [Textbook in chemistry] Förra delen. Stockholm. 483 p.

Beskow, G. (1929) Tjälproblemets grundfrågor. [Basics on the frost in soil problem] *Svenska Vägföreningens Tidskrift*, 16, 12–26.

Beskow, G. (1930) Om jordarternas kapillaritet. En ny metod för bestämning av kapillärkraften (eller kapillära stighöjden). [On the capillarity of soils. A new method how to determine the capillarity] SGU C356, Stockholm. 65 p.

Beskow, G. (1935) Tjälbildningen och tjällyftningen med särskild hänsyn till vägar och järnvägar. [Swedish version of Beskow 1947] Statens väginstitut, Medd. 48. Stockholm. 242 p.

Beskow, G. (1947) Soil freezing and frost heaving with special application to roads and rail-roads. With a special supplement for the English translation of progress from 1935 to 1946. Publ. by the Technological Institute, Northwestern University, Evanston, Illinois. The same paper, retyped (1991) In: Black, P.B. & Hardenberg, M.J. (eds.) Historical Perspectives in Frost Heave Research. The Early Works of Taber, S. & Beskow, G. US Army Corps of Engineers. Cold Regions Research & Engineering Laboratory, Special Report 91–23, Hanover, NH, USA.

BYGG (1947) Handbok för hus-, väg. och vattenbyggnad. [Handbook for house-, roads- and water constructors] In: Wåhlin, E. (ed.) 1st ed. 4 Vols. Stockholm.

Carlsson, L. (1974) Groundwater with high content of chloride in a quaternary marine clay area in western Sweden. *Nordisk hydrologisk konferens, Aalborg*, II, 433–457. København.

De Geer, J. (1979) Några anmärkningar rörande källor i Sverige. [Some remarks on springs in Sweden] *Vannet i Norden*, 12, 107–114.

Ekström, G. (1927) Klassifikation av svenska åkerjordar. [Classification of Swedish agricultural soils] Diss. SGU C345. Stockholm.

Ekström, G. & Flodkvist, H. (1926) Hydrologiska undersökningar av åkerjord inom Örebro län. [Hydrogeological examinations on agricultural soils in County of Örebro] SGU C334. Stockholm. 48s.

Ekström, G. (1938) Preliminärt nordiskt förslag till jordvattnets terminologi. [Preliminary Nordic proposal on teminology of water in the ground] *Nordisk Jordbrugsforskning*, 20, 299–313. Köpenhamn.

Engqvist, P. & Olsson, T. (1974) Försök att i återhämtningskurvor beräkna moränbrunnars kapacitet. [An attempt to calculate capacity of wells in till from recovery tests] Preliminär rapport. Vannet i Norden, Nr. 4, pp. 21–34. Oslo.

Eriksson, E. & Khunakasem, V. (1968) The chemistry of ground waters. In: Eriksson, E., Gustafsson, Y. & Nilsson, K. (eds.) *Ground Water Problems*. Oxford, Pergamon Press. pp. 110–146.

von Feilitzen, R. (1954) Några synpunkter på Kristianstadsslättens grundvattenproblem. [Some points of view on the groundwater problems in the Kristianstad Plain] *Vattenhygien*, 10, 123–132.

Flodkvist, H. (1931) Kulturtechnische Grundwasserforschungen. [Agricultural groundwater research] SGU C371, Stockholm. 317 p.

Frängsmyr, T. (1969) Geologi och skapelsetro. Föreställningar om jordens historia från Hiärne till Bergman. [Geology and the Doctrine of the Creation. Ideas on the History of the Earth from Hiärne to Bergman] Diss. Lychnos bibliotek 26. Stockholm, Almqvist & Wiksell. 372 p.

Gandahl, R. & Bergau, W. (1957) Two methods for measuring the frozen zone in soil. *Proceedings of the Fourth International Conference on Soil Mechanics and Foundation Engineering, London, 12–24 August 1957*, I. pp. 32–34.

Grundvatten (2000) Nr. 2, SGU, Uppsala.

Gustafson, G. (1974) A method of calculating the hydraulic properties of esker aquifers. *Nordisk hydrologisk konferens, Aalborg*, II, 525–543. København.

Gustafson, G. (1978) A method of calculating the hydraulic properties of leaky esker-aquifer systems. Diss. Chalmers tekniska högskola, Geologiska institutionen, A23, Göteborg.

Gustafsson, O. (1972) Beskrivning till hydrogeologiska kartan Trelleborg NV/Malmö SV.[Description of the hydrogeological map Trelleborg NV/Malmö SV] SGU Ag 4, Stockholm.

Hållén, K. (1920) Anläggning för konstgjort grundvatten vid Örebro stads vattenledningsverk, Skåmsta. [The works for artificial groundwater at the waterworks at Skåmsta in Örebro] *Svenska kommunal-tekniska föreningens handlingar*, 1919 (9), 1–23.

Hansen, F.V. (1903) Statistisk sammanställning rörande vatten- och afloppsledningar i Sveriges städer och samhällen. [Statistics on water- and waste water pipes in cities and municipalities of Sweden] Stockholms stad, Drätselnämndens tryckta handlingar, år 1903 Nr. 35.

Hamberg, N.P. (1868) Kemisk undersökning af vattnet i åtskilliga brunnar i Stockholm. [Chemical examination of the water in several wells in Stockholm] *KVA Översikt*, 25, 159–193.

Hellgren, A. (1959) Skadliga grundvattenändringar. [Harmful groundwater level changes] Stadsbyggnad. pp. 117–123. Stockholm.

Hiärne, U. (1679) Een kort berättelse om the nys upfundne surbrunnar wid Medewij uthi Östergöthland, huru the bäst skole brukas, och på hwad sätt i medler tijdh diæten anställes : them til underrättelse, som ther aff sigh willia betjäna. [A short narrative about the recently found springs with "carbonated water" at Medevi in Östergötland, how they should be used ....] Stockholm. 122 p.

Hiärne, U. (1683) Den lilla wattuprofwaren hwarigenom de rätta och hälsosamma Suurbrunnar ifrån de falska eller gemene Jernwatn, som här och där i wårt K. Fädernesland finnas, igenkännas och åthskillies. [The small water tester....] Stockholm. 93 p.

Hiärne, U. (1702–06) Den korta anledningen, til åtskillige malm och bergarters, mineraliers och jordeslags etc. efterspörjande och angifwande, beswarad och förklarad, jämte deras natur, födelse och i jorden tilwerkade, samt uplösning och anatomie, i giörligaste måtto beskriven. [The short reason, to the inquiry on considerable ores and rocks, minerals and soils etc, answered and explained, together with their nature.... .] Stockholm. 2 Vol., 416 p.

Hiärne, U. (1916) Urban Hjärnes självbiografi. [Autobiography of Urban Hiärne] Uppsala universitets årsskrift 1916, Program 4:2, Äldre svenska biografier 5. Uppsala.

Houtkamp, H. & Jacks, G. (1972) Geohydrologic well-logging. *Nordic Hydrology*, 3, 165–182

Hult, A. (1991) Källan till vattnet. Sanningar och sägner om vattnet i marken. [The origin of water. Truth and myths of water in the ground.] *Gidlunds bokförlag*, 198 p.

Jacks, G. (1972) Chemistry of groundwater in igneous rock at Angered, Gothenburg. *Nordic hydrology*, 3, 140–164.

Jansa, V. (1952) Artificial replenishment of underground water. In: Int. Water Supply Association, 2nd Congress, Paris, June 9–13, 149-191.

Jansa, O.V. (1954) Artificial ground-water supplies in Sweden (a second report). In: Union Geod. Geophys. Internat., Association Internat. Hydrologie Sci. Assemblée Gen., Rome, Vol. 2, Publ. 37. pp. 269–275.

Jönsson, J. (1889) Berättelse öfver undersökningen af området med artesiskt vatten mellan Malmö och Romele klint. [Narrative of examination of the area with artesian water between Malmö and the horst of Romele] Malmö Stadsfullmäktiges protokoll 1897, Bilaga 3 (dated 1889). Malmö.

Johansson, S. (1914) Die Festigkeit der Bodenarten bei verschiedenem Wassergehalt nebst Vorschlag zu einer Klassifikation. [The solidness of soils with different water contents, and a proposal for a classification] Diss. SGU C256, Stockholm.

Johansson, S. (1916) Agrogeologiska undersökning av Ultuna egendom. [Agricultural examinations at the estate of Ultuna] SGU C 271. Stockholm. 95 p.

Johansson, S. (1927) Tjälproblemet och programmet för dess lösande. [The frost problem and the programme for its solution] *Svenska vägföreningens tidskrift*, 4, 315–322.

Knutsson, G. (1966) Grundvatten i moränmark. [Groundwater in till soils] Statens naturvetenskapliga forskningsråd, Svensk naturvetenskap. pp. 236–249. Stockholm.

Knutsson G (1967) Tracing ground-water flow in sand and gravel using radioactive isotopes. *Steirische Beiträge zur Hydrogeologie, Jahrgang* 1966/67, 13–31

Knutsson, G. (1971) Studies of ground-water flow in till soils. *Geologiska föreningens i Stockholm förhandlingar*, 93, 553–573.

Knutsson, G., Ljunggren, K. & Forsberg, H.G. (1963) Field and laboratory tests of chromium-51-EDTA and tritium water as a double tracer for groundwater flow. In: IAEA, *Radioisotopes in Hydrology*. pp. 347–363. Vienna.

Knutsson, G. & Morfeldt, C.-O. (1973) Vatten i jord och berg. [Water in earth and bedrock] Ingenjörsförlaget, Stockholm. 172 p.

Larsson, I. (1959) Tektoniska förutsättningar för grundvatten i urberget. [Tectonic conditions for groundwater in the bedrock] *Svensk geografisk årsbok*, 35, 169–183.

Larsson, I. (1962) Studies on ground water in the quaternary deposits of the Kristianstad Plain. *Lund Studies in Geography*, A19, Lund. 55 p.

Leander, B. (1971) Alnarpsströmmen. [The Alnarp Groundwater Flow] In: *Grundvattenförekomst i sydvästra Skåne. Symposium vid Lunds tekniska högskola 7–8 juni 1971*. pp. 39–60. Lund.

Lindroth, S. (1967) Kungl. Svenska Vetenskapsakademiens Historia 1739–1818 [History of the Royal Swedish Academy of Sciences], del 1:1. KVA, Stockholm.

Lindh, G. & Falkenmark, M. (eds.) (1973) Hydrologi. En inledning till vattenresursläran. [Hydrology. An introduction to the science of water resources] Studentlitteratur, Lund. 465 p.

Lindskoug, N.-E. & Nilsson, L.-Y. (1974) Grundvatten och byggande. STEGAs arbete 1966–73. [Groundwater and civil engineering. The works of STEGA 1966-73]]Byggforskningen R20:1974. Stockholm. 163 p.

Meier, O. & Sund, B. (1952) Geologins betydelse för vattenborrningen i Sverige. [The significance of geology for the water borings in Sweden] *Vattenhygien*, 8, 1–11.

Möller, Å., Engqvist, P. & Müllern, C.-F. (1971) Beskrivning till hydrogeologiska kartan Örebro NV. [Description to the hydrogeological map of Örebro NV] SGU Ag3, Stockholm.

Nilsson, K. (1966) Geological data from the Kristianstad Plain, Southern Sweden. SGU C605, Stockholm. 32 p.

Nordberg, L. & Modig, S. (1974) Investigation of effective porosity of till by means of a combined soil moisture/density gauge. In: *Isotope Techniques in Groundwater Hydrology 1974, Proc. of a Symp. Wienna, 11–15 March 1974*, II. Wien, IAEA. pp. 313–340.

Nordberg, L. & Persson, G. (1974) The national groundwater network of Sweden. SGU Ca 48, Stockholm. 160 p.

Nordendahl, E. (1926) Protokoll från det av Svenska väginstitutet anordnade diskussionsmötet i tjälfrågan i Luleå den 5 och 6 okt. 1925. [The protocoll from the meeting on frost problems in Luleå Oct. 5–6, 1925] *Svenska vägföreningens tidskrift.* 27 p.

Nordenskiöld, A.E. (1896) Om borrningar efter vatten i urberget. [On the borings for water in the bedrock] *Geologiska föreningens i Stockholm förhandlingar*, 18, 269–284

Olsson, H. (1971) Kemiens historia i Sverige intill år 1800. [History of Chemistry in Sweden up to the year 1800] *Lychnos bibliotek 17:4.* Stockholm, Almqvist & Wiksell. 384 p.

Persson, G. (1973) Tritiumdata från Kristianstadsslätten 1963–72. [Tritium data from the Kristianstad Plain 1963–1972] *Vannet i Norden*, vol. 6 h. 2. pp. 3–17.

Pousette, J., Müllern, C.-F., Engquist, P. & Knutsson, G. (1981) Hydrogeologisk karta över Kalmar län med beskrivning och bilagor. [Description to the hydrogeological map of Kalmar County] SGU Ah 1. Uppsala.

Ramberg, L. (1912) Kemisk undersökning av Malmö stads vattenfattning vid Torreberga. [Chemical examination of the waterworks of Malmö at Torreberga] Kongl. *Fysiografiska sällskapets handlingar*, NF Vol. 23, Nr. 5. Lund. 13 p.

Richert, J.G. (1891) Om grundvattens förekomst och användning. Berättelse från en studieresa år 1889. [On the occurence and use of groundwater. Narrative of a study tour in 1889] Göteborg. 79 p.

Richert, J.G. (1893) Hydrografiska undersökningar för Malmö vattenledning. [Hydrologic examinations for the waterworks of Malmö] Drätselkammarens Handlingar Bil. 10. 32 p.

Richert, J.G. (1895) Göteborgs nya vattenverk. [The new waterworks in Göteborg] Berättelse. Göteborg. 82 p.

Richert, J.G. (1897) Användning af konstgjordt grundvatten för Göteborgs vattenledning. [The use of artificial groundwater in the waterworks of Göteborg] Tekniska samfundets i Göteborg handlingar, 1896 Nr. 14, Bil. 2, 7–57.

Richert, J.G. (1898) Om ökad vattentillgång för stadens vattenverk. [About increased water supply for Malmö watervorks] Malmö Stadsfullm. Protokoll, Bih. 108. 1–14, Malmö.

Richert, J.G. (1900) On Artificial Underground Water. Stockholm, Fritzes. 33 p.

Richert, J.G. (1902) Artificial infiltration basins. *The Public Health Engineer*, Oct. 18, 371–372.

Richert, J.G. (1911) Om Sveriges grundvattenförhållanden. [On the groundwater conditions in Sweden] Stockholm, CE Fritzes. 108 p.

Rothelius, E. & Sund, B. (1955) Borrning efter grundvatten i Sverige. [Borings for groundwater in Sweden] Tidskrift för värme-, ventilations- och sanitetsteknik, 26, 251–257.

Runeberg, E.O. (1765) Anmärkningar om någre förändringar på jordytan i allmänhet, och under de kalla Climat i synnerhet. [Notes about some changes on the Earth's surface, and especially during the cold climates] *KVA Handlingar*, XXVI, 81–115.

Sahlbom, N. (1915) Om radioaktiviteten hos svenska källvatten och dess samband med de geologiska förhållandena. [On radioactivity in Swedish spring waters and its relation to the geology environment] *Arkiv för kemi, mineralogi och geologi*, 6 (3), 1–51.

Sandegren, R., Högbom, A. & Svenonius, F. (1922) Beskrivning till kartbladet Väse. [Description to the geological map of Väse] SGU Aa151, Stockholm.

SBL: Svenskt bibliografiskt lexicon. [Bibliographic Lexicon of Sweden]. Index at http://home.swipnet.se/sbl/.

Scheelhaase, F. (1911) Beitrag zur Fragen der Erzeugung künstlichen Grundwassers aus Fluss-wasser. [Contributions to the issue of obtaining artificial groundwater from river water] *Journal für Gasbeleuchtung und Wasserversorgung*, 54, 665–675.

Schütz, F. (1961) Knut Lindmark – trafikbyggare i Stockholm. [Knut Lindmark – a traffic builder in Stockholm] *Väg-och vattenbyggaren*, 7, 247–248.

Sjögren, H. & Sahlbom, N. (1907) Undersökningar av radioaktiviteten hos svenska källvatten. [Examination of radioactivity in Swedish spring waters] Arkiv för kemi, mineralogi och geologi, 3 (2), 1–28.

Statens naturvårdsverk, Viak, A.B. (1969) Gotlands vattenförsörjning. Statens naturvårdsverk, Publ. 1969:7. Solna. 120 p.

Svenska vatten-och avloppsverksföreningen (1976) VA-verk 1975. Statistiska uppgifter över kommunala vatten- och avloppsverk. [Statistics on water- and sewerage works 1975, from the Swedish Association of water- and sewerage works] VAV S75, Stockholm.

TNC 86. Geologisk ordlista [Glossary of Geology, in 8 languages]. Tekniska nomenklaturcentralen, Stockholm. 482 p.

Torell, O. (1889) Utlåtande rörande artesiskt vatten till Malmö stad. Bihang till Malmö stadsfullmäktiges protokoll 1897 Nr. 35, Bil. 2. Malmö.

Torulf, N. (1962) Allmänna ingenjörsbyrån AB – AIB. [About a company] Stockholm. 87 p.

Troedsson, T. & Nykvist, N. (1973) Marklära och markvård. [Textbook on soils and soil conservation] AW Läromedel, Stockholm. 403 p.

Trofast, J. (1992) Jacobus Berzelius akademiska avhandlingar. [The academic dissertations of Jacobus Berzelius] Berzeliussällskapets skriftserie 1. Wiken, Höganäs. 95 p.

Tullström, H. (1955) Hydrogeologiska förhållanden inom Slite köping på Gotland. [Hydrogeological investigations in municipality of Slite at Gotland] SGU C538, Stockholm.

Tullström, H. (1960) Sveriges geologiska undersöknings hydrogeologiska sektion och brunnsarkiv. [The groundwater section and well archive at Geological Survey of Sweden] Vattenhygien, 16, 46–49.

Vattenbyggnadsbyrån (1969) Alnarpsströmmen. Utredning rörande vattentillgång och dess lämpliga utnyttjande. Utförd på uppdrag av samarbetskommittén för Alnarpsströmmen av VBB 1969. [The Alnarp groundwater flow. Investigations on water accessibility and appropriate utilization. Made on request from the Alnarp groundwater flow co-operation committee by VBB. A consultants report] 56 p, 7 app, 21 pl.

Wallerius, N. (1746) Rön, hwarigenom åtskillliga naturens lagar, angående vattnets och andra flytande materiers utdunstande, frambringas. [Observations, through which several laws of nature, concerning the vaporization of water and other liquid compounds, are bringed forth] KVA Handlingar, VII, 1–21.

Wallerius, J.G. (1748) Hydrologia, eller Watturiket, indelt och beskrifvit, jämte anledning til vattuprofvers anställande. [Hydrologia or the water kingdom, subdivided and described, and a reason why analyze the water] Stockholm. 150 p.

Westerberg, N. (1916) Om vattenledningar. Kortfattad lärobok. [On waterworks. A short textbook] Norstedts, Stockholm. 150 p.

Wickström, L. (1994) 60 år med KM. Kjessler & Mannerstråle – Rapsodi 1984-1994. [60 years with KM.] 24 p.

Winqvist, G. (1968) Artificial replenishment of ground water. In: Eriksson, E., Gustafsson, Y. & Nilsson, K. (eds) Ground Water Problems. Oxford, Pergamon Press. pp. 197–211.

Wimmerstedt, A. (1866) Chemisk undersökning af Medevi helsovatten. [Chemical examinations of the spa water at Medevi] Diss. Lund. 28 p.

The progress of hydrogeology in Britain: 1600 to 2000

J.D. Mather

Department of Geology, Royal Holloway University of London, Egham, Surrey, United Kingdom

Arguably the beginnings of modern scientific hydrogeology in Britain can be traced to the end of the reign of Elizabeth 1 (1558–1603) coinciding with a renaissance in the use of mineral springs for therapeutic purposes. Some of these springs had been used in Roman times but as the Roman Empire declined their use diminished and was later actively discouraged. During the Elizabethan period the Protestant government became concerned that discouraging the use of springs in Britain was providing an excuse for wealthy Catholics to congregate at Spa in Belgium, then under the control of Spain (Hembry, 1990). So for reasons of political expediency, and because people were reluctant to abandon their use of mineral waters, an accommodation was reached and the policy of prohibition abandoned.

The interest in mineral waters which ensued led writers to question accepted ideas of the origin of springs and their dissolved constituents which were based largely on the writings of the Greek philosophers and biblical teachings. Thus Edward Jorden (d.1632), a Bath physician and chemist who was notable for his rejection of the supernatural as a cause in nature, recognised that the heat and composition of springs, such as that at Bath, were derived from a subterranean source and explored the nature of the minerals involved (Jorden, 1631). Lodwick Rowzee, a physician practicing at Ashford in Kent, recognised that "... most water retaineth some favour of the ground through which it runneth ...." (Rowzee, 1632, p. 8). However, their discourses were largely philosophical, raised questions rather than provided answers, and it was to be later in the century before significant advances were made.

THE SCIENTIFIC REVOLUTIONARIES

The latter part of the seventeenth century in Britain was a time of unparalleled scientific and technological discovery. Here scientific free-thinking was able to flourish unchecked by the dead hand of the Catholic Church, which curtailed scientific experiment and observation in much of continental Europe. This culminated in 1660 in the foundation of the Royal Society which promoted the use of experiments to test hypotheses rather than debating the merits of ideas in philosophical terms. Over the next 140 years Fellows of the Society published a range of papers in which many ingenious ideas were presented and many of the principal concepts of hydrogeology were foreshadowed.

In the early years of the Royal Society Robert Hooke (1635–1703), their first curator of experiments, carried out a series of microscopic observations. One series of experiments involved small glass canes [capillaries] from which he speculated that the height to which water rose was proportional to "the bigness of the holes of the pipes" (Hooke, 1665, p. 11). He also carried out experiments with fresh and salt water in a U-tube, using his results to explain the flow of water in mountain springs. A decade later, the Society published an English translation of the section of Pierre Perrault's book on the origin of fountains (Perrault, 1674), which demonstrated experimentally that rainfall is more than adequate to account for the flow of rivers and springs (Anon, 1675). The concept of the hydrological cycle was further developed by the Society's clerk, Edmond Halley (1656–1742), who showed that water evaporated from the oceans and returned as rainfall (Halley, 1687). Also at this time the foundations of hydrogeochemistry were laid by the chemist and natural philosopher Robert Boyle (1627–1691) who formalised a scheme for the examination of mineral waters, particularly ferruginous waters, using colour indicators and other qualitative tests (Boyle, 1685).

Many Fellows of the Society found it difficult to accept the idea that springs owed their origin solely to rain and dew. Robert Plot (1640–1696), first keeper of the Ashmolean Museum at Oxford and onetime secretary of the Society, produced a detailed critique, first in Latin (Plot, 1685) and then in English (in Plot, 1686) in which he came out in support of a sea water origin. In this he was supported by John Woodward (1665–1728), Professor of Physick at Gresham College, (Woodward, 1695) and the clergyman and naturalist William Derham (1665–1735) who stated "That springs have their origin from the sea, and not from rains and vapours ... I conclude from the perennity of divers springs; which always afford the same quantity of water." (Derham, 1716, p. 50). In the opposite camp was another clergyman and naturalist John Ray (1627–1705) who stated clearly that all rivers and fountains "proceed from rainwater" (Ray, 1693, p. 38/39).

Over the next 100 years a number of groundwater related papers appeared in the "*Philosophical Transactions*" of the Society. Some of these reported solely on the strata intersected by wells or on the characteristics of mineral waters. The phenomenon of intermittent springs aroused considerable interest and complex systems of reservoirs and siphons were erected to explain their flow patterns (Atwell, 1732). The physician John Hunter (1754–1809) made observations on groundwater temperatures, including some around Brighton in Sussex. One well was near the beach and the water level rose and fell with the tide. He noted that it "does not correspond exactly with the tides, but follows them with an interval of about three hours" (Hunter, 1788, p. 61). In 1783, the Derbyshire physician Erasmus Darwin (1731–1802), grandfather of Charles, used his knowledge of inclined strata to improve the water quality in an old well in his garden (Darwin, 1785). Although his conceptual model was wrong he was one of the first to erect such a model and test it by sinking a well (Mather, 2006).

The sinking of deep wells in Britain begins to be recorded from the late eighteenth century and it was the military that undertook one of the first successful projects. Fresh water supplies for coastal garrisons were of concern as drinking water could not be supplied from external sources at times of siege. Although success was thought unlikely, permission was granted to sink a deep well within the confines of the dockyard at Sheerness, on the south side of the Thames estuary in Kent. Shallow wells in superficial

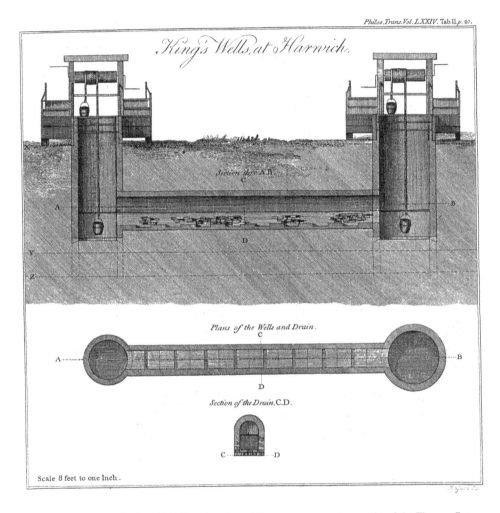

Philos. Trans. Vol. LXXIV. Tab.II.p. 20.

King's Wells, at Harwich.

Figure 1   Drawing of the King's Wells at Landguard-Fort, on the northern side of the Thames Estuary (from Page 1784, Plate 2). The distance between the centres of the two dug wells is 40 ft (12.2 m). X indicates the elevation of the high water mark, Y the low water mark and Z low water at spring tides.

deposits yielded salt water and over a thirteen month period in 1781/2 a deep well was dug through these deposits, and underlying hard clays, to quicksand at 330 feet (101 m). As the well was dug the upper salt water was cut off using an arrangement of wooden frames, brick steining and rammed clay. The quicksand yielded an excellent supply of water and the construction of the well was a significant achievement (Page, 1784). Following a visit by the King during construction the well became known as the King's Well.

Other deep wells were less successful. In areas of London south of the River Thames running sands at the base of the clay proved impenetrable as they boiled up and forced abandonment of the wells. In some cases a boring was made from the bottom of the well but without success (Barrow and Wills, 1913). This problem was not solved until 1794 when Benjamin Vulliamy (1747–1811), a watchmaker of Swiss descent, succeeded in sinking and completing a well into the sands beneath the London Clay near Notting Hill. He reported his achievement in some detail to the Royal Society (Vulliamy, 1797).

The military engineer in charge of the Sheerness work, Sir Thomas Hyde Page (1746–1821), was also ordered to improve the supply to a fort on the northern side of the Thames estuary on the Essex/Suffolk border. He started to sink another deep well similar to that at Sheerness but quickly found that fresh water lay above salt water in the sands on which the garrison was built. Two shallow wells were dug with a brick drain between them to collect the water (Figure 1). He was unable to account for the body of freshwater overlying salt water but speculated "Whether the greater specific gravity of the salt-water is sufficient to prevent a mixture with the fresh upon a higher line, I cannot venture to say; but the fact of there being a separation is beyond a doubt, and the depths may be ascertained to a degree of great accuracy" (Page, 1784, p. 18).

THE LAND DRAINERS

As well as the scientific gentlemen who reported their work through the pages of the Royal Society's "*Philosophical Transactions*", there was another group of men who had begun to develop an appreciation of the principles of hydrogeology. These were farmers, surveyors and engineers who were involved in land drainage. This had assumed increasing importance towards the middle of the eighteenth century with the need to feed a steadily increasing population. Most attention has been given to the work of the Warwickshire farmer Joseph Elkington (1740–1806) who was credited with discovering an improved method of drainage in 1764 (Figure 2 from Johnstone, 1797). However, there were other drainers active and, in his tour of the north of England in 1768, the agricultural writer, Arthur Young (1741–1820), records several cases where agricultural land had been improved by drainage. He is particularly enthusiastic about work on the estate of Thomas, Marquis of Rockingham, at Wentworth House in the West Riding of Yorkshire (Young, 1770), where covered drains had been introduced. From 1795 counter-claims of originality against Elkington were made by the Scottish agriculturist James Anderson (1739–1808) who was spurred into action when Elkington was awarded a premium of £1,000 to divulge his methods (Anderson, 1797). Anderson contended that the only way in which Elkington's mode of draining differed from that practised in general throughout Britain, consisted in what he called, tapping for springs, a mode of draining which he, Anderson, had been practising for over 30 years and which he had described in 1775 (Anderson, 1775).

Whatever the truth of the various claims and counterclaims it is clear that these land drainers had a sound understanding of the hydrological cycle and how water moves through the subsurface. They recognised that some groundwater was perched and separated from underlying saturated strata by low permeability material (Figure 3). They understood the nature of confined conditions and, in his description of Elkington's methodology, Johnstone (1797) uses the word "confined" to describe a reservoir "pent up between two impervious strata". This may be the first published use of the word "confined" in a hydrogeological context (Stephens and Ankeny, 2004).

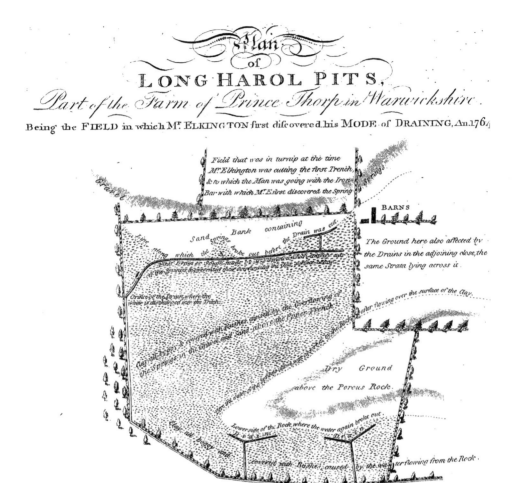

Figure 2 Plan of Long Herol Pits on Princethorpe Farm in Warwickshire, where Joseph Elkington pioneered his improved method of drainage in 1764 (from Johnstone, 1797 Figure 2). The explanation reads: "A in the plan represents the place where the clay pointed up on the surface above the line and below the bottom of the Trench, the depth of which not reaching the spring, induced Mr Elkington to push down the Iron Bar; which at 4 feet below the bottom of the Drain, caused the water to burst up and this was the first means that led him to think of applying the Auger as a more proper instrument in such cases, where the depth of the Drain does not reach that of the spring, and upon this all his future practise has been grounded".

It has been argued that these land drainers, working in the latter half of the eighteenth century, "played a key role in the evolution of the field of hydrogeology" (Stephens and Stephens, 2006, p. 1376). However, their understanding of hydrogeology was at the scale of a hillside and it was to be another surveyor/engineer, William Smith (1769–1839), who himself began draining land in 1796 (Torrens, 2004), whose appreciation of stratigraphy brought order to the search for underground water supplies needed to support the industrial revolution of the nineteenth century.

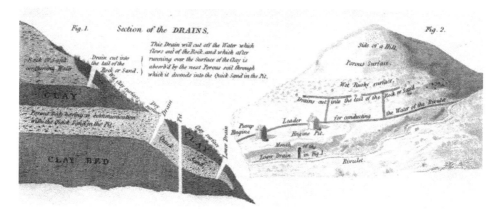

Figure 3 The application of Elkington's method of drainage. Adapted from Plan 10 in Johnstone (1797) and published by Longmans in 1816.

THE EARLY 19TH CENTURY

Smith's first known order of strata, drawn up in 1797, shows a keen awareness of the occurrence of spring lines and his better known 1799 version shows springs tabulated against five separate formations (Torrens, 2004). He had a good understanding of the stratigraphic control of spring lines, as well as the significance of hydraulic head, using this knowledge to advise landowners and canal companies on water supplies (Phillips, 1844; Sheppard, 1920). However, his only published work on water supply was a paper, read to the Yorkshire Philosophical Society (Smith, 1827), in which he discussed a method for supplementing the supply of Scarborough in the summer by controlling the discharge from an overflowing borehole made some years previously for draining the land. It was left to John Farey (1766–1826), Smith's pupil, to publicise his work and its relevance to water supply, particularly to the sinking of deep wells in the Thames Basin (Farey, 1807).

The gentlemen of the Geological Society of London, although not interested in the practical applications of geology, were involved in the collection of facts by experiment or observation in order to provide a basis for geological theories (Rudwick, 1963). In 1818 one of their members, William Phillips (1775–1828), summarised published strata descriptions to provide an outline of the geology of England and Wales (Phillips, 1818). Although this contained little concerning groundwater, the second edition, to which William Daniel Conybeare (1787–1857) added much original material, includes some hydrogeological information (Conybeare and Phillips, 1822). Within each stratum description, following discussion of the character, extent, thickness, etc, is a section entitled "phenomena of water and springs". Although some of these sections are only 2 or 3 lines in length others, such as those referring to the London Clay and Chalk, run to several pages. Terms, such as "porous", "porosity" and "impervious to water" are used and the poor quality of water in the London Clay is ascribed to the decomposition of pyrite. Water levels in four wells over a distance of 15 miles from the River Thames to Epping are plotted on a diagram which is probably the earliest British hydrogeological cross-section. A brave attempt is made by Conybeare to explain the variations in the water levels observed and he concluded that "The only general rule that can be deduced is, – that the water of wells can in no case rise to a higher level, than the highest point of the strata collecting them ..." (Conybeare and Phillips, 1822, p. 36). In contrast the *"Principles of Geology"*, the influential book by Charles Lyell (1797–1855), contains almost nothing on groundwater. Chapter 12 of Volume 1 is concerned with springs but only in their role as a means of transferring material in solution from depth and not in their role as a source of potable water (Lyell, 1830–1833).

In the early part of the nineteenth century significant advances in the study of groundwater were being made in France largely as a result of the drilling of artesian wells. The term artesian was not used by Conybeare and Phillips (1822) although they described the phenomenon of overflowing wells. The term seems to have first appeared in British journals around 1823 the year in which John Farey called for the translation of relevant material published abroad (Farey, 1823). Over the next decade the practical memoir of Garnier (1822) and the more theoretical work of Héricart de Thury (1829) were used widely in the absence of any comparable text in English, the first of which did not appear until 1849 (Swindell, 1849). The theory behind the occurrence of artesian flow was well understood by William Buckland (1784–1856) who produced excellent explanatory sections in his Bridgewater Treatise (Figure 4 from Buckland, 1836) based largely on French texts available in English in the 1830s (Héricart de Thury, 1830; Arago, 1835).

From the 1820s onwards papers related to groundwater began to appear in the *"Transactions* and *Proceedings of the Geological Society"* and from 1836 in the *"Transactions of the Institution of Civil Engineers"*. Many of the published items are merely descriptions of wells or boreholes together with the strata intersected (e.g. Yeats, 1826; Donkin, 1836). However, in 1831 the *"Proceedings of the Geological Society"* record that a letter was read "On the Influence of Season over the Depth of Water in Wells" (Bland, 1831). The work was later published in the Philosophical Magazine (Bland, 1832) and described a series of monthly observations made in a well near Sittingbourne in Kent from January 1819 to June 1831. These observations represented the first British systematic record of fluctuations in groundwater level and demonstrated that levels varied seasonally with the greatest depth of water "at and about the longest day" and the least depth "at or about the shortest day". Bland also measured the water levels along three traverses across the Chalk and was able to demonstrate that the height to which water rose in wells correlated with the rise and fall of the hills (Bland, 1832). At about the same time, the mining geologist William Jory Henwood (1805–1875) began to relate the amount of water pumped out of the Cornish Mines with rainfall, mine depth and whether the mine was in granite or slate (Henwood, 1831 and 1843).

By 1840 the construction of artesian wells and boreholes was generally understood and many had been sunk or drilled (Mylne, 1840). However, the concept did not meet with universal approval amongst engineers and there were some who felt that boring should be the last method of resort for the purpose of supplying a large town (Seward, 1836). Many boreholes failed because they were drilled by engineers with no geological background who assumed that you could drill almost anywhere

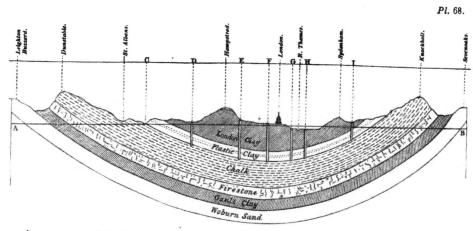

Pl. 68.

Section shewing the cause of the rise of water in Artesian Wells in the basin of London.

Figure 4 Section from Buckland (1836, Plate 68) showing the basin-shaped disposition of strata in the
London Basin and illustrating the rise of water in artesian wells. According to the explanation
in the text, water from rainfall accumulates in the joints and crevices of the Chalk and Plastic
Clay formations to the line A B at which height it overflows by springs in valleys such as that
represented in the section by C. The horizontal line A B represents the level to which water
would rise by hydrostatic pressure, in any perforations through the London Clay into sands
of the Plastic Clay or into the Chalk, such as those represented at D, E, F, G, H and I. If the
perforation is made at G or H, where the surface of the country is below the line A B, the
water will rise in a perpetually flowing artesian fountain.

and find artesian water (see Farey, 1822). Although Britain lagged behind in the development of drilling technology, pumping technology was well advanced largely because
of equipment inherited from the mining industry (Younger, 2004).

URBAN WATER SUPPLY

By the beginning of the 19th century, traditional sources of supply to metropolitan
London were proving inadequate (Figure 5) and, in 1828, a Royal Commission was
appointed to enquire into the matter. The commissioners recommended the filtration of
river water through sand but this failed to quieten public anxiety and the distinguished
engineer Thomas Telford was engaged to suggest a practicable plan (Telford, 1834).
However, nothing came of this and in 1840 a new scheme was put forward to sink
wells into the Chalk adjacent to the River Colne, near Watford, north of London. The
railway engineer Robert Stephenson (1803–1859) was asked to advise on the scheme
which he supported (Stephenson, 1840). His proposals were based on the premise
that recharge was so rapid that there was little evaporation and most of the rainfall
accumulated in the lower part of the Chalk forming an enormous natural reservoir
from which water could be abstracted without affecting surface springs and streams.

Figure 5 View of the conduit at Bayswater published on the 5th October 1796 by L. Stockdale, Piccadilly. The water that supplied the conduit came from the Westbourne which arose by several small streams from the west side of Hampstead Heath to the north. South of Bayswater, in Hyde Park, its valley is occupied by the Serpentine.

The scheme was vigorously opposed by a number of local landowners and mill-owners. Pamphlets and counter-pamphlets were issued and the debate became extremely acrimonious (Preene, 2004; Mather, 2008). John Dickinson and Co owned a number of paper mills along local rivers and, since 1835, had used a percolation gauge to measure the amount of rainfall which percolated through the soil to a lower

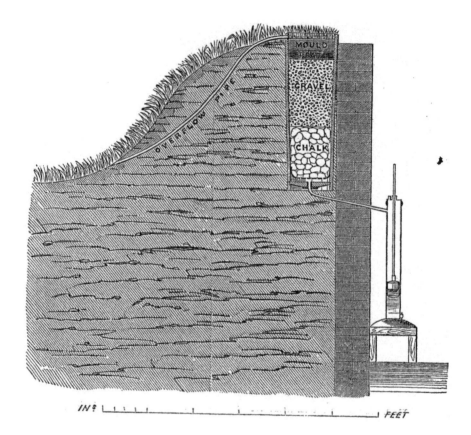

Figure 6 Sectional view of the percolation gauge used by the papermakers, John Dickinson and Co,
at Apsley Mill adjacent to the River Gade in Hertfordshire between 1835 and 1853, as figured
by Homersham (1855).

level and which was available for the supply of springs and rivers (Figure 6). The
company's early measurements showed that only one third of the rainfall reached the
Chalk in contrast to Stephenson's view that this was almost 100% (Mather, 2008).
Another objector was the Reverend James Charles Clutterbuck, whose family owned
property adjacent to the River Colne. Clutterbuck made a series of detailed systematic
observations of groundwater levels the conclusions from which were reported in a
series of presentations to the *Institution of Civil Engineers* (Clutterbuck, 1842, 1843
and 1850). His measurements suggested that very little surplus water was available
for abstraction from the Chalk when that carried off by the river was subtracted from
that which percolated to the water table.

The scheme did not go ahead but the debate which took place is important because
it demonstrates how significantly knowledge of groundwater had advanced since 1800.
Stephenson (1841) accurately described the shape of the cone of depression around
a pumping well. The observations made by both Dickenson and Co and Clutterbuck enabled them to recognise the intimate relationship between surface water and

groundwater. Clutterbuck made systematic observations of groundwater levels, which he applied in a practical and innovative way to estimate aquifer recharge.

Joseph Prestwich (1812–1896), one of the gentleman geologists of the Geological Society, also became interested in the problems of London's water supply. In 1850 he read a paper to the Royal Institute of British Architects suggesting that the Cretaceous Upper and Lower Greensands beneath London might provide a suitable auxiliary source of supply (Prestwich, 1850). This paper was later expanded into a book (Prestwich, 1851) which became widely quoted and used, so much so that it was reissued shortly before his death (Prestwich, 1895). Although the Lower Greensand had a more limited range beneath London than he envisaged, and hence never yielded the volume of water anticipated, the book is significant for a number of reasons. In it Prestwich presents a review of the geology of the country around London followed by an appraisal of the hydrogeological properties of individual stratigraphic units and likely safe yields; it is in effect the first British hydrogeological memoir. Prestwich provides a map and sections which divide the strata according to their permeability (Figure 7 from Prestwich, 1851) and the map is probably the first British geological map to show hydrogeological information. Later Prestwich was a member of the Royal Commission on Water Supply, which reported in 1869, and became a widely respected authority on the application of geology to water supplies.

Working at about the same time as Prestwich, David Thomas Ansted (1814–1880) began experiments on the absorption of water by Chalk (Ansted, 1850) demonstrating that at least one third of the bulk of fully saturated chalk consisted of water. Data on the absorbent power of various rock types had been collected previously by the Commissioners appointed to select the building stone for the new Houses of Parliament, one of whom was William Smith (Barry et al., 1839). Numbers of deep boreholes in the London area increased rapidly and by 1844 government offices around Westminster together with the fountains in Trafalgar Square were supplied by wells sunk to the Chalk (Amos, 1860). By 1849 a section across the London Basin illustrating the origin of London's artesian wells was widely circulated (Figure 8 from Morris, 1849).

It was not only in south-east England that geologists and engineers became interested in groundwater. Whilst working in the coalfields of Lancashire and Cheshire, Edward Hull (1829–1917) recognised the value of the Permo-Triassic sandstones as a source of water to supply local towns. He recommended the line of a fault in these rocks as the best site for a well as it was certain to draw in water from a long distance (Hull, 1865). The Permo-Triassic rocks of the Liverpool and Manchester area were also the subject of considerable research activity from the mid-1800s. Most workers were amateur members of the local geological societies (Tellam, 2004) but Robert Stephenson also contributed (Preene, 2004).

By the middle of the 19th century boreholes were being drilled to considerable depths but it was soon recognised that water obtained from such wells was often unfit for domestic use. A borehole at Chichester in Sussex, completed in 1844 at 1054 ft (321 m) in the Cretaceous Upper Greensand, was chalybeate and strongly impregnated with hydrogen sulphide (Whitaker and Reid, 1899). An even deeper hole at Southampton, begun in 1842 and eventually abandoned in the Chalk at 1323.5 ft (403 m) some years later, yielded water containing 898 mg/l NaCl (Whitaker, 1910). At a depth of 1000 ft (305 m) the progress made did not exceed 3 ft (0.9 m) a week even under the most favourable conditions. By 1864 "... a heap of stones, a solitary sapling, and the

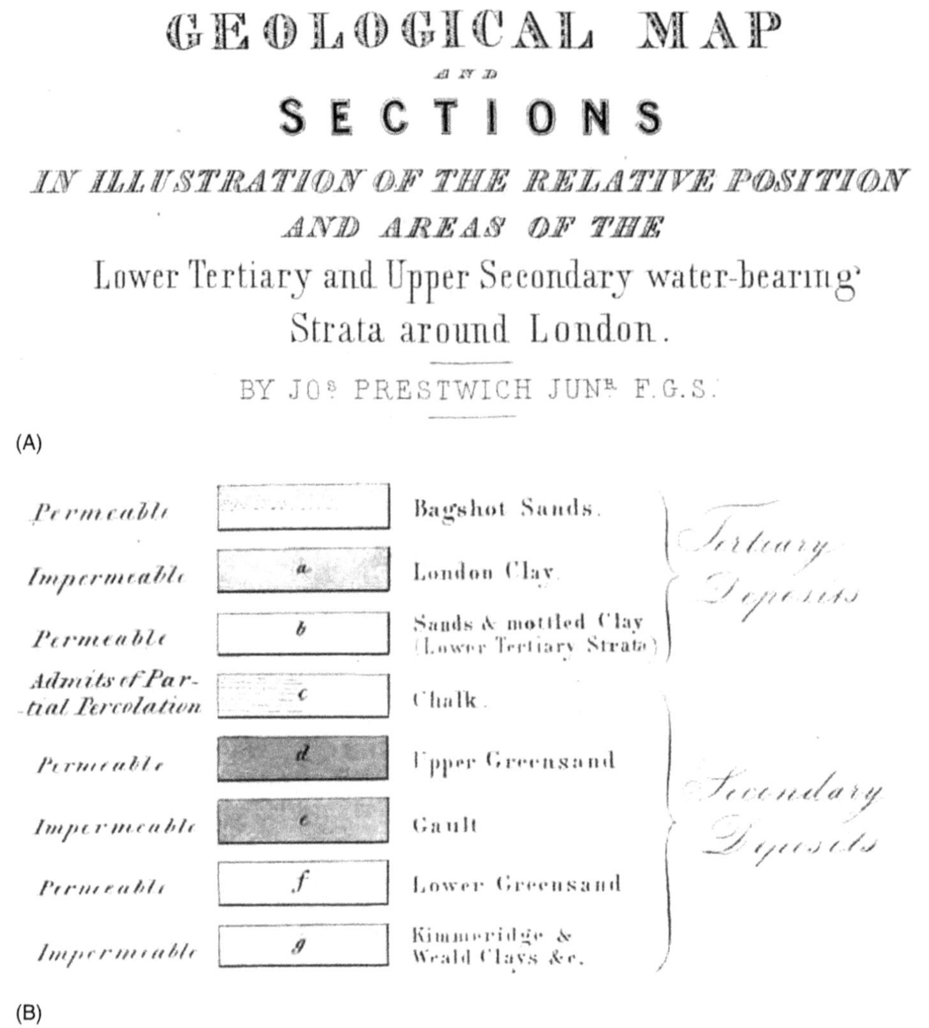

GEOLOGICAL MAP
AND
SECTIONS
IN ILLUSTRATION OF THE RELATIVE POSITION
AND AREAS OF THE
Lower Tertiary and Upper Secondary water-bearing
Strata around London.

BY JO⁹ PRESTWICH JUN⁸ F.G.S.

(A)

Permeable		Bagshot Sands.	Tertiary Deposits
Impermeable	*a*	London Clay.	
Permeable	*b*	Sands & mottled Clay (Lower Tertiary Strata)	
Admits of Partial Percolation	*c*	Chalk.	
Permeable	*d*	Upper Greensand	Secondary Deposits
Impermeable	*e*	Gault	
Permeable	*f*	Lower Greensand	
Impermeable	*g*	Kimmeridge & Weald Clays &c.	

(B)

Figure 7 Title (A) and Key (B) to the map and sections illustrating 'The water-bearing strata of the country around London" (Prestwich, 1851).

interest of the money expended, [were] the only relics left to record the fate of one of the most costly and interesting experiments in well-making ever perpetuated" (Bond, 1864, p. 209, his italics). The experience gained from these costly failures enabled the conditions which are necessary for the success of an artesian well to be set out. These conditions were (Bond, 1864, p. 207);

1   the existence of a porous stratum having a sufficient outcrop on the surface to collect an adequate amount of rainfall, and passing down between two impermeable strata;

Figure 8 Section of the London Basin Artesian Wells published in London by James Reynolds in 1849.

2   the level of the outcropping portion of the porous stratum must be above that of the orifice of the well, so as to give a sufficient rise to the water;

3   there must be no outlet in the porous stratum by which its drainage can leak out, either in the shape of a dislocation, by which it can pass into lower strata, or a natural vent, by which it can rise to the surface at a lower level than the well.

These conditions cover the seven prerequisites for artesian flow supposedly first presented by the American, Thomas Chrowder Chamberlin (1843–1928) in 1885, by over 20 years (cf. Chamberlin, 1885, p. 134–135).

Water quality became an issue early in the century but it was not until 1854 that John Snow (1813–1858), working in Westminster, demonstrated beyond doubt that cholera was spread by contamination of drinking water (Price, 2004). Subsequent research showed that sewage effluent derived from a local cesspool was to blame. Snow's study represents one of the first studies of an incident of groundwater contamination in Britain. By now analysts had become more confident in their determination of the constituents of water. Mineral springs such as those at Bath (Edmunds, 2004) became field sites for testing developments in analytical chemistry, but doubts about the true composition of such waters remained (see discussion in Hamlin, 1990).

As early as 1850 Lyon Playfair (1819–1898), at that time the chemist at the School of Mines and later a noted administrator and educational reformer, provided an excellent qualitative description of cation exchange in Chalk groundwaters (Playfair in discussion of Clutterbuck, 1850, p. 160). However, most of the early papers published by chemists were reports on the species dissolved in groundwater although some speculated on the source of groundwater and how it achieved its composition (Campbell, 1857).

THE GEOLOGICAL SURVEYORS

Prior to 1870 the Geological Survey of Great Britain made little contribution to the application of geology to water supply. As early as 1850, some 15 years after its

HYDROGEOLOGY:

ONE OF THE DEVELOPMENTS OF

MODERN PRACTICAL GEOLOGY.

By JOSEPH LUCAS, F.G.S.,

Read at the Ordinary General Meeting of the INSTITUTION OF SURVEYORS
February 26th, 1877.

THE PRESIDENT IN THE CHAIR.

Figure 9 The heading from the paper by Joseph Lucas (1877a) in which he defined hydrogeology as follows: "Hydrogeology takes up the history of rain water from the time that it touches the soil, and follows it through the various rocks which it subsequently percolates".

formation and under its first Director Henry Thomas de la Beche (1796–1855), complaints were already being made that mapping work should be transferred from North Wales, where no need existed for early geological information, to the metropolitan districts to investigate the deep water-bearing strata (Clutterbuck, 1850). The second Director of the Survey, Roderick Impey Murchison (1792–1871), was a distinguished gentleman geologist and Fellow of the Royal Society, who had little interest in the economic applications of geology (Flett, 1937). However, three surveyors whom he recruited were to make an impact in the last 30 years of the nineteenth century.

According to the bibliography prepared by Whitaker (1888), Hull's memoir on the geology of the country around Bolton-le-Moors in Lancashire was the first to include well sections (Hull, 1862). However, after Murchison's death in 1871 the number of memoir pages devoted to well sections increased enormously covering some 141 pages in the London Basin Memoir (Whitaker, 1872). Subsequently all memoirs relating to south east England had lists of well sections and information concerning water supply. Eventually these lists came to dominate the geological memoirs such that in 1899 the first Water Supply Memoir, on Sussex, was produced (Whitaker and Reid, 1899).

Many of the memoirs were authored by William Whitaker (1836–1925) who had joined the Survey in 1857 (George, 2004). Whitaker was an assiduous collector of records of well sections and temporary exposures and he also compiled numerous bibliographies. However, he made little use of the data he collected, preparing lists of information rather than using this information to understand hydrogeological processes. According to Wilson (1985, p. 125) "There is no doubt ... that Whitaker was the father of English hydrogeology" although Mather (1998) has characterised his contribution as "worthy".

Joseph Lucas (1846–1926) joined the Geological Survey in 1867 and spent 9 years mapping in Yorkshire before being forced to resign for a disciplinary offence (Mather,

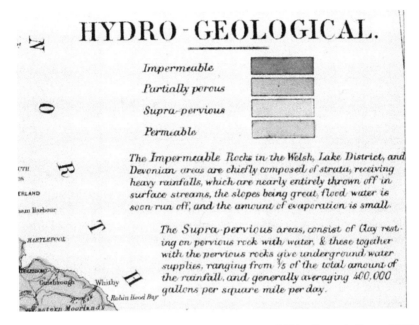

HYDRO - GEOLOGICAL.

Impermeable

Partially porous

Supra-pervious

Permeable

The Impermeable Rocks in the Welsh, Lake District, and Devonian areas are chiefly composed of strata, receiving heavy rainfalls, which are nearly entirely thrown off in surface streams, the slopes being great, flood water is soon run off, and the amount of evaporation is small.

The Supra-pervious areas, consist of Clay resting on pervious rock with water, & these together with the pervious rocks give underground water supplies, ranging from ⅕ of the total amount of the rainfall, and generally averaging 400,000 gallons per square mile per day.

Figure 10   Key to the hydrogeological map of England and Wales (from de Rance, 1882).

et al., 2004). In 1874, whilst still with the Survey, he was the first person to use the term "hydrogeology" in its modern context (Lucas, 1874; Mather, 2001) and defined this new subject in a series of papers in the 1870s (Figure 9 from Lucas, 1877a). He used water level data to draw the first British maps showing groundwater contours and described how to carry out a hydrogeological survey (Lucas, 1874, 1877b and 1878). His innovative work was immediately taken up by other workers and the term hydrogeology was widely adopted (e.g. Prestwich, 1876; Inglis, 1877). For the next ten years of his life Lucas lobbied for hydrogeological surveys to be carried out over the whole country but his ideas were ignored.

The third geologist in the Survey to make an impact was Charles Eugene de Rance (1847–1906) who joined in 1868, the year after Lucas. On behalf of the Survey he assisted the Rivers Pollution Commission in the preparation of its sixth report on domestic water supply (Rivers Pollution Commission, 1875). The Commissioners concluded that "spring waters" and "deep well waters" were the best sources of supply providing wholesome and palatable water for drinking and cooking. This testimonial provided a boost to the development of groundwater. In 1874 he was appointed the secretary of a Committee appointed by the British Association for the Advancement of Science for the purpose of investigating the circulation of underground waters. Edward Hull was the Chairman of this Committee and although it sat for 20 years it achieved little of substance. De Rance's work on these two bodies together with his own research enabled him to produce a 600 page volume on the water supply of England and Wales (de Rance, 1882). The book is an account of the water supplied to each town and urban sanitary authority in England and Wales and includes a small hydrogeological map (Figure 10), on which the strata are divided in terms of their permeability.

HYDROGEOLOGY AS A SUB-DISCIPLINE

In the latter part of the 19th century the involvement of the Geological Survey was but one part of a trend which saw the study of the geological aspects of water supply becoming respectable. Joseph Prestwich was elected President of the Geological Society in 1870 and in 1872 chose as the subject of his Presidential Address "Our Springs and Water-supply and Our Coal-measures and Coal-supply" (Prestwich, 1872). In 1876 John Evans (1823–1908) also chose "Water Supply" as one of the topics of his second Presidential Address (Evans, 1876) and in 1898 Whitaker was elected President choosing "Water-supply and Sanitation" as his first topic (Whitaker, 1899).

Throughout the country groundwater was now being exploited for both urban and rural supplies and geologists, chemists and engineers were involved in exploration and development. Numerous papers and reports appeared, many emphasising the vastness of the underground water supplies available. The debate about supplying London from the Chalk continued with geologists emphasising the interdependence of groundwater and surface water and the dangers of overexploitation (e.g. Evans, 1876; Hopkinson, 1891) whilst most engineers still considered that rivers took their supply "from the skin of the chalk formation" with a much larger quantity of water available for extraction from "the great body of the Chalk below" (Harrison, 1891, p. 21).

Many papers were descriptive but some contributed towards an understanding of hydrogeological processes. In 1884 an earthquake caused much damage to buildings in Essex and had a significant impact on groundwater levels causing rises of up to 7 feet in some wells. Numerous reasons for the rise were advanced the most widely accepted being put forward by de Rance who suggested that the shock caused a widening of the fissures in the Chalk increasing the flow which led in turn to a rise in water level (de Rance, 1884). Whitaker was involved as a consultant in an early study of groundwater pollution (Whitaker, 1886). A brewery well in Brentford, west London, was found to be polluted from another well 297 ft (90.5 m) away which had been turned into the drainage for a privy belonging to a printing works. The connection between the wells was proved by the use of lithium chloride as a tracer. The case, Ballard v. Tomlinson established the principle that no owner had the right to pollute a source of water common to his own and other wells.

The agricultural chemist, Robert Warington (1838–1907) brought together existing information to trace the connection between the composition of rain, drainage and deep well waters (Warrington, 1887). To detect contamination from sewage, he recognised that it was necessary to know the normal concentration of chloride and nitrate in uncontaminated supplies. He measured concentrations of 11 ppm chloride and 4.4 ppm nitrate in uncontaminated Chalk wells north of London. He compared the concentration of chloride in outcrop samples with that found in chalk waters beneath the Tertiary cover, concluding that a portion of the chlorides in the deep strata was probably "derived from a residue of sea salt remaining in the rock" (Warington, 1887, p. 551).

In a report (Figure 11) on the amount of water available from the Chalk downs near Worthing in Sussex, the engineer and meteorologist Baldwin Latham (1836–1917) described the manner in which water moved through the unsaturated zone above the water table (called by Latham the water line). "On examining the chalk formation above the water line, it is found to be completely saturated with water,

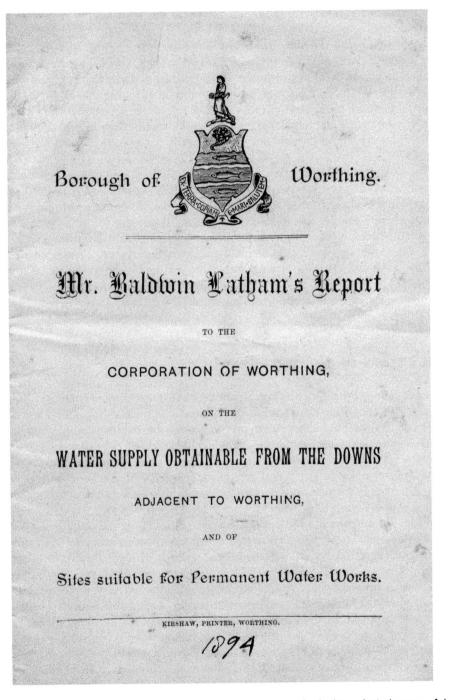

Figure 11   Cover of the report of Mr Baldwin Latham reporting on his hydrogeological survey of that part of the South Downs which was likely to afford a supply of water to the Sussex town of Worthing (Latham, 1894). In this report he accurately estimated the rate at which water moved through the unsaturated Chalk matrix to the water table.

and when the balance of the rainfall, after evaporation, enters the earth near the surface, it displaces a corresponding quantity that has been held in the chalk at the bottom or near the water line, and the amount of water actually pushed out from the strata varies with the amount of rain that enters the ground; as a rule, the rain water in sound chalk would only descend on an average about a yard every year, so that the number of yards in depth of a chalk well, from the surface of the ground to the water line, would roughly indicate the number of years the rain of any particular year would require to descend to the water line" (Latham, 1894, p. 13). It was to be over 70 years before the important role played by matrix flow in unsaturated chalk was "rediscovered" and the rate of movement deduced by Latham confirmed (Smith et al., 1970).

The influence of barometric pressure on the discharge of springs was described (Latham, 1882) and the dual permeability of the Chalk was recognised. Adits or headings were dug laterally from shafts to intersect as many fissures as possible and these reached a length of over two miles (3.2 km) at the Ramsgate Waterworks in Kent. It was recognised that vast amounts of water were available in the body of the Chalk but that "... the failure of wells, sunk in chalk free from fissures and cavities, proves that capillary water does not travel with sufficient swiftness to be available" (Dawkins, 1898, p. 262). Nearly all the work was concentrated in England largely because of the abundance of good quality surface waters in Wales, Scotland and Ireland (Robins, et al., 2004).

EARLY YEARS OF THE 20TH CENTURY

At the beginning of the 20th century attention focussed on the ownership of groundwater and the "Kent Water Preservation Association" was formed to conserve the waters of Kent for its own inhabitants and to prevent them being exploited to supply London. The association was later extended to include other counties, the name altered to "The Underground Water Preservation Association" and a pamphlet issued (Beadle, 1902). It was now generally admitted that groundwater levels beneath London were falling as a result of over-abstraction (Beadle, 1903) and this was demonstrated conclusively by Barrow and Wills (1913).

In the years between 1900 and 1938 the Geological Survey published 27 Water Supply Memoirs. These consisted mostly of records of wells and boreholes many of which could be characterised as "unchecked records of wells that have been sunk through ill-defined strata" (Bailey, 1952, p. 202). Prior to 1940 the approach of the Survey was extremely conservative and progress compared very unfavourably with that made by the United States Geological Survey during this period. The general descriptive approach of the British Survey is well illustrated by the textbook on the "Geology of Water Supply" written by Horace Bolingbroke Woodward (1848–1914) a former senior member of the Directorate (Woodward, 1910).

A number of workers knew of the pioneering work going on in the USA and Europe and were able to use it to examine groundwater flow in British aquifers (e.g. Baldwin-Wiseman, 1907). Later Norman Savage Boulton (1899–1984), in work which predated that of Theis in the USA, examined the time-variant flow to a pumped well in a confined aquifer (Downing et al., 2004). However, to his great disappointment his manuscript was rejected for publication, a sad reflection of the conservative views of his peers in the UK in the early 1930s.

Some innovative work was carried out by consultants such as Charles Lapworth (1842–1920), the first professor of geology at Mason College (subsequently Birmingham University), who advised widely on water supply. His work for the brewing industry (Lapworth, 1901) was carried forward by successive professors of geology at Birmingham (Boulton, 1932; Shotton, 1952; Lloyd, 1986). Lapworth's son Herbert (1875–1933) established his own consultancy practice in 1910 (de Freitas and Rosenbaum, 2008) and, amongst other topics, specialised in the geology of water supply. One of his colleagues, Rupert Cavendish Skyring Walters (1888–1980), summarised the hydrogeology of the Chalk (Walters, 1929), compiling groundwater contour maps for the Chalk throughout England (Figure 12). Later he completed a comparable study of the Jurassic Oolitic Limestone (Walters, 1936a) and wrote a well received book on water supply (Walters, 1936b). Arthur Beeby-Thompson (1873–1968) was another well known consultant who was still active in his 80s (Beeby Thompson, 1969).

Many geologists worked for the Geological Surveys of Britain's overseas colonies and protectorates where they became involved in the provision of water supplies. The use of geophysical techniques in borehole siting was pioneered in West Africa (Hazell, 2004) and Southern Rhodesia, now Zimbabwe (Barker, 2004). The British colonial geologists Frank Dixey (1892–1982) and Cyril S. Fox (d.1952) wrote practical textbooks which were widely used overseas (Dixey, 1931; Fox, 1949).

In 1901 the origin of alkaline waters in the chalk was explained by the gradual dissolution of sodium carbonate (Fisher, 1901). Later John Clough Thresh (1850–1932) showed that the rocks themselves possessed "the power of softening hard water by substituting sodium salts for those of calcium and magnesium", describing the process of cation exchange (Thresh, 1912, p. 43).

Groundwater pollution, particularly outbreaks of typhoid fever resulting from sewage leaking directly into wells, received considerable publicity. However the inherent safety of well waters was recognised and experience showed that "water in slowly percolating through a few feet of compact soil cannot carry with it the microbe causing typhoid fever" (Thresh, 1908, p. 109). Sections on contamination and the risk involved were included in some of the longer Geological Survey Water Supply Memoirs (e.g. Whitaker, 1912). Parliament empowered many water boards to make bye-laws for protecting land around wells and in the Margate Act of 1902 the water board was given the power to control drains, closets, cesspools etc. over an area of 1500 yards (1372 m) from any well or adit (Thresh and Beale, 1925).

During 1932–1934 there was a severe drought in southern and central England. This led to the appointment of an Inland Water Survey Committee in 1935 and the eventual formation of a Water Unit within the Geological Survey in 1937 (Downing, 2004a). The first geologist assigned to this Unit, Francis Hereward Edmunds (1893–1960), began to introduce order to the Survey's water records. The state of knowledge in the Survey at that time was summarised by the then Director, Bernard Smith (1881–1936), in his Cantor lectures to the Society of Arts (Smith, 1935). A Survey handbook on the "Water Supply of England and Wales from Underground Sources" was in preparation at the outbreak of war in 1939 and some of the subject matter was subsequently published (Edmunds, 1941). A permeable

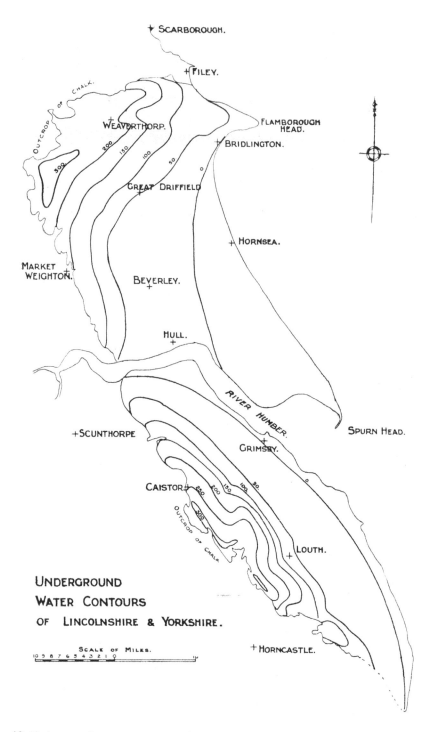

Figure 12 Underground water contours in the chalk of Lincolnshire and Yorkshire (Fig. 1 in Walters, 1929).

rock was defined as one which, "when maintained at saturation point, ... can supply a well of ordinary dimensions with a minimum daily yield of 75 gallons (341 litres) and so meet the requirements of a small private house" (Edmunds, 1941, p. 25). In discussion this definition was described as being "as novel as it is original" and did not meet with universal approval (Morton in discussion of Edmunds, 1941, p. 88).

During the war many staff were diverted to water supply work and contributed to a build up of expertise within the Survey which was to prove valuable in post-war Britain. The urgent need for additional and emergency supplies both at home and overseas also involved academics such as William Bernard Robinson King (1889–1963), then Professor of Geology at University College, London, and Frederick William Shotton (1906–1990), a lecturer at Cambridge, both of whom advised the army (Rose, 2004) and Percy George Hamnall Boswell (1886–1960), Professor of Geology at Imperial College London, who advised the Metropolitan Water Board (Boswell, 1949). In addition consultants continued to make a major input (e.g. Herbert Lapworth Partners, 1946).

CENTRALISATION AND EXPANSION

The Water Act of 1945 was the first piece of major legislation affecting water supply in the UK for almost 100 years and was part of the social revolution which followed the end of the Second World War (Downing, 1993). The period between 1945 and 1963 has been described as an era of resource assessment by Downing and Headworth (1990) and the 1945 Act provided the framework within which these assessments were made. Information on the aerial distribution of rainfall was already well established in the UK and the work of Penman (1948) enabled evaporation to be estimated. Thus there was the opportunity to assess groundwater resources with far greater accuracy than in the past.

The Geological Survey took the lead in this work as, following the 1945 Act, they became advisors to the Government on technical aspects of groundwater development (Downing, 2004a). Under the leadership of Stevenson Buchan (1907–1996), the Survey recruited a group of staff who came to consider themselves as hydrogeologists rather than geologists with a peripheral interest in groundwater (Gray and Mather, 2004). They were primarily engaged on regional hydrogeological surveys to assess groundwater resources but they were also responsible for testing all new major public supply and industrial wells in order to assess appropriate sustainable yields before abstraction licences were issued by the Government.

Using data from many of the wells tested by the Survey, Jack Ineson (1917–1970) began to apply methods developed in the USA to heterogeneous British aquifers such as the Chalk (Downing and Gray, 2004). Other staff pioneered the use of downhole logging techniques adapted from the oil industry, analysed the river/groundwater interface and synthesised the well records using the data to compile hydrogeological maps (Downing, 2004a; Gray and Mather, 2004). In the late 1950s and early 1960s, the Ministry of Housing and Local Government, in cooperation with the Survey, made a series of hydrological surveys to identify the availability and use of water resources in England and Wales.

Outside the Survey, Boulton was working on the delayed yield observed when unconfined aquifers are pumped under non-steady state conditions (Boulton, 1954; Downing *et al.*, 2004). The potential of artificial recharge began to be taken seriously in the London Basin (Boniface, 1959) and natural radioactivity in groundwaters became an issue (Turner, *et al.*, 1961).

Unfortunately the 1945 Water Act did not recognise the close links between groundwater and surface water which led to difficulties, as aquifer development and river management were the responsibility of different organisations. The reduction in stream flows as a consequence of increasing groundwater development, particularly in the Chalk, became a major factor leading to the Water Resources Act of 1963 (Downing, 2004b). This created the Water Resources Board (WRB), to plan water resources development on a national scale, and 29 catchment-based River Authorities to control abstraction, prevent pollution, drain land and protect fisheries. The period which ensued from 1963 to 1974 has been called an era of groundwater management by Downing and Headworth (1990).

The WRB made three major regional studies of water resources and, in 1973, proposed a national water strategy (Downing, 2004b). The core of its Geology Division was made up of seven staff transferred from the Geological Survey, led by Ineson. It was enthusiastic in its support for the development of groundwater within integrated water resource systems, in particular the use of groundwater for river regulation, its use in conjunction with surface water and artificial recharge. In association with Ken Rushton, of Birmingham University, it built some of the first electrical analogue and then mathematical models used in Britain (Downing, 2004b). Many of the studies carried out by WRB relied on the cooperation of the water supply and river authorities some of which began to recruit their own geological staff (Headworth, 2004).

The 1963 Act removed both experienced staff and the groundwater advisory service from the Geological Survey. However, with the support of successive Directors, the Survey increased its research role, undertaking both fundamental and applied research projects in the UK and overseas (Gray, 2004). Laboratories were commissioned for research into groundwater modelling, hydrogeochemistry and the measurement of aquifer physical properties. Production of hydrogeological maps was continued using data held by the Survey and groundwater contamination, particularly from landfill leachates and intensive agriculture received significant support from central Government (Gray, 2004).

At the University of Bath, John Napier Andrews (1930–1994) studied the release of radon from rock matrices and its entry into groundwater (Andrews and Wood, 1972). He subsequently became a pioneer in the application of noble gases to problems in hydrogeology (e.g. Andrews and Lee, 1979). Analysis of the effects of the severe drought of 1975–6 failed to discern any long-term adverse effects on British aquifers with respect to either water levels or water quality (Day and Rodda, 1978).

The years between 1963 and 1975 were probably the most significant in the history of British hydrogeology. The number of individuals involved in groundwater work, excluding those within consulting engineering companies, rose from less than 20 to around 150. This increasing demand resulted in the setting up of Masters Courses in Hydrogeology at University College London in 1965 and Birmingham University in 1971. The need for a discussion forum led to the formation of the Hydrogeological Group of the Geological Society in 1974 and the Sub-Committee for Hydrogeology of the British National Committee for Geology, serviced by the Royal Society, in 1975. During this period hydrogeology changed from a fringe subject to a mainstream branch of geology in the UK.

REGIONAL DEVELOPMENT AND QUALITY CONSIDERATIONS

Since the mid-1970s the development of hydrogeology in Britain has been intimately connected with changes in legislation. A weakness of the Water Resources Act was that there was inadequate provision for the coordination of water resources development and water quality control which were in the hands of separate organisations (Downing, 1993). By the mid-1970s there was also a move to devolve power to the regions rather than concentrate it centrally and the Water Act of 1973 was part of the Government's reorganisation of local government to achieve this aim. The WRB was disbanded and ten regional water authorities created whose areas of operation were defined by river catchment boundaries so that the whole of the water cycle in a particular area, including water supply and sewage treatment, was under the control of a single body (Brassington, 2004). A further restructuring took place in 1989 when the ten water authorities were privatised to become water supply and sewage utility companies with their regulatory function transferred to a new body, the National Rivers Authority. In 1996 the latter became part of the new Environment Agency. The greater emphasis on quality from 1975 onwards prompted Downing and Headworth (1990) to define this period as an era of groundwater quality.

In Scotland and Northern Ireland groundwater received little attention before the mid-1970s when the Geological Survey appointed dedicated hydrogeologists in Edinburgh and Belfast (Robins *et al.*, 2004; Gray, 2004). The water industry in Scotland was not privatised in the same way as that in England and Wales and became the responsibility of one multi-functional authority – Scottish Water. Since 1996 environmental regulation, including groundwater protection has been the responsibility of the Scottish Environmental Protection Agency. In Northern Ireland responsibilities for water supply and regulation are vested in the Department of the Environment for Northern Ireland.

The need of regional authorities for personnel to staff water resources and regulatory sections meant an increasing demand for hydrogeologists. Since the changes of 1989 in England and Wales some reduction in numbers has taken place as both the utilities and regulators increased their reliance on consultants. However, with the increase in contract work, the consultants themselves have a requirement for hydrogeologists which many have found difficult to satisfy. Both the Geological Survey and the Water Research Centre (formed in 1974 from the research arm of WRB and the Water Research Association, the research organisation funded by the water supply industry) have continued with research and contract work. The interest in hydrogeology is now such that the Hydrogeological Group of the Geological Society has a membership of around 1050. Some of the work carried out by this large cohort has been described by Headworth (2004) and Brassington (2004).

One major change over the last 30 years has been the diversification of the scope of the work now carried out by the hydrogeological community in Britain. The traditional fields of water supply and water quality continue to dominate but hydrogeologists have also made major contributions to studies on the disposal of radioactive wastes (Chaplow, 1996); geothermal energy (Barker *et al.*, 2000); the problem of rising groundwater levels (Brassington, 1990); mineralisation (Barker *et al.*, 1999) and climate change (Edmunds and Milne, 2001).

DISCUSSION AND CONCLUSIONS

A conscious decision has been taken not to call this chapter the history of hydrogeology in Britain. Although a wealth of information has been accumulated, and collated into a coherent story, this remains a series of snapshots and reviews of the advances which were made during particular periods and some of the individuals involved. Much information remains hidden away in obscure journals and forgotten pamphlets from the 18th and 19th centuries, which are available only in a few libraries. As one delves further into this material it becomes apparent that particular individuals, to whom certain advances are attributed, were in fact merely expressing the scientific thought of the day.

During the last 400 years a series of "drivers" can be identified which resulted in advances in our knowledge of hydrogeology. The first of these is the increasing interest in mineral springs in the early 17th Century which resulted in the questioning of traditional ideas on the origin of springs. 1660 saw the foundation of the Royal Society and the use of experiments to test hypotheses and explain natural phenomena. Scientific enquiry continued to be a driving force into the 18th century but the need to bring more land into agricultural production, to feed an expanding population, encouraged a generation of land drainers to build up a broad understanding of the movement of shallow groundwater and its control.

From the beginning of the 19th century the search for water supplies to support the industrial revolution became a priority. Technological advances in France and the stratigraphical work of William Smith enabled wells to be sunk and/or bored with a much greater chance of success. By the middle of the century traditional sources of supply to many conurbations were proving inadequate and the conflicts which arose between water engineers and early hydrogeologists led to measurements of percolation and estimates of recharge to groundwater reservoirs. In the London area, Clutterbuck made systematic observations on groundwater levels, applying these in a practical and innovative way, Prestwich wrote the first hydrogeological memoir and Snow demonstrated that cholera was spread by contaminated well water.

The increasing demand for water supplies led to an interest within the fledgling Geological Survey. William Whitaker was an avid collector of records of well sections and, although most of his work was descriptive, its sheer volume enhanced the profile of groundwater as a source for water supply. His colleague Joseph Lucas used the term hydrogeology in its modern context, drew the first British maps to show groundwater contours and was probably the first person to call himself a hydrogeologist. Charles de Rance produced the first hydrogeological map of the whole of England and Wales.

By 1900 the study of underground water had become an accepted branch of geology but the following 30 years was a period of relative inactivity in Britain. The few practitioners were active in developing groundwater supplies but contributed little to the science. Despite the clear need for a hydrogeological survey of the whole country no Government staff were appointed to work on underground water until the 1930s and perhaps this was the reason why the important advances in hydrogeology took place in mainland Europe and North America rather than in Britain.

The position changed in the early 1930s when a severe drought put a strain on water resources. The setting up of an Inland Water Survey Committee led to the formation of a Water Unit within the Geological Survey in 1937 and, shortly afterwards, at the beginning of the Second World War, many staff were diverted to groundwater work. At the end of the War, the 1945 Water Act provided, for the first time, a framework within which groundwater resources could be assessed and heralded a new era for hydrogeology in which the Geological Survey, joined after 1965 by the Water Resources Board, led the way. Although initially development took place in a piecemeal fashion with the emphasis on local problems and solutions, the Water Act of 1945 saw the beginnings of a quantitative approach to hydrogeology and the recognition that sectional interests needed to be subordinate to the national interest.

The period between 1945 and 1990 has been divided into 3 phases of (1) resource assessment (1945–1963); (2) groundwater management (1963–1974) and (3) groundwater quality (1974–1990) (Downing and Headworth, 1990). The period between 1963 and 1974 was probably the most significant in the development of British hydrogeology and saw the subject develop into a mainstream branch of geology, taught at University level, and represented within the Learned Societies. The period from 1990–2006 has been one of increasing integration with other earth and environmental sciences such that a 4th phase of "groundwater in integrated environmental management" has been suggested (Skinner, 2008). These developments have seen a continual increase in the number of practising hydrogeologists and a gradual change in the emphasis of their work, a trend which is likely to continue into the future.

REFERENCES

Amos, C.E. (1860) On the government waterworks in Trafalgar Square. *Proceedings of the Institution of Civil Engineers*, 19, 21–52.

[Anderson, J.] A farmer (1775) *Essays Relating to Agriculture and Rural Affairs; in Two Parts, Illustrated with Copperplates*. Edinburgh, William Creech.

Anderson, J. (1797) *A Practical Treatise on Draining Bogs and Swampy Grounds, Illustrated by Figures; with Cursory Remarks upon the Originality of Mr Elkington's Mode of Drainage*. London, G. G. and J. Robinson.

Andrews, J. N. & Wood, D.F. (1972) Mechanism of radon release into rock matrices and entry into groundwaters. *Transactions of the Institute of Mining and Metallurgy*, B81, 198-209

Andrews, J.N. & Lee, D.J. (1979) Inert gases in groundwater from the Bunter Sandstone of England as indicators of age and palaeoclimatic trends. *Journal of Hydrology*, 41, 233–252.

Anon. (1675) A particular account, given by an anonymous French Author in his book of the Origin of Fountains, printed 1674 at Paris; to shew, that the Rain and Snow-waters are sufficient to make Fountains and Rivers run perpetually. *Philosophical Transactions of Royal Society of London*, 10, 447–450.

Ansted, D.T. (1850) On the absorbent power of Chalk and its water contents under different conditions. *Proceedings of the Institution of Civil Engineers*, 9, 360–375

Arago, D.F.J. (1835) On springs, artesian wells, and spouting fountains. *Edinburgh New Philosophical Journal*, 18, 205–246.

Atwell, J. (1732) Conjectures upon the nature of intermitting and reciprocating springs. *Philosophical Transactions of the Royal Society of London*, 37 (for 1731–32), 301–316.

Bailey, E.B. (1952) *Geological Survey of Great Britain*. London, Thomas Murby and Co.. 278p.

Baldwin-Wiseman, W.R. (1907) The influence of pressure and porosity on the motion of sub-surface water. *Quarterly Journal of the Geological Society of London*, 63, 80–105.

Barker, J. A., Downing, R.A., Gray, D.A., Findley, J., Kellaway, G.A., Parker, R.H. & Rollin, K.E. (2000) Hydrogeothermal studies in the United Kingdom. *Quarterly Journal of Engineering Geology and Hydrogeology*, 33, 41–58.

Barker, J.A., Downing, R.A., Holliday, D.W. & Kitching, R. (1999) Hydrogeology. In: Plant, J.A. & Jones, D.G. (eds.) Development of regional exploration criteria for buried carbonate-hosted mineral deposits: a multidisciplinary study in Northern England. *British Geological Survey, Technical Report*, WP/91/1, 119–126.

Barker, R.D. (2004) The first use of geophysics in borehole siting in hardrock areas of Africa. In: Mather, J.D. (ed.) *200 Years of British Hydrogeology*. Geological Society, London, Special Publications, 225, 263–269.

Barrow, G. & Wills, L. J. (1913) *Records of London Wells*. Memoirs of the Geological Survey. England and Wales. London, HMSO, London.

Barry, C., de la Beche, H.T., Smith, W. & Smith, C.H. (1839) *Report with Reference to the Selection of Stone for Building the New Houses of Parliament*. London, Commissioners of Her Majesty's Woods, Forests, Land Revenues, Works and Buildings.

Beadle, C. (1902) The abstraction of underground water and its local effects. *Journal of the Sanitary Institute*, 23, 467–474.

Beadle, C. (1903) Evidence as to the cause and effect of the lowering of the permanent water levels in the London water basin. *Journal of the Sanitary Institute*, 24, 400–403.

Beeby-Thompson, A. (1969) *Exploring for Water*. London, Villiers Publications.

Bland, W. (1831) On the influence of season over the depth of water in wells. *Proceedings of the Geological Society of London*, 21, 339–340.

Bland, W. (1832) Letter from William Bland, Jun. Esq., of New Place in the Parish of Hartlip, near Sittingbourne, Kent, to Dr Buckland; recording a series of observations made by himself, on the rise and fall of water in wells in the County of Kent. *Philosophical Magazine and Annals of Philosophy, New Series*, 11, 88–96.

Bond, F.T. (1864) The geology of mineral springs. *Popular Science Reviews*, 4, 203–213

Boniface, E.S. (1959) Some experiments in artificial recharge in the lower Lee Valley. *Proceedings of the Institution of Civil Engineers*, 14, 325–338.

Boswell, P.G.H. (1949) *A review of the resources and consumption of water in the Greater London area*. London, Metropolitan Water Board.

Boulton, N.S. (1954) The drawdown of the water table under non-steady conditions near a pumped well in an unconsolidated formation. *Proceedings of the Institution of Civil Engineers*, 3, 564–579.

Boulton, W.S. (1932) Underground water supplies in the Midlands for brewery purposes. *Journal of the Institute of Brewing*, 38, 353–360.

Boyle, R. (1685).*Short memoirs for the natural experimental history of mineral waters*. London, Samuel Smith.

Brassington, F.C. (1990) A review of rising groundwater levels in the United Kingdom. *Proceedings of the Institution of Civil Engineers, Part 1*, 88, 1037–1057.

Brassington, F.C. (2004) Developments in UK hydrogeology since (1974) In: Mather, J.D. (ed.) *200 years of British Hydrogeology*. Geological Society, London, Special Publications, 225, 363–385.

Buckland, Rev. W. (1836) *Geology and mineralogy considered with reference to natural theology*. Bridgewater Treatise 6. London, William Pickering.

Campbell, D. (1857) On the source of the water of the deep wells in the Chalk under London. *Quarterly Journal of the Chemical Society*, 9, 21–27.

Chamberlin, T.C. (1885) Requisite and qualifying conditions of artesian wells. *US Geological Survey, Annual Report*, 5, 131–175.

Chaplow, R. (1996) The geology and hydrogeology of Sellafield: an overview. *Quarterly Journal of Engineering Geology*, 29, S1–S12.

Clutterbuck, J.C. (1842) Observations on the periodical drainage and replenishment of the subterraneous reservoir in the Chalk Basin of London. *Proceedings of the Institution of Civil Engineers*, 2, 155–160.

Clutterbuck, J.C. (1843) Observations on the periodical drainage and replenishment of the subterraneous reservoir in the Chalk basin of London – continuation of the paper read at the Institution, May 31st 1842. *Proceedings of the Institution of Civil Engineers*, 3, 156–165.

Clutterbuck, J.C. (1850) On the periodical alternations, and progressive permanent depression, of the Chalk water level under London. *Proceedings of the Institution of Civil Engineers*, 9, 151–180.

Conybeare, W.D. & Phillips, W. (1822) *Outlines of the Geology of England and Wales, with an Introductory Compendium of the General Principles of that Science, and Comparative Views of the Structure of Foreign Countries. Part 1.* London, William Phillips.

Darwin, E. (1975) An account of an artificial spring of water. *Philosophical Transactions of the Royal Society of London*, 75, 1–7.

Dawkins, W.B. (1898) On the relation of geology to engineering (The James Forrest Lecture).*Proceedings of the Institution of Civil Engineers*, 134, 254–277.

Day, J.B.W. and Rodda, J.C. (1978) The effects of the 1975–76 drought on groundwater and aquifers. *Proceedings of the Royal Society of London, A. Mathematical and Physical Sciences*, 363, 55–68.

Derham, W. (1716) *Physico-Theology or a Demonstration of the Being and Attributes of God, from His Works of Creation. Being the Substance of 16 Sermons Preached in St Mary le Bow Church, London at the Honourable Mr Boyles Lectures in the Years 1711 and 1712.* London, W. Innys.

Dixey, F. (1931) *A Practical Handbook of Water Supply.* London, Thomas Murby.

Donkin, J. (1836) Some accounts of borings for water in London and its vicinity. *Transactions of the Institution of Civil Engineers*, 1, 155–156.

Downing, R. A. (1993) Groundwater resources, their development and management in the UK: an historical perspective. *Quarterly Journal of Engineering Geology*, 26, 335–358.

Downing, R. A. 2004a. The development of groundwater in the UK between 1935 and 1965 – the role of the Geological Survey of Great Britain. In: Mather, J.D. (ed.) *200 years of British Hydrogeology.* Geological Society, London, Special Publications, 225, 271–282.

Downing, R. A. 2004b. Groundwater in a national water strategy, 1964–79. In: Mather, J.D. (ed.) *200 years of British Hydrogeology.* Geological Society, London, Special Publications, 225, 323–338.

Downing, R. A., Eastwood, W. and Rushton, K. R. (2004) Norman Savage Boulton (1899–1984): civil engineer and groundwater hydrologist. In: Mather, J.D. (ed.) *200 years of British Hydrogeology.* Geological Society, London, Special Publications, 225, 319–322.

Downing, R. A. and Gray, D. A. (2004) Jack Ineson (1917–1970). The instigator of quantitative hydrogeology in Britain. In: Mather, J.D. (ed.) *200 years of British Hydrogeology.* Geological Society, London, Special Publications, 225, 283–286.

Downing, R. A. and Headworth, H. G. (1990) The hydrogeology of the Chalk in the UK; the evolution of our understanding. In: *Chalk: Proceedings of the International Chalk Symposium held at Brighton Polytechnic 1989.* London, Thomas Telford. pp. 555–570.

Edmunds, W. M. (2004) Bath thermal waters: 400 years in the history of geochemistry and hydrogeology. In: Mather, J.D. (ed.) *200 years of British Hydrogeology*. Geological Society, London, Special Publications, 225, 193–199.

Edmunds, F.H. (1941) Outlines of underground water supply in England and Wales. *Transactions of the Institution of Water Engineers*, 46, 15–104.

Edmunds, W. M. and Milne, C. J. (eds.) (2001) *Palaeowaters in Coastal Europe: Evolution of Groundwater since the Late Pleistocene*. Geological Society of London, Special Publications, 189.

Evans, J. (1876) The Anniversary Address of the President. *Quarterly Journal of the Geological Society of London*, 32, Proceedings 91–121.

Farey, J. (1807) On the means of obtaining water. *Monthly Magazine and British Register*, 23, 211–212.

Farey, J. (1822) On overflowing wells and bore-holes. *Monthly Magazine and British Register*, 54, 35–37.

Farey, J. (1823) On artesian wells and boreholes. *Monthly Magazine and British Register*, 56, 309.

Fisher, W. W. (1901) On alkaline waters from the chalk. *Analyst*, 26, 202–213

Flett, J, S. (1937) *The First Hundred Years of the Geological Survey of Great Britain*. London, HMSO.

Fox, Sir C. R. (1949) *The Geology of Water Supply*. London, The Technical Press Ltd. de Freitas, M. H. and Rosenbaum, M. S. 2008. Engineering geology at Imperial College London; 1907–2007. *Quarterly Journal of Engineering Geology and Hydrogeology*, 41, 223–228.

Garnier, F. A. J. (1822)*De l'art du fontenier sondeur, et des puits artésiens*. Paris.

George, W. H. (2004) William Whitaker (1836–1925) – geologist, bibliographer and a pioneer of British hydrogeology. In: Mather, J.D. (ed.) *200 years of British Hydrogeology*. Geological Society, London, Special Publications, 225, 51–65.

Gray, D. A. (2004) Groundwater studies in the Institute of Geological Sciences between 1965 and 1977. In: Mather, J.D. (ed.) *200 years of British Hydrogeology*. Geological Society, London, Special Publications, 225, 295–318.

Gray, D. A. and Mather, J. D. (2004) Stevenson Buchan (1907–1996): field geologist, hydrogeologist and administrator. In: Mather, J.D. (ed.) *200 years of British Hydrogeology*. Geological Society, London, Special Publications, 225, 287–293.

Halley, E. (1687) An estimate of the quantity of vapour raised out of the sea by the warmth of the sun; derived from an experiment shown before the Royal Society, at one of their late meetings. *Philosophical Transactions of the Royal Society of London*, 16 (for 1686–1692), 366–370

Hamlin, C. (1990) *A Science of Impurity, Water Analysis in Nineteenth Century Britain*. Bristol, Adam Hilger.

Harrison, J. H. (1891) On the subterranean water in the chalk formation of the Upper Thames, and its relation to the supply of London. *Proceedings of the Institution of Civil Engineers*, 105, 2–25.

Hazell, R. (2004) British hydrogeologists in West Africa – an historical evaluation of their role and contribution. In: Mather, J.D. (ed.) *200 years of British Hydrogeology*. Geological Society, London, Special Publications, 225, 229–237.

Headworth, H. G. (2004) Recollections of a golden age: the groundwater schemes of Southern Water 1970 to 1990. In: Mather, J.D. (ed.) *200 years of British Hydrogeology*. Geological Society, London, Special Publications, 225, 339–362.

Hembry, P. (1990) *The English Spa 1560–1815*. London, Athlone Press.

Henwood, W. J. (1831) Facts bearing on the theory of the formation of springs, and their intensity at various periods of the year.*Philosophical Magazine and Annals of Philosophy, New Series*, 9, 170–177.

Henwood, W. J. (1843) On the quantities of water which enter the Cornish mines. *Transactions of the Royal Geological Society of Cornwall*, 5, 411–444.

Héricart de Thury, L. É. F. (1829) *Considérations géologiques et physiques sur la cause du jaillissement des eaux des puits forés ou fontaines artificielles et récherches sur l'origine ou l'invention de la sonde, l'état de l'art du fontenier–sondeur, et le degré de probabilité du succès des puits forés.* Paris, Bachelier.

Héricart de Thury, L. É. F. (1830) Observations on the cause of the spouting of overflowing wells or artesian fountains. *Edinburgh New Philosophical Journal*, 9, 157–165.

Homersham, S. C. (1855) In discussion of P. W. Barlow. On some peculiar features of the water-bearing strata of the London Basin. *Proceedings of the Institution of Civil Engineers*, 14, 42–95.

Hooke, R. (1665) *Micrographia: or some physiological descriptions of minute bodies made by magnifying glasses with observations and inquiries thereupon.* London, Royal Society.

Hopkinson, J. (1891) Water and water supply with special reference to the supply of London from the chalk of Hertfordshire. *Transactions of the Hertfordshire Natural History Society*, 6, 129–161.

Hull, E. (1862) *The Geology of the Country around Bolton-le-Moors, Lancashire.* Memoirs of the Geological Survey. England and Wales. London, HMSO.

Hull, E. (1865) On the New Red Sandstone as a source of water supply for the central towns of England. *Quarterly Journal of Science*, 2, 418–429.

Hunter, J. (1788) Some observations on the heat of wells and springs in the Island of Jamaica, and on the temperature of the earth below the surface in different climates. *Philosophical Transactions of the Royal Society of London*, 78, 53–65.

Inglis, J. C. (1877) On the hydrogeology of the Plymouth District.*Annual Report and Transactions of the Plymouth Institution and Devon and Cornwall Natural History Society*, 6, 105–121.

Johnstone, J. (1797) *An Account of the Most Approved Mode of Draining Land; According to the System Practised by Mr. Joseph Elkington, Late of PRINCETHORP, in the County of Warwick: with an Appendix, Containing Hints for the Farther Improvement of Bogs and Other Marshy Ground, After Draining; Together with Observations on Hollow and Surface Draining in General. The Whole Illustrated with Explanatory Engravings.* Edinburgh, Mundell and Son.

Jorden, E. (1631) *A Discourse of Naturall Bathes and Minerall Waters. Wherein First the Originall of Fountains in Generall is Declared. Then the Nature and Differences of Minerals, with Examples of Particular Bathes from Most of Them. Next the Generation of Minerals in the Earth, from Whence Both the Actuall Heate of Bathes, and Their Vertues are Proved to Proceede. Also by What Meanes Minerall Waters are to be Examined and Discovered. And Lastly, of the Nature and Uses of Bathes, but Especially of our Bathes at Bathe in Sommersetshire.* London, Thomas Harper.

Lapworth, C. (1901) Underground water supply in relation to brewing. *Journal of the Federated Institutes of Brewing*, 7, 443–458.

Lapworth Partners, Herbert, (1946) *A hydro-geological survey of Kent.* London, The Advisory Committee on Water Supplies for Kent.

Latham, B. (1894) *Report to the Corporation of Worthing on the Water Supply Obtainable from the Downs Adjacent to Worthing and of Sites Suitable for Permanent Water Works.* Worthing, Kirshaw.

Latham, B. (1882) On the influence of barometric pressure on the discharge of water from springs.*Report of the 51st Meeting of the British Association for the Advancement of Science, York, August/September 1881.* London, John Murray. p. 614.

Lloyd, J. W. (1986) Hydrogeology and Beer. *Proceedings of the Geologists' Association*, 97, 213–219.

Lucas, J. (1874) *Horizontal Wells. A new application of geological principles to effect the solution of the problem of supplying London with pure water.* London, Edward Stanford.

Lucas, J. (1877a) Hydrogeology: one of the developments of modern practical geology. *Transactions of the Institution of Surveyors*, 9, 153–184.

Lucas, J. (1877b) *Hydrogeological Survey, Sheet 1.* London, Edward Stanford.

Lucas, J. (1878) *Hydrogeological Survey, Sheet 2.* London, Edward Stanford.

Lyell, C. 1830–33.*Principles of Geology, 3 vols.* London, John Murray.

Mather, J. D. (1998) From William Smith to William Whitaker: the development of British hydrogeology in the nineteenth century. In: Blundell, D.J. & Scott, A.C. (eds.)*Lyell: the Past is the Key to the Present.* Geological Society, London, Special Publications, 143, 183–196.

Mather, J. D. (2001) Joseph Lucas and the term "hydrogeology". *Hydrogeology Journal*, 9, 413–415.

Mather, J. D. (2006) Erasmus Darwin (1731–1802) and the principle of the overflowing well. *Quarterly Journal of Engineering Geology and Hydrogeology*, 39, 313–320.

Mather, J. D. (2008) The hydrogeological work of Sir John Evans: his role in the battle between geologists and engineers for the water of the Chilterns. In: Arthur MacGregor (ed.) *Sir John Evans 1823–1908. Antiquity, commerce and natural science in the age of Darwin.* Oxford, The Ashmolean.

Mather, J. D., Torrens, H. S. and Lucas, K. J. (2004) Joseph Lucas (1846–1926) – Victorian polymath and a key figure in the development of British hydrogeology. In: Mather, J.D. (ed.) *200 years of British Hydrogeology.* Geological Society, London, Special Publications, 225, 67–88,

Morris, J. (1849) *A Series of Large Geological Diagrams, Illustrating the Principles of this Important and Practical Science. With Explanatory Notes upon Each Diagram.* London, J. Reynolds.

Mylne, R. W. (1840) On the supply of water from artesian wells in the London Basin, with an account [by W. C. Mylne] of the sinking of the well at the reservoir of the New River Company, in the Hampstead Road. *Transactions of the Institution of Civil Engineers*, 3, 229–244.

Page, T. H. (1784) Descriptions of the King's Wells at Sheerness, Languard–Fort and Harwich. *Philosophical Transactions of the Royal Society of London.* 74, 6–19.

Penman, H. L. (1948) Natural evaporation from open water, bare soil and grass. *Proceedings of the Royal Society of London*, 193, 120–146.

Perrault, P. (1674) *De l'origine des fontaines. (On the origin of springs, trans. A. la Roque, 1967.* Hafner Publishing Co., New York).

Phillips, J. (1844) *Memoirs of William Smith, LL. D., author of the "Map of the strata of England and Wales".* London, John Murray.

Phillips, W. (1818) *Selection of Facts from the Best Authorities, Arranged so as to form an Outline of the Geology of England and Wales.* London, William Phillips.

Plot, R. (1685) *De origine fontium, tentamen philosophicum. In praelectione habita coram Societate Philosophica nuper Oxonii instituta ad scientiam naturalem promovendam.* Oxford, Sheldonian Theatre.

Plot, R. (1686) *The Natural History of Staffordshire.* Oxford, The Theatre.

Preene, M. (2004) Robert Stephenson (1803–1859) – The first groundwater engineer. In: Mather, J.D. (ed.) *200 years of British Hydrogeology.* Geological Society, London, Special Publications, 225, 107–119.

Prestwich, J. (1850) On the geological conditions which determine the relative value of the water–bearing strata of the Tertiary and Cretaceous Series, and on the probability of finding in the lower members of the latter beneath London fresh and large sources of water supply. *Proceedings Royal Institute of British Architects*, 8 July, 1850, 14p.

Prestwich, J. (1851)*A Geological Inquiry Respecting the Water-Bearing Strata of the Country around London, with Reference Especially to the Water-Supply of the Metropolis; and Including Some Remarks on Springs.* London, Van Vorst.

Prestwich, J. (1872) The Anniversary Address of the President. *Quarterly Journal of the Geological Society*, 28, liii–xc.

Prestwich, J. (1876) *On the Geological Conditions Affecting the Water Supply to Houses and Towns, with Special Reference to the Modes of Supplying Oxford: Being a Lecture Given on October 22, 1875, with Additional Notes.* Oxford and London, James Parker.

Prestwich, J. (1895) *A Geological Inquiry Respecting the Water-Bearing Strata of the Country Around London, with Reference Especially to the Water-Supply of the Metropolis; and Including some Remarks on Springs. Re-Issue, with Additions by the Author.* London, Gurney and Jackson.

Price, M. (2004) Dr John Snow and an early investigation of groundwater contamination. In: Mather, J.D. (ed.) *200 years of British Hydrogeology.* Geological Society, London, Special Publications, 225, 31–49.

de Rance, C. E. (1882) *The Water Supply of England and Wales; its Geology, Underground Circulation, Surface Distribution, and Statistics.* London, Edward Stanford.

de Rance, C. E. (1884) The recent earthquake. *Nature*, May 8 1884, 31.

Ray, J. (1693) *Three Physico-Theological Discourses Concerning 1. The Primitive Chaos and Creation of the World, 2. The General Deluge, its Causes and Effects, 3. The Dissolution of the World and Future Conflagration, Wherein are Largely Discussed the Production and Use of Mountains, the Original of Fountains, of Formed Stones, and Sea-Fishes Bones and Shells Found in the Earth, the Effects of Particular Floods and Inundations of the Sea, the Eruptions of Vulcanos; the Nature and Causes of Earthquakes: with an Historical Account of those Two Late Remarkable Ones in Jamaica and England. With Practical Inferences. The 2nd Edition Corrected, very much Enlarged and Illustrated with Copper-Plates.* London, Samuel Smith.

Rivers Pollution Commission, (1875) *Sixth Report of the Commissioners Appointed in 1868 to Inquire into the Best Means of Preventing the Pollution of Rivers. The Domestic Water Supply of Great Britain.* London, HMSO.

Robins, N. S, Bennett, J. R. P. and Cullen, K. (2004) Groundwater versus surface water in Scotland and Ireland – the formative years. In: Mather, J.D. (ed.) *200 years of British Hydrogeology.* Geological Society, London, Special Publications, 225, 183–191.

Rose, E. P. F. (2004) The contribution of geologists to the development of emergency groundwater supplies by the British army. In: Mather, J.D. (ed.) *200 years of British Hydrogeology.* Geological Society, London, Special Publications, 225, 159–182.

Rowzee, L. (1632) *The Queenes Welles. That is, a Treatise of the Nature and Vertues of Tunbridge Water. Together with an Enumeration of the Chiefest Diseases, which it is Good for, and Against which it may be Used, and the Manner and Order of Taking it.* London, John Dawson.

Rudwick, M. J. S. (1963) The foundation of the Geological Society of London: its scheme for co-operative research and its struggle for independence. *British Journal for the History of Science*, 1, 325–355.

Seward, J. (1836) On procuring supplies of water for cities and towns, by boring. *Transactions of the Institution of Civil Engineers*, 1, 145–150.

Sheppard, T. (1920) *William Smith: His Maps and Memoirs.* Hull, A. Brown and Sons.

Shotton, F. W. (1952) Underground water supply of Midland breweries. *Journal of the Institute of Brewing*, 58, 449–456

Skinner, A. C. (2008) Groundwater: still out of sight but less out of mind. *Quarterly Journal of Engineering Geology and Hydrogeology*, 41, 5–19

Smith, B. (1935) *Geological Aspects of Underground Water Supplies.* Cantor Lectures November/December, 1935. London, Royal Society of Arts. 55p.

Smith, D. B., Wearn, P. L., Richards, H. J. and Rowe, P. C. (1970) Water movement in the unsaturated zone of high and low permeability strata using natural tritium. In:*Isotope Hydrology 1970.* Vienna, International Atomic Energy Authority. pp. 73–87.

Smith, W. (1827) On retaining water in rocks for summer use. *Philosophical Magazine, New Series*, 1, 415–417.

Stephens, D. B. and Ankeny, M. D. (2004) A missing link in the historical development of hydrogeology. *Ground Water*, 42, 304–309.

Stephens, D. B. and Stephens, D. A. (2006) British land drainers: their place among pre-Darcy forefathers of applied hydrogeology. *Hydrogeology Journal*, 14, 1367–1376.

Stephenson, R. (1840) *Report to the Provisional Committee of the London and Westminster Water-Works, Etc., Etc.* Reproduced in the Morning Advertiser of December 29, 1840.

Stephenson, R. (1841) *London, Westminster and Metropolitan Water Company. Second Report to the Directors.* London, Atkinson.

Swindell, J.G. (1849) *Rudimentary Treatise on Well-Digging, Boring, and Pump Work with Illustrations.* London, John Weale.

Telford, T. (1834) *Metropolis Water Supply. Report of Thomas Telford, Civil Engineer, February 1834, on the Means of supplying the Metropolis with pure Water.* London, House of Commons.

Tellam, J. H. (2004) 19th Century studies of the hydrogeology of the Permo-Triassic Sandstones of the Northern Cheshire Basin, England. In: Mather, J.D. (ed.) *200 years of British Hydrogeology.* Geological Society, London, Special Publications, 225, 89–105.

Thresh, J. C. (1908) The detection of pollution in underground waters, and methods of tracing the source thereof. *Transactions British Association Waterworks Engineers*, 12, 108–137.

Thresh, J. C. (1912) The alkaline waters of the London Basin. *Chemical News*, 106, 25–27 and 40–44.

Thresh, J. C. and Beale, J. F. (1925) *The Examination of Waters and Water Supplies. 3rd Edition.* J & A Churchill, London.

Torrens, H. S. (2004) The water-related work of William Smith (1769–1839). In: Mather, J.D. (ed.) *200 years of British Hydrogeology.* Geological Society, London, Special Publications, 225, 15–30.

Turner, R. C., Radley, J. M. and Mayneord, W. V. (1961) Naturally occurring alpha-activity of drinking waters. *Nature*, 189, 348–352.

Vulliamy, B. (1797) An Account of the Means employed to obtain an overflowing Well. In a Letter to the Right Honourable Sir Joseph Banks, Bart. K.B. P.R S. *Philosophical Transactions of the Royal Society of London.* 87, 325–331.

Walters, R. C. S. (1929) The hydro-geology of the Chalk of England. *Transactions of the Institution of Water Engineers*, 34, 79–110.

Walters, R. C. S. (1936a) The hydro-geology of the Lower Oolite rocks of England. *Transactions of the Institution Water Engineers*, 41, 134–158.

Walters, R. C. S. (1936b) *The Nation's Water Supply.* Ivor Nicholson and Watson, London.

Warington, R. (1887) A contribution to the study of well waters. *Journal of the Chemical Society*, 51, 500–552.

Whitaker, W. (1872) *The geology of the London Basin. Part 1 – The Chalk and the Eocene Beds of the Southern and Western tracts.* Memoirs of the Geological Survey. England and Wales, 4, HMSO, London.

Whitaker, W. (1886) On a recent legal decision of importance in connection with water supply from wells. *Geological Magazine*, 3, 111–114.

Whitaker, W. (1888) Chronological list of works referring to underground water, England and Wales. Appendix in 13th Report of the British Association Committee appointed for the purpose of investigating the circulation of underground waters in the permeable formations of England and Wales and the quantity and character of water supplied to various towns and districts from these formations. *In; Report of the 57th Meeting of the British Association, Manchester, August/September 1887*, John Murray, London, 384–414.

Whitaker, W. (1899) The Anniversary Address of the President. *Quarterly Journal of the Geological Society of London*, 55, liii–lxxxiii.

Whitaker, W. (1910) *The water supply of Hampshire (including the Isle of Wight), with records of Sinkings and Borings*. Memoirs of the Geological Survey. England and Wales. HMSO, London.

Whitaker, W. (1912) *The water supply of Surrey, from underground sources, with records of sinkings and borings*. Memoirs of the Geological Survey. England and Wales. HMSO, London.

Whitaker, W. and Reid, C. (1899) *The Water Supply of Sussex from Underground Sources*. Memoirs of the Geological Survey. England and Wales. London, HMSO.

Wilson, H. E. (1985) *Down to earth: One Hundred and Fifty years of the British Geological Survey*. Edinburgh, Scottish Academic Press. 189p.

Woodward, H. B. (1910) *The Geology of Water-Supply*. London, Edward Arnold.

Woodward, J. (1695) *An Essay Toward a Natural History of the Earth and Terrestrial Bodies, Especially Minerals as also of the Sea, Rivers and Springs. With an Account of the Universal Deluge and of the Effects that It had upon the Earth*. London, Richard Wilkin.

Yeats, T. (1826) Section of a well sunk at Streatham Common, in the county of Surrey. In a letter addressed to – Brown Esq. secretary to the Westminster Fire-Office; and by him communicated to the Geological Society. *Transactions of the Geological Society of London*, series.2, 2, 135–136.

Young, A. (1770) *A six months tour through the North of England containing, an Account of the present State of Agriculture, Manufactures and Population, in several Counties of this Kingdom*. Vol. 1. P. Dublin, Wilson and others.

Younger, P. L. (2004) "Making water": The hydrogeological adventures of Britain's early mining engineers. In: Mather, J.D. (ed.) *200 years of British Hydrogeology*. Geological Society, London, Special Publications, 225, 121–157.

A history of hydrogeology in the United States Geological Survey 1850–1990

Joseph S. Rosenshein[1] & John E. Moore[2]

[1] Bear Trail Virginia Beach, VA, USA
[2] Grape St. Denver, CO, USA

ABSTRACT

Prior to 1900, hydrogeological studies were carried out by geologists who were self-taught in groundwater science. From 1900 to 1930, the principal interest in hydrogeology was resource investigations. By the 1930s, agricultural irrigation was becoming more important to the economy, and this need drove hydrogeology in a more quantitative direction. The focus was on well hydraulics. The 1960s saw a shift towards regional aquifer system analysis as analog model technology advanced, and then in the 1970s by digital computer model technology. In the 1970s, emphasis slowly shifted to contaminant hydrogeology. In the 1980s, societal recognition of groundwater contamination as a serious environmental problem led to a major increase in the employment of hydrogeologists and advancements in the science and understanding of solute transport and multi-phase flow and transport of organic chemicals.

INTRODUCTION

This chapter focuses on the major scientific and technical accomplishments in hydrogeology during much of the 20th century, as well as the state of hydrogeology in the years leading up to this time. Aspects of the history of hydrogeology in the U.S. have been addressed Rosenshein et al., editors (1986)); Moore and Hanshaw (1987); Davis and Davis (2005); Meyer et al. (1988); Narasimhan (2005) and Reilly (2004). Table 1 lists the major milestones in the history of hydrogeology in the United States from 1879 to 1990.

EARLY HISTORY AND THE MEINZER ERA

Prior to 1879, the science of hydrogeology in the U.S. was rudimentary. Before this time, much hydrogeologic interest was focused on understanding development of spring and artesian water supplies and developing well drilling capabilities to the point that the artesian supplies could readily be utilised. Although about 14 state geological surveys had been established by 1860, the real pressure on them was to address geological mapping and mineral resources. Even so, many of the early hydrogeologic studies were carried out as much by state geological surveys as academia. For example, Harper (1857) includes some aspects of hydrogeology in his report on the geology

Table I Major Milestones in the History of Hydrogeology in the United States (1879–1990).

1879	U.S. Geological Survey is established. The objective is to apply science and prompt publication of findings.
1885	T.C. Chamberlin's report on artesian wells is published.
1896	The first USGS Water Supply Paper *Pumping Water for Irrigation is published.* G.K. Gilbert's report on the Arkansas Valley of Colorado is published.
1896	Darton reports on the artesian Dakota Sandstone in South Dakota. He described vertical leakage through low permeability shales.
1899	King's classic report on groundwater principles is published.
1906	Veatch complete one of the early quantitative groundwater investigations.
1907	Deep well turbine pumps developed for irrigation in California.
1909	Ellis completes a study of occurrence of water in crystalline rock in Connecticut.
1912	O.E. Meinzer became chief of what would eventually be designated as the Ground Water Branch.
1912	Lee publishes the results of a study of the Owens Valley California which included tank experiments of rates of water use.
1916	Mendenhall published a report on the San Joaquin Valley California.
1923	Meinzer reports on the Occurrence of Ground-Water in the United States with a discussion of principles.
1925	Brown studies saltwater encroachment along the Atlantic Coast.
1928	Meinzer paper on the compressibility of artesian aquifers is published.
1932	Meinzer prepares a paper outlining methods to estimate groundwater supplies.
1935	Theis publishes an equation to describe non-steady groundwater flow to a well.
1936	Stringfield prepares a potentiometric map of Florida. One of the first maps prepared in the United States.
1939	Meinzer publishes a report Ground Water in the United States.
1940	Theis publishes a paper on the source of water to wells.
1940	Hubbert publishes a report on the general theory of groundwater motion.
1941	Theis presents a method to estimate the effect of groundwater withdrawal from a well on a nearby stream.
1943	Jacob develops an analytic solution for recession of the potentiometric surface when recharge ceases.
1946	Jacob develops a physical basis for the Theis equation.
1947	Carslaw & Jaeger publish a classic book on conduction of heat in a solid.
1948	Ferris applies the Theis equation and the method of images for locating hydrogeologic boundaries.
1950	Foster demonstrates by laboratory experiments that the occurrences of high bicarbonate groundwater in the Atlantic Coastal Plain are caused by ion-exchange.
1951	Thomas publishes a book *The Conservation of Ground Water.*
1952	Bennett completes a report on the Baltimore area presenting an early use of flow net analysis.
1955	Hantush & Jacob develop mathematical models to analyze leaky aquifer problems.
1959	Cooper reports on the dynamic balance of freshwater and saltwater in a coastal aquifer.
1959	Hem publishers *Study and interpretation of the chemical characteristics of natural waters.*
1959	Todd publishes his book *Ground Water Hydrology.*
1960	Garrels publishes his text on mineral equilibria.
1960	Hantush publishes his modified artesian aquifer equation to include storage in the confining layer.
1962	Ferris publishes *Theory of Aquifer Tests.*
1963	McGuiness prepares a report on groundwater conditions in each state.
1963	Stallman points out that groundwater flow is an efficient mechanism for the transport of heat.

(Continued)

Table 1 Continued.

1963	Toth receives first Meinzer Award from the Geological Society of America for his paper on regional groundwater flow.
1963	Cooper & Rorabaugh develop methodology for estimating changes in bank storage and groundwater contribution to streamflow.
1963	Walton & Prickett prepare one of the first papers demonstrating the use of analog models in hydrogeology.
1964	Chow publishes *Advances in Hydroscience*.
1965	LeGrand publishes a paper on groundwater contamination.
1965	Domenico & Mifflin publish an analytical method to evaluate land subsidence.
1967	Freeze & Witherspoon publish a numerical solution of groundwater flow in complex hydrogeologic environments.
1968	Pinder & Bredehoeft publish a paper describing the use of a digital model in aquifer evaluation.
1969	Poland and Davis prepare a report on subsidence due to fluid withdrawal.
1969	Neuman & Witherspoon publish a paper that summarizes various methods for solutions to leaky aquifer situations.
1970	Bredehoeft & Young prepare an article that demonstrates the interrelation of groundwater, surface water, and water use.
1974	Zohdy and others publish a technical manual on application of surface geophysics to groundwater.
1979	Freeze & Cherry publish a textbook *Ground Water*.
1980	Fetter publishes a textbook *Applied Hydrogeology*.
1988	McDonald & Harbaugh develop the ModFlow digital groundwater model.

of Mississippi; Norton (1897) discusses artesian wells in Iowa; and McCallie (1898) reported on Georgia's artesian well system. With the establishment of the U.S. Geological Survey (USGS) in 1879, the foundation was laid for the systematic development of hydrogeology as a science. John Wesley Powell, the second Director of the Survey, recognised the importance of systematic collection of hydrological data. Prior to 1903, most USGS groundwater studies were either carried out by the Geological Branch as part of the Branch's geological efforts or by Branch geologists on loan to the Division of Hydrography. By 1903, the Geological Survey had established the Division of Hydrology as part of the Hydrographic Branch to carry out studies on the occurrence of groundwater. By 1906, the Division of Hydrology had evolved into the Groundwater Division.

Among the early contributors to hydrogeological science were Chamberlin (1885) who published a treatise on artesian wells; Gilbert (1896) who investigated groundwater in the Arkansas River Valley in Colorado; Darton (1896) who described and mapped the classic artesian aquifer, the Dakota Sandstone; and King (1899) who reported on laboratory and field evaluations of groundwater storage, flow and directional permeability. Ellis, Newell, Slichter, Fuller, Lindgren, Lee, Mendenhall (fifth Director of the USGS), and Siebenthal all made notable contributions prior to 1910.

In 1912, a change took place in the USGS that would result in 34 years of noteworthy advances in the science of hydrogeology in the United States. This change eventually had an affect on the development of the science of hydrogeology outside the United States. O.E. Meinzer became chief of what would eventually be designated as the Ground Water Branch. During his career, his scientific contributions and leadership changed hydrogeology into a fully-fledged science and he is recognised, at least in the US, as the father of modern day groundwater hydrology. G.B. Maxey (1986) reported that Meinzer recognised

- resource evaluation studies must be as quantitative as possible,
- a real need existed to bring together knowledge and methodology and to develop new information in both categories,
- hydrogeology was a multidisciplinary field requiring the combined talents of geologists, physicists, chemists, and engineers to develop basic principles of the science and to apply those principles, and
- a systematic approach to the solution of groundwater problems should be made by an organised team of experts.

Meinzer's publications were extensive and many underpinned hydrogeology as practiced today. In 1943, Meinzer received the Bowie Award of the American Geophysical Union for his contribution to science.

The first study on the affects of phreatophytes on groundwater was carried out by Lee in 1922. The affects of discharge of groundwater on water loving vegetation in western U.S. would eventually become a major area of concern. The first study of groundwater and surface-water interaction was made by Mendenhall and others (1916) in a report on the San Joaquin Valley, California, where they used streamflow records for an evaluation of groundwater conditions. They recognised that the aquifers in the valley were directly connected to the streams, and they used records of paired gauging stations to determine channel loss. In a study of saltwater encroachment along the Atlantic Coast, Brown (1925) made one of the early applications in the U.S. of the Ghyben-Herzberg principle. The basic concepts developed in his study eventually led to further applications of the principle in other studies of saltwater encroachment in the United States.

During the Meinzer Era, most of the hydrogeologic investigations in the United States were carried out by the hydrogeologists from the USGS (Figure 1). The results of these investigations were published by various state agencies or in the USGS Water-Supply Paper series). Maxey (1985) reported that by 1930 more than 100 resource investigations had been carried out in all parts of the U.S covering aspects of groundwater occurrence, availability, movement, and water quality. Among the significant studies carried out during this era were those by Fiedler & Nye (1933), who described the effects of large-scale groundwater development on the Roswell artesian basin in New Mexico; Stringfield (1936), who mapped the potentiometric surface of the Floridan aquifer, thereby contributing to the understanding of groundwater flow in a regionally important aquifer; and Sayre and Bennett (1942) whose work showed that the Edwards Limestone aquifer is part of a major aquifer system that extends across much of central Texas.

Prior to 1935, the method for testing the hydraulic properties of aquifers was the Theim steady state or equilibrium method. In 1935, Theis published a paper in which he derived the non-equilibrium formula to describe nonsteady flow to a pumping well. This formula was derived using the analogy of heat flow in solids. He was the first to introduce the concept of time to the mathematics of groundwater hydraulics and the importance of aquifer storage. This work is one of the greatest contributions to the

Division of Ground Water
U. S. Geological Survey
July 1, 1932

Figure 1   Members of USGS Division of Ground Water 1932 (USGS).
In 1906, what is now the Office of Ground Water was called the Ground Water Division.
In 1932, the Office of Ground Water was called the Division of Ground Water.

science of groundwater hydraulics in the 20th Century and opened the door to modern quantitative investigation.

Figure 2   C.V. Theis (USGS).

In 1940, Theis followed this work with a paper on the source of water derived from wells, the significance of the cone of depression created by pumping, and the key factors that control the response of an aquifer to development. In 1941, he presented a method to estimate the effects on a stream from pumping from a nearby well. Among other important contributions to hydrogeology during the latter stage of the Meinzer Era were the contributions of: Hubbert (1940), who developed a general theory of groundwater motion and described the influence of groundwater flow on the accumulation of hydrocarbons; Jacob who published a series of papers in which he (1) defined the transient-flow equations using flow hydraulic rates from the heat-flow analogy (1940); (2) developed an analytical solution for the recession of the potentiometric surface when recharge ceases (1945); (3) analysed radial flow in a leaky artesian aquifer and identified the influence of leakage from confining beds (1946): and Wenzel (1942) who published an extensive report on the methods for determining permeability of water-bearing materials which included laboratory and discharging well methods.

With the end of the Meinzer Era, the ground work had been established for the application of quantitative methods to define aquifer properties by field tests. In the 33 years that followed, 1947 to 1980, the emphasis remained on the water-supply aspects of groundwater. This emphasis slowly became shared with other national and local issues such as defining regional aquifer systems, underground storage of liquid and solid wastes, groundwater contamination, water rights, land subsidence, numerical simulation of aquifer systems, and water management.

POST-MEINZER ERA

Quantitative methods

The development and refinement of quantitative methods continued to play an important role in hydrogeology in the post-Meinzer Era. Aquifer test methods for determining the hydraulic characteristics of aquifers were expanded. Ferris wrote a chapter on ground water for inclusion in a text book on hydrology (Wisler & Brater, 1948) that provided an understanding of the hydrogeologic aspects of the quantitave methods available at that time. This chapter became a standard reference for hydrogeologists. Jacob wrote a chapter on groundwater for inclusion in Hunter Rouse, Engineering Hydraulics (1950) in which he recognised that the coefficient of storage could be determined if the hydraulic system had reached a steady-state condition. Hantush and Jacob (1955) addressed the problem of nonsteady flow in an infinite leaky aquifer. This paper was followed by a series of papers by Hantush on analysis of aquifer hydraulics; one of his most important was his modification (1960) of the theory of leaky confined aquifers where storage of water in the semipervious confining bed(s) is taken into account. The modified method provided a way to determine the hydraulic conductivity of the semipervious confining layer(s). In 1968, Hantush received the Meinzer Award from the Geological Society of America for his paper on hydraulics of wells (1964). *The Theory of Aquifer Tests* (Ferris *et al.* 1962) is still used as a reference for aquifer tests. The report included not only the methods for determining aquifer characteristics from aquifer tests but also areal methods such as numerical analysis (Stallman, 1962), and flow-net analysis (Bennett, 1962); and image theory and hydrologic boundary analysis.

In 1962, the Illinios State Water Survey published a report on analytical methods for well and aquifer evaluation (Walton, 1962). From 1961 to 1963, the USGS published a series of reports dealing with aquifer tests; of special note are the reports on shortcuts and special problems in aquifer tests and methods of determining permeability, transmissibility and drawdown. From 1956 through 1966, Stallman published a number of papers dealing with quantitative aspects of hydrogeology. In 1967, Stallman received the Meinzer Award for his publication on multiphase fluids in porous media (Stallman, 1964).

Numerical methods

Stallman presented an method for evaluating aquifer hydrology in Ferris *et al.* (1962), in which he demonstrated how numerical analysis can be used to determine aquifer characteristics. This followed various papers on numerical analysis of regional water levels to define aquifer hydrology (Stallman, 1956a) and use of numerical methods for analysing data on groundwater levels (Stallman, 1956b). The application of his numerical method was hampered by the complex nature of the calculations, a problem only resolved with the advent of widespread affordable computers in the 1970s.

In the pre-computer age of the 1950s, Skibitzke and others began investigating the electric analog computer technology for the solution of transient groundwater flow equations. This adaptation was aided by Karplus (1958) in his book on analog simulation. The adaptation was based on the analogy with flow of electricity in a resistor-capacitor network. In 1960, Skibitzke published a paper on electronic computers as an aid to the analysis of hydrologic problems. By 1960, the USGS had established an analog modelling unit. The ability to simulate the hydrology of complex, multi-aquifer systems had a significant impact on regional areal evaluation of aquifer systems and their response to stress. Analog computer models were eventually replaced by digital computer models; about 100 analog models were constructed and analysed

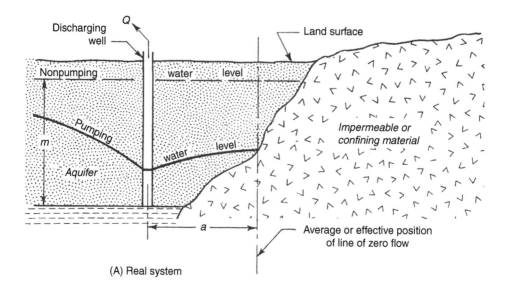

(A) Real system

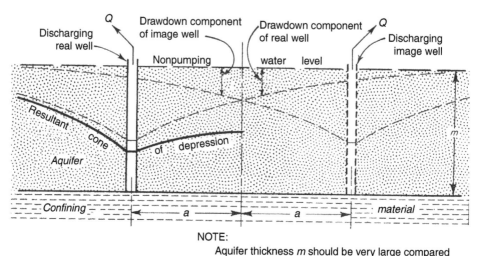

NOTE:
Aquifer thickness *m* should be very large compared
to resultant drawdown near real well

(B) Hydraulic counterpart of real system

Figure 3 Idealised section views of a discharging well in a semi-infinite aquifer bounded by an
impermeable formation, and of the equivalent hydraulic system in an infinite aquifer (USGS).

by the USGS. One of the first applications of an analog model was the evaluation of
the effect of large scale groundwater withdrawals in the Houston, Texas area (Wood
& Gabrysch, 1965).

Walton & Prickett (1963) used the work of Skibitzke to describe the construction
and application of electric analog models in hydrogeology. This paper facilitated the
use of analog modelling to simulate aquifer hydrology.

Figure 4 H.E. Skibitzke (USGS).

By the late 1960s, the development and availability of the main frame digital computer had reached a point whereby the digital computer could profoundly influence modelling and data storage. The main frame computer has since been replaced for many uses first by the minicomputer and then the personal desktop computer.

The USGS played a significant role in the development of digital groundwater flow models (Appel & Bredehoeft, 1976). The Illinois State Water Survey also released a public-domain code that became widely used (Pricket & Lonnquist, 1971). Bredehoeft & Pinder (1973) described a quasi three-dimensional digital model to analyse areal flow in multiaquifer groundwater systems. Trescott (1973) reported on an iterative model for aquifer evaluation which simulated two-and three-dimensional groundwater flow and Konikow & Bredehoeft (1973) presented a method to model chemical quality variations in a groundwater system by applying mass transport equations. In 1978, they published a computer model of solute transport and dispersion in ground water. By 1976, the USGS had developed twelve operational digital models and were in the process of testing 17 other models. These models covered groundwater flow, land subsidence, coupled groundwater and stream systems, coupled flow and transport of chemical constituents, coupled flow and transport of heat, and coupled flow and transport of conservative constituents and heat. McDonald & Harbaugh (1988) introduced the MODFLOW code, which is now the most widely used groundwater flow model in the world.

By 1990, some short courses in computer modelling were being taught associated with professional society meetings and stand alone short courses were available to the practicing hydrogeologist. Courses in computer modelling were being taught in universities and colleges, and education and training in modelling was becoming readily available. In some universities, work was being done on development of stochastic groundwater flow models. Other major advances were being made in inverse (parameter-estimation) modelling. These advances had the advantage of automating the model calibration process while minimising errors in fitting the observed data. These approaches improved overall modelling efficiency and also enabled assessments of how good the models were by calculating goodness of model fit, predictive errors, confidence intervals, and uncertainties in parameter estimates.

Regional aquifer systems

Many hydrogeologists have long recognised a need to define and evaluate the regional sized aquifers and to treat them as a system. As early as 1896, Darton showed that the Dakota Sandstone covered a large multi-state area. Progress toward regional-scale analysis was made when, in 1923, (1) Meinzer divided the United States into groundwater provinces in part based on geology, and (2) Thomas (1952) modified the groundwater regions (Figure 5). In 1984, Heath divided the United States, Puerto Rico, and the Virgin Islands into 15 areally extensive regions of relatively similar groundwater conditions. Heath (1988) divided North America into 31 areally extensive hydrogeologic regions and provided a map of the aquifers in these regions (Figure 6).

Theis (1940) demonstrated the importance of understanding the sources of water to an aquifer on a system-wide basis. The quantitative methods and technology (flow-net analysis and computer modelling) necessary to carry out this type of evaluation did not become available until the 1950s and later. By 1960, sufficient information was available to permit McGuiness to publish a comprehensive review of groundwater conditions in each state with a map of the productive aquifers in the United States. The map showed that many of the aquifers were areally extensive and covered parts of more than one state. All of the above efforts laid the foundation for a regional approach to the aquifer systems underlying the United States. By the 1970s, this need was exacerbated by the increased demand for groundwater, the conflicting interest of states and local governments for groundwater, and the growing importance of the management of the groundwater resources. In late 1978 the USGS defined 28 regions in the United States, the Caribbean Islands, and Hawaii for analysis. This analysis defined the hydrogeological framework for each important regional aquifer system, accessed the available hydrochemical data, characterise the groundwater flow system, and defined the effects of development on the flow system. Digital modelling was heavily used in the characterisation of the flow systems. By 1990, reports on many of these systems had been published. (See Sun and Johnston (1994) for a detailed summary and reference list.)

ROLE OF PROFESSIONAL SOCIETIES

Professional societies have provided a forum for exchange of ideas, served as a vehicle for attracting students to the science. The key societies are the American Geophysical Union (AGU), American Institute of Hydrology (AIH), Geological Society of America (GSA), International Association of Hydrogeologists (IAH), and the National Ground Water Association (NGWA). These organisations recognise important contributions through awards given for groundwater science.

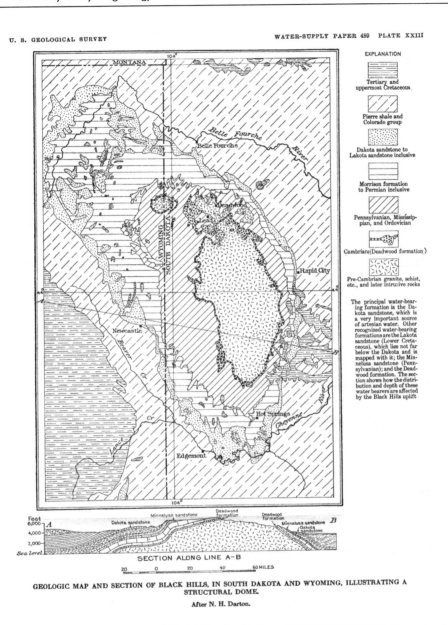

U. S. GEOLOGICAL SURVEY

WATER-SUPPLY PAPER 489   PLATE XXIII

EXPLANATION

Tertiary and
uppermost Cretaceous

Pierre shale and
Colorado group

Dakota sandstone to
Lakota sandstone inclusive

Morrison formation
to Permian inclusive

Pennsylvanian, Mississip-
pian, and Ordovician

Cambrian (Deadwood formation)

Pre-Cambrian granite, schist,
etc., and later intrusive rocks

The principal water-bear-
ing formation is the Da-
kota sandstone, which is
a very important source
of artesian water. Other
recognized water-bearing
formations are the Lakota
sandstone (Lower Creta-
ceous), which lies not far
below the Dakota and is
mapped with it; the Min-
nelusa sandstone (Penn-
sylvanian); and the Dead-
wood formation. The sec-
tion shows how the distribution and depth of these
water bearers are affected
by the Black Hills uplift

SECTION ALONG LINE A–B

GEOLOGIC MAP AND SECTION OF BLACK HILLS, IN SOUTH DAKOTA AND WYOMING, ILLUSTRATING A
STRUCTURAL DOME.

After N. H. Darton.

Figure 5 Geologic map and section of Black Hills after Meinzer, 1923, USGS WSP 489. Ground-water provinces of Meinzer (USGS).

AGU: The AGU was established in 1909 by the National Research Council. Many of the early papers on quantitative aspects of hydrogeology such as those by Jacob (1940) were published in Transactions of the AGU and others in the Journal of Geophysical Research. Since 1965, AGU's journal Water Resources Research has served as an outlet for important research oriented papers related to hydrogeology. In the early 1960s, the AGU Hydrology Section established a Ground Water Committee. This

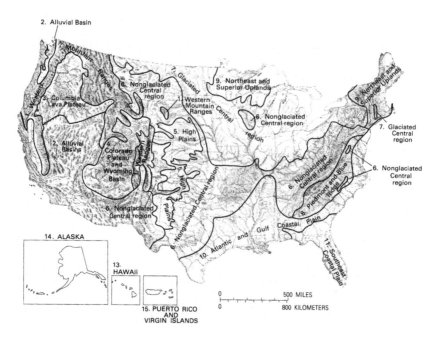

Figure 6 Groundwater regions of Heath (USGS).

Committee has served as a focal point for organising symposia and poster sessions on groundwater science at the annual meetings, which offer the opportunity for students and experienced hydrogeologists to present papers and gain exposure in the science. A symposium at the 1981 annual meeting honoring John Ferris for his contributions to quantitative aspects of hydrogeology resulted in an AGU Water Resources Monograph on Ground Water Hydraulics (Rosenshein & Bennett, 1984). Additional Water Resources Monographs cover other aspects related to hydrogeology such as the use of numerical models for groundwater management (Bachmat *et al.*, 1980).

AIH: The American Institute of Hydrology was formed as a scientific and educational organisation in 1981. Until AIH was established, no uniform way existed to determine or evaluate the competence of practicing hydrologists or hydrogeologists. The AIH is the only nationwide organisation to offer certification to professionals in all fields of hydrology. The AIH maintains a registry of hydrologists and hydrogeologists that have met educational, professional experience, publications, and conduct requirements and have passed a written test. A few states have established registration of hydrologists and hydrogeologists based on the requirements of AIH for certification.

GSA: The Geological Society of America has played a strong and focused role in hydrogeology. In 1959, the GSA established the Hydrogeology Division. The Division was founded principally through the efforts of G. Burke Maxey and Phillip E. LaMoreaux; with G. Burke Maxey as acting Chairman. This Division became one of the Society's largest. Many members of the Hydrogeology Division are fellows of the Society. Many prominent hydrogeologists have served as Chairman of the Division. (See http://gsahydrodiv.unl.edu/general/chairs.htm.)

To mark the Centennial of the Geological Society of America (1888–1988), the Society undertook a project called the Decade of North American Geology (DNAG). As part of this project the Hydrogeology Division prepared a volume on the *Hydrogeology of North America*. This volume includes the work of about 100 North American hydrogeologists. The volume covers aspects of regional and comparative hydrogeology, groundwater and geological processes, and scientific and societal problems.

IAH: The U.S. National Chapter was organised in 1972 as a Subcommittee of the U.S. National Committee on Geology (a Subcommittee of the National Research Council). The U.S. National Chapter represents the U.S. hydrogeological community's interests in the hydrogeologic activities and programs of the International Geological Congress and the Congresses of the IAH. The U.S. National Committee has held joint meeting with other U.S. professional societies and in 1975, hosted the 12th IAH Congress, on the theme Karst Hydrogeology, held in Huntsville, Alabama and in 1985, the 17th Congress, Hydrology of Low Permeability Rock, in Tucson, Arizona.

NGWA: The National Ground Water Association was originally founded as the National Water Well Association (NWWA) in 1948 and changed its name in 1991. Early in its history, the NWWA established a Technology Division. In part, because of the many contacts between hydrogeologists conducting groundwater resources investigations and well drillers, the hydrogeological membership in the Technology Division grew. In 1991, the Technology Division became the Association of Ground Water Scientists and Engineers (AGWSE). In 1963, the NWWA started publication of the journal Ground Water. Prior to this time, hydrogeologists had very limited outlets for publication of papers. The decision to begin publication of the journal by NWWA was in part due to the efforts of W.C. Walton, who at that time was Chief of the Ground Water Section of the Illinois State Water Survey. Walton served as the first editor of the journal, followed by Jay Lehr, and subsequently John Bredehoeft.

EDUCATION AND TRAINING IN HYDROGEOLOGY

Maxey in his paper on the Meinzer Era of hydrogeology points out that in 1976 most of the hydrogeologists in the U.S. had received training from the USGS, from ex-members of the USGS, or from students of ex-members. This is a reflection of the status of education and training available to a hydrogeologist in academia before the mid 1960s. Universities and colleges then placed emphasis in geological education on traditional areas of geology. The first text book on groundwater was published in 1937 by Tolman (of Stanford University). The USGS and (beginning in 1930) state agencies with cooperative programs with the USGS, were the principal sources for supplemental education and training. It was not until the 1960s that this situation began to change. Early in the Meinzer Era, the Ground Water Branch began providing additional education and training to USGS staff. The Ground Water Branch carried out informal education through in-house publications such as groundwater notes. Many of these notes were published in the water-supply paper series on shortcuts and special problems in aquifer hydraulics. In addition, the Branch was responsible for the publication of groundwater related books and chapters of the techniques of water-resources investigations of the USGS and publications such as Heath's (1987) report on basic groundwater hydrology.

Figure 7 John G. Ferris (USGS).

About 1950, the Ground Water Branch formalised some of the training by holding two-week Basic Ground-Water Short Courses. John Ferris and Stan Lohman played key roles in teaching these concentrated courses. In the 1960s, the Ground Water Branch began teaching an Advanced Ground-Water Short Course. By the 1960s, the need for a National Training Centre became apparent and this was established in 1968 in the Denver area. Although the USGS has continued to provide training and education for its staff, its influence on the education and training of hydrogeologists has decreased as the universities and colleges increased their education programs in hydrogeology in the late 1960s through the 1980s.

By the 1950s, universities and colleges slowly began to expand their teaching efforts in hydrogeology. In 1955, G. Burke Maxey became Chief of the Ground Water Section of the Illinois Geological Survey. He also taught courses in hydrogeology at the University of Illinois and was responsible for number of doctoral theses in geology with a heavy emphasis in hydrogeology, including that by John Bredehoeft. In 1962, he became Director of the Desert Research Institute, a part of the University of Nevada. He continued to turn out (mentor) a number of students with doctoral degrees that have become leaders and educators in the field of hydrogeology. Many current leaders in the field can trace the roots of their hydrogeological education back to Prof. Maxey (who was very proud of his many "grand-students" and "great grand-students").

In 1957, the USGS assigned John Ferris to the University of Arizona to serve as an adviser and principal lecturer for a new curriculum in hydrology initiated jointly by the university through the efforts of John Harshbarger and others and the USGS. Ferris stayed in this post until 1965. This program issued the first doctoral degree in hydrology with an emphasis on groundwater science.

In the second half of the twentieth century numerous text and reference books on groundwater science were published. Among those are Todd (1959) *Ground Water*

Hydrology, Chow, editor (1964) *Advances in Hydroscience,* Freeze & Cherry (1979) *Ground Water,* Fetter (1980) *Applied Hydrogeology,* and Domenico & Schwartz (1990), *Physical and Chemical Hydrogeology.*

Some universities began awarding advanced degrees in groundwater science. Among these schools were the Universities of Arizona, California at Berkley, and Illinois, New Mexico Institute of Mining and Technology, the Pennsylvania State University, and Stanford University.

Marked decreases in employment opportunities in petroleum industry and new legislation concerned with safe drinking water and environmental contamination combined to increase both the supply and demand for hydrogeologists. The growth of the Hydrogeology Division of the GSA into one of the largest Divisions also highlighted the need for serious changes in geologic education to meet the new demand. By 1990 the number of universities and colleges offering undergraduate and graduate level courses in hydrogeology and interdisciplinary degrees in hydrogeology had begun a significant increase. Between 1980 and 1990, short courses in aspects of hydrogeology became common as part of some of the professional society meetings. Stand alone one-week short courses in new areas of concern in hydrogeology such were also offered annually. The ability of hydrogeologists to remain abreast of new applications in hydrogeology significantly increased as the availability of these continuing education courses increased.

GROUNDWATER QUALITY AND CHEMICAL HYDROGEOLOGY

Many large cities and small communities used public wells for water supply. Because these wells served as community wells and were generally located in populated areas, they were often subject to contamination. Springs, in parts of the eastern U.S., were developed as spas and used for their supposed therapeutic affects. Analysis of water from springs and artesian wells was frequently performed by medical doctors who had interest in the quality of the water and often also in geology. Among these were William Meade and John Honeywood Steel who published a number of reports on quality of water of springs between 1817 to 1832. The constituents that were determined were generally rudimentary.

Much of the early groundwater quality work focused on the natural quality of groundwater. One of the earliest reports on geochemical interpretation of analyses of natural water was published by Palmer (1911). As ability to analyse the chemical quality of groundwater improved, the number of constituents determined increased. Methods were developed to increase the accuracy of the determinations of constituents as well as to detect ever smaller concentrations. Collins (1923 presented a graphical method to aid in the interpretation of analyses of groundwater. Some methods were developed to make partial analysis more useful, such as Collins' (1928) method for estimating total dissolved solids from constituents generally determined by partial analysis of water. Piper (1944) presented a graphic procedure for the geochemical interpretation of water analysis. Stiff (1951) presented a graphical method to interpret water analysis by means of patterns. Both graphical methods are still in use. The study and interpretation of the chemical characteristics of natural water was described by Hem (1959). Hem's report was updated in 1970 and in 1980 and has continued to serve as a standard reference and as a text book on the subject. In 1960, Garrels published a book on mineral equilibria that provided the tools of chemical thermodynamics to hydrogeologists. Much of the chemical hydrogeologic work since 1960 has been based on aspects of his work. By the 1980s, computer programs (Appel & Reilly, 1988) had been developed to calculate (1) chemical equilibrium of natural waters (WATEQ), (2) solution-mineral equilibrium (SOLMNEQ), (3) mass transfer for geochemical reactions in ground water (BALANCE), and (4) geochemical calculations (PHREEQE). By the 1980s, advanced laboratory methods had been developed to determine trace concentrations of a wide variety of organic and inorganic constituents in groundwater and protocols had been developed for analyses.

The papers cover various aspects of chemical hydrogeologic development including the evolutional period, occurrence and geochemical significance of salt water, the equilibrium approach, isotopes in ground water, and heat and mass transport. Among the papers included are Back (1960) on the origin of hydrochemical facies of ground water in the Atlantic Coastal Plain; Kohout (1960) on the cyclic flow of saltwater in the Biscayne aquifer of Southeastern Florida; Hanshaw, Back, and Rubin (1965) on carbonate equilibria and radiocarbon distribution related to ground-water flow in the Floridan limestone aquifer; Bricker and Garrels (1967) on mineralogic factors in natural water equilibria; Pearson and White (1967) on ages and flow rates of water in the Carrizo Sand, Atascosa County, Texas; and Anderson (1979) on using models to simulate the movement of contaminants through ground-water flow systems.

Groundwater contamination

Before about 1970, most hydrogeological interest in groundwater contamination was centered on contamination of aquifers by non-organic constituents. Discharge of industrial wastes and sewage to streams was a major source of contamination of aquifer-stream systems in southern New England, and elsewhere. The Clean Water Act of 1961 led to action to address this problem. Local discharge of sewage into former coal mines, sinkholes, and wells also contributed to the direct contamination of aquifers in parts of the U.S.

McGuinness (1963) cited a number of significant sources of groundwater contamination. These included contamination of aquifers by industrial wastes with arsenic, cadmium, chromate, nitrate, sulfate, chlorinated solvents, and potentially radioactive materials; by agricultural chemicals such as fertilizers, insecticides, herbicides, and defoliants; leachates from landfills; and effluents from septic tanks that included household detergents, nitrate, and pathogens. McGuinness noted also the contamination of aquifers by hydrocarbons leaking from storage tanks. Serious attention to this type of contamination would wait until the 1970s and 1980s. Before the 1970s, contamination of aquifers by salt-water intrusion was probably the most serious groundwater contamination concern in the U.S. and was studied by the USGS, State, and local agencies.

The concern about groundwater contamination took a major change in direction in the 1970s with the focus of media attention on the significance of industrial contamination of groundwater at places like Love Canal in the State of New York. The United States realized the need to take effective steps to address the problems of severe groundwater contamination with the passage of the Resource Conservation and Recovery Act (RCRA) (1976) and the Superfund Act (Comprehensive Environmental

Figure 8   C.L. McGuinness (USGS).

Response, Compensation, and Liability Act) (1980). The Acts resulted in serous consideration of groundwater contamination from a wide range of industrial sources. Organic compounds were of concern with base-neutral and acid extractables, volatiles, and pesticides at the top of the list. In the late 1970s through the 1980s, the United States increased its efforts to address groundwater contamination at military facilities and its National Laboratories. By the 1980s, the country had established a site for the underground storage of its radioactive military wastes and was addressing groundwater contamination at military facilities through the Department of Defense Environmental Restoration Program. The United States started investigating the feasibility of storage of power-plant radioactive wastes at the Department of Energy's Yucca Mountain site in Nevada. Extensive hydrogeological and geological studies were carried out at the site aimed at addressing the regulatory agencies requirements and the concerns of the State of Nevada, local agencies, and the public.

The USGS began a long-term program on Toxic Waste-Ground Water Contamination in 1982 which was expanded in 1985 to become the Toxic Substances Hydrology Program. In 1986, the USGS began a program on National Water Quality Assessment (NAWQA). Prior to 1990, emphasis of studies in the Toxic Substances Hydrology Program as related to hydrogeology was on subsurface point-source contamination, contaminant occurrence, contaminant plumes, natural attenuation of contaminants, and to some extent the unsaturated zone and the use of tracers. Many of these investigations involved field site and laboratory studies, and where appropriate modelling.

THE ROLE OF PUBLIC AGENCIES IN HYDROGEOLOGY IN THE UNITED STATES

The needs of Federal, State, and local agencies for information and hydrogeological analysis have played a major role in developing the science. Although the USGS has had a long history of involvement in hydrogeology (1880–1990), other Federal agencies have also influenced the science's development. For example, the Agricultural Research Centre, Department of Agriculture, has been involved in studies of the unsaturated zone, the Bureau of Land Management in studies of hydrogeology on the Federal lands that it administers, Bureau of Mines in hydrogeological aspects of mining, and the Department of Energy and its laboratories in the storage of radioactive wastes.

In 1930s the USGS began establishing cooperative groundwater program with State and local agencies. Under the cooperative program, the USGS frequently carried out the activities for the cooperating state or local agency or the activities were carried out jointly. Before the 1970s, the hydrogeological reports of the cooperative program were frequently published by the cooperator in its publication series. Since then, many of the reports produced by the cooperative program have been published in the USGS Water-Resources Investigation and Water-Supply Paper Series.

A wide range of State and local agencies have long-term cooperative programs with the USGS. These include offices of state engineers, state departments of environment, state departments of water resources, state water development boards, state geological surveys, and within states such as Florida, water-management district administrations. Many of these agencies also carry out hydrogeological investigations with their own groundwater staff.

A number of state geological surveys have been involved with hydrogeology including the Alabama, Kansas, Maryland, Missouri, Wisconsin, and Illinois Geological Surveys. The Geological Surveys of Illinios and Kansas have had Ground Water Sections as part of the Surveys for at least 60 years. In Illinois, groundwater investigations are carried out by the Illinois Geological Survey and the Illinois State Water Survey. Many of the state surveys are responsible for collection and storage of well records and maintain the well records.

SPECIAL AREAS OF HYDROGEOLOGY
OF HISTORICAL SIGNIFICANCE

A number of areas of hydrogeology have received special interest in the United States. These areas include aquifer-stream systems, salt-water encroachment, and underground storage of wastes.

Aquifer-stream interrelations: Alluvial valley aquifer-stream systems have historically been important sources of water supply. Discharge of groundwater from these systems provides the base flow of streams during dry weather. Because of the interrelations of the aquifer and associated stream, aquifer hydrology is impacted by changes in the stream regime and changes that adversely affect aquifer hydrology also affect stream flow. Frequently, man's activities have had an impact on both systems and these can take place over a short time period.

Because of their importance, these systems have received considerable attention in the technical literature. One of Theis' important early papers (1940) dealt with the effects of pumping of a well on a nearby stream. Other papers of note include Rorabaugh's (1956) evaluation of factors affecting induced infiltration in his landmark paper on groundwater in northeastern Louisville, Kentucky; Ferris (1951) method for determining aquifer transmissivity from fluctuations of stream stage and the transmission of these fluctuations into the associated aquifer; Pinder *et al.* (1969) paper on the use of aquifer response to river stage fluctuations to determine aquifer diffusivity; Rorabaugh (1964) on estimating changes in bank storage and groundwater contribution to stream flow; and Bredehoeft & Young (1970) hydrological model that demonstrated the interrelation between aquifer, stream, and water use.

Alluvial valley aquifer-stream systems have been sources of large water supplies. Special methods such as drilling horizontal wells into the aquifer adjacent to the stream bed and into the stream bed have been used to induce infiltration from streams into an aquifer. In places, the induced infiltration has resulted in marked decrease in stream flow. In the more arid parts of the United States, heavy pumping from aquifers adjacent to streams has in places resulted in dry steam beds. For example, since at least the 1970s, such impact on stream flow has been a source of conflict between the State of Kansas and the State of Colorado concerning flow in the Arkansas River. For more than 16 km from the Colorado-Kansas State line, the Arkansas River in Kansas has no flow for long periods. Even in the humid parts of the United States, large groundwater withdrawals from alluvial valley aquifers have affected the flow of associated streams.

Salt-water encroachment: For almost 100 years, the salt-water encroachment of fresh-water aquifers has been of concern along the Atlantic Coast of the United States. Marine water encroachment has also been a concern along the Gulf Coast of the United State and California. Before the 1970s, marine encroachment was the most serious groundwater contamination concern in the United States. Brown's 1925 study was one of the earliest, but numerous papers and reports on this subject have followed. The practice of constructing drainage canals inland from the coast to create water-front residential property has markedly contributed to the problem in the south Florida area elsewhere along the Florida Gulf Coast. Drainage canals from inland areas of Florida to the coast have lowered the fresh-water head and served as conduits for movement of salt water inland. Kohout (1961) published a case history of saltwater encroachment caused by a drainage canal in the Miami area, Florida, and local water managers have since installed control structures on most canals to minimise saltwater intrusion. Lusczynski & Swarzenski (1966) presented another example of marine water encroachment in southern Nassau and southeastern Queens Counties, Long Island, New York.

Cooper (1959) presented a hypothesis on the dynamic balance of fresh water and salt water in a coastal aquifer. Kohout (1960) published a paper on the cyclic flow of salt water in the Biscayne aquifer in southeastern Florida based on a 20 year historical record of the movement of the fresh-water salt-water front. These observations showed that the salt water flows in a cycle from the floor of the sea into a zone of diffusion and back to the sea. This cycle lessens the extent to which the salt water intrudes the aquifer. Additional publications of note are Banks & Richter (1953) on sea-water intrusion into groundwater basins along the California coast; Glover (1964) on patterns of fresh-water flow in a coastal aquifer where fresh-water discharge takes place below sea level; Collins & Gelhar (1971) on seawater intrusion in layered aquifers; and Pinder and Cooper (1970) on techniques for simulating the transient position of the salt-water front.

Underground storage of wastes: McGuinness (1963) investigated the disposal of sewage and industrial wastes by wells into aquifers. Hickey & Vecchioli (1986) summarised various aspects of subsurface injection of liquid wastes in with emphasis on injection practices in Florida. In some states such as Kansas, oil-field brines have been disposed of in the subsurface since 1935. The use of the deeper saline parts of aquifers for disposal or storage of industrial and sewage treatment plant effluent increased markedly during the 1950s through the 1980s. During this period, the interest in subsurface storage of non-industrial fluids was in part motivated by the problems faced with discharge of treated effluent to surface-water bodies and the potential for recovery of the stored fluid (treated sewage plant effluent or stream flow) for reuse.

Florida's history of storage and recovery of sewage treatment plant effluent and storm water provides a good example of the status and application through 1990. The storage of sewage treatment plant effluent in the saline part of the Floridan aquifer has been of interest since the 1950s. Part of this interest was motivated by environmental regulations that required the effluent to be treated before discharge to surface-water bodies. Since 1959 a number of injection well fields have been constructed along the east coast to depths of 1000 m and the effluent disposed into the highly permeable boulder zone of the Floridan aquifer. According to Meyers (1989), south Florida had nine systems in operation. Because of this zone's high transmissivity, the likelihood of recovery of the injected fluid was small. Although a number of recovery tests have been run on injected treatment plant waste in shallower and less transmissive zones in the Floridan aquifer, as of 1990, no operating recovery systems have been installed.

In central Florida, a large part of the fresh water pumped from the Floridan aquifer is used is to irrigate green belts, golf courses, and residential lawns. Since 1977, Pinellas County and the City of St. Petersburg, along the west central Florida Gulf Coast, have considered the aspects of storing advanced treated effluent and storm water in the saline part of the Floridan aquifer for future recovery for non-potable use. In the late 1970s through most of the 1980s, detailed investigations of the hydraulic properties of four permeable saline-water zones in the Floridan aquifer were carried out to determine their suitability for injection, storage, and recovery of treated effluent and storm water. Hickey (1982) carried out four injection tests which ranged from 3 to 91 days in duration. His tests show that the transmissivity of the most permeable zone was too high to be used for recovery of stored effluent or storm water.

Groundwater mining

The mining of groundwater is occurring in many areas due to large increases in demand for irrigation and municipal supplies. The U.S. Geological Survey has noted that increased groundwater use mining is a significant issue in almost every state. The development of groundwater has led to declining water levels and artesian head in a number of states. In several areas, groundwater levels have declined by 90m or more in the past 10 years. Five areas where groundwater mining has had a major impact are the Ogallala aquifer of Texas, New Mexico, Kansas, and Colorado; Denver, Colorado; Tucson, Arizona; Chicago, Illinois; Central Valley of California; and Las Vegas, Nevada. In these areas, groundwater mining has caused serious environmental problems including stream flow depletion, land subsidence, saltwater intrusion, increased cost to deepen wells, increased costs to pump groundwater, decrease in well yield, drying up of shallow wells and springs and degradation of water quality.

The US Geological Survey published the first in a series of hydrogeologic atlases in 1990 (Oklahoma and Texas). This report series present a comprehensive summary of the Nation's groundwater resources and are a basic reference for the location, geography, geology, and hydraulic character of the major aquifers. The main source of information was the USGS Regional Aquifer Analysis Program. The Atlas series includes 13 chapters each representing regional areas that collectively cover 50 states and Puerto Rico. The final chapter (Alaska, Hawaii and Puerto Rico) was published in 1999.

International activities

Hydrogeologists in the United States have had a long-term interest in international hydrogeology. The professional societies, as well as individual hydrogeologists, have played an active role in UNESCO's International Hydrological Decade and its follow up activity the International Hydrological Program. U.S. universities and colleges have also been active in international hydrogeological activities.

The USGS has probably had the longest history in the international arena. Taylor (1976) that between 1940 and 1970, some 340 water-resources projects were completed in 80 host countries. For example from, 1957 to 1976, projects were undertaken in Argentina, Australia, Brazil, Cambodia, Chile, Egypt, Ghana, Guyana, India, Iran, Israel, Jordan, Korea, Libya, Nigeria, Pakistan, Panama, Peru, Sudan, Thailand, Tunisia, Turkey, Vietnam, and Zimbabwe. More than 425 hydrogeologist from 60 countries received training at USGS facilities. From 1976 through 1990, the activities with various countries have been at a decreased level.

SUMMARY AND CONCLUSIONS

Great progress has been made in the last 140 years in the development of the science of hydrogeology. In the United States, this development has been driven by the ever changing needs of society, and for the most part, research and applications of the science have kept abreast with societal needs. The science developed into a multidisciplinary science much as Meinzer visualised.

The major professional societies have played an important role in the development of the science and have in part fostered the involvement of United States hydrogeologists in international activities.

From 1879 to about 1970, the U, S, Geological Survey played a major role in the development of the science and training of hydrogeologists. The establishment in 1930 of the cooperative program between the U.S. Geological Survey and State and local agencies broadened the role of other agencies in the development of the science. Since the 1970s, universities have played a significant role in the education of hydrogeologists and in hydrogeologic research.

Hydrogeology has a strong and broad historical foundation in the United States. This foundation will markedly aid future development of this country's hydrogeological science.

ACKNOWLEDGMENTS

The authors wish to express their sincere gratitude to L.F. Konikow and P.E. LaMoreaux for their review of this paper. Their excellent review comments and suggested changes to the text markedly improved this paper.

REFERENCES

Appel, C.A. & Bredehoeft, J.D. (1976) *Status of ground-water modelling in the U.S. Geological Survey*. U.S. Geological Survey Circular 737: 9 p.

Appel, C.A. & Reilly, T.E. (1988) *Selected reports that include computer programs produced by the U.S. Geological survey for simulation of ground-water flow and quality*. U.S. Geological Survey Water Resources Investigations Report 87-4271: 64 p.

Bachmat, Bredehoeft, J.D., Andrews, Holtz & Sebastian (1980) Use of numerical models for ground water management. Groundwater management: the use of numerical models. *American Geophysical Union Water Resources Monograph*, 5, 127.

Back, W. & Freeze, R.A. (eds.) (1983) *Chemical hydrogeology, benchmark papers in geology*, Volume 73. Hutchinson Ross Publication Company. 416 p.

Banks, H.O. & Richter, R.C. (1953) Sea water intrusion into groundwater basins bordering the California coast and inland bays. *American Geophysical Union Transactions*, 34, 575–582.

Bennett, R.R. (1962) Flow-net analysis, pp. 139–144. In: Ferris, J.G., Knowles, D.B., Brown, R.H. & Stallman, R.W. (eds.) *Theory of aquifer tests: U.S. Geological Survey Water-Supply Paper* 1536-E. p. 174.

Bennett, R.R. & Meyer, R.R. (1952) Geology and ground water resources of the Baltimore area. *Maryland Department of Geology, Mines and Water Resources Bulletin*, 4, 573.

Bredehoeft, J.D. & Pinder, G.F. (1973) Mass transport in flowing groundwater. *Water Resources Research*, 9 (1), 194–210.

Bredehoeft, J.D. & Young, R.A. (1970) The temporal allocation of ground water – a simulation approach. *Water Resources Research*, 6 (1), 3–21.

Brown, J.S. (1925) *A study of coastal ground water, with special reference to Connecticut*. U.S. Geological Survey Water-Supply Paper 537: 101 p.

Carslaw, H.S. & Jaeger, J.C. (1947) *Conduction of heat in solids*. Oxford Press. 510 p.

Chamberlin, T.C. (1885) *The requisite and qualifying conditions of artesian wells*. U.S. Geological Survey 5th Annual Report. pp. 125–173.

Chow, V.T. (1964) *Advances in hydroscience*. Volume 1. New York, Academic Press, 462 p.

Collins, M.A. & Gelhar, L.W. (1971) Seawater intrusion in layered aquifers. *Water Resources Research*, 7, 971–979.

Collins, W.D. (1923) Graphic representation of water analysis: p. 25, In: Back, W. & Freeze, R.A. (eds.) (1983) *Chemical hydrogeology, benchmark papers in geology*. Volume 73: Stroudsburg, Pennsylvania, Hutchinson Ross Publishing Company.

Collins, W.D. (1928) *Notes on practical water analysis*. U.S. Geological Survey Water-Supply Paper 596-H.

Cooper, H.H. (1959) A hypothesis concerning the dynamic balance of fresh water and sea water in a coastal aquifer. *Journal of Geophysical Research*, 64, 461–467.

Cooper, H.H. & Rorabaugh, M.I. (1963) *Ground-water measurements and bank storage due to flood stages in surface streams*. U.S. Geological Survey Water-Supply Paper 1536-J: 343–346.

Darton, H.H. (1896) *Preliminary report on artesian waters of a portion of the Dakotas*. U.S. Geological Survey Annual Report 17, pt. 2: 603–694.

Davis, S.N. & Davis, A.G. (2005) *Hydrology in the United States 1780–1950*. Tucson, Arizona, University of Arizona Department of Hydrology and Water Resources HWR No. 05-02.

Domenico, P.A. & Mifflin, M.D. (1965) Water from low-permeability sediments and land subsidence. *Water Resources Research*, 1, 562–576.

Domenico, P.A. & Schwartz, F.W. (1990) *Physical and chemical hydrogeology*. New York, John Wiley and Sons. 824 p.

Ellis, E.E. (1909) A study of the occurrence of water in crystalline rocks. In: Gregory, H.E. (ed.) *Underground water resources of Connecticut*. U.S. Geological Survey Water-Supply Paper 232. 200 p.

Ferris, J.G. (1948) Ground water, chapter 6, pp. 127–191. In: Wisler and Brater (eds.) *Hydrology*. New York, Wiley and Sons.

Ferris, J.G. (1951) *Cyclic fluctuations of water level as basis for determining aquifer transmissivity*. International Union of Geodesy and Geophysics, International Association of Scientific Hydrology Assembly, Brussels. Volume. 2, pp. 148–155.

Ferris J.G., Knowles, D.B., Brown, R.H. & Stallman, R.W. (1962) *Theory of aquifer tests*. U.S. Geological Survey Water-Supply Paper 1536-E. 174 p.

Fetter, C.W. (1980) *Applied hydrogeology*. Columbus, Ohio, Charles E. Merrill Publishing Company. 488 p.

Fiedler, A.G. & Nye, S.S. (1933) *Geology and ground-water resources of the Roswell Artesian Basin in New Mexico*. U.S. Geological Survey Water-Supply Paper 639. 372 p.

Foster, M.D. (1950) The origin of high sodium bicarbonate waters in the Atlantic and Gulf Coastal Plains. *Geochimica Acta*, 1 (1), 33–48.

Freeze, R.A. and Cherry, J.A. (1979) *Ground water*. New Jersey, Prentice-Hall, Inc. 604 p.

Freeze, R.A. & Witherspoon, P.A. (1967) Theoretical analysis of regional groundwater flow. *Water Resources Research*, 3, 623–634.

Garrels, H.M. (1960) *Mineral equilibria at low temperature and pressure*. New York, Harper Brothers, 254 p.

Gilbert, G.K. (1896) *The underground water of the Arkansas Valley in eastern Colorado*. U.S. Geological Survey 17th Annual Report, pt.11: 551–601.

Glover, R.E. (1964) The pattern of fresh-water flow in a coastal aquifer. In: *Sea Water in Coastal Aquifers*, U.S. Geological Survey Water-Supply Paper 1613-C. pp. 32–35.

Hantush, M.S. (1960) Modification of the theory of leaky aquifers. *Journal of Geophysical Research*, 65 (11), 3713–3725.

Hantush, M.S. (1964) Hydraulics of wells, In: Chow, V.T. (ed.) *Advances in hydroscience*. New York, Academic Press. Volume. 1. pp. 281–432.

Hantush, M.S. & Jacob, C.E. (1955) Nonsteady radial flow in an infinite leaky aquifer. *American Geophysical Union Transactions*, 36 (1), 95–100.

Harper, L. (1857) *Preliminary report on the geology and agriculture of the State of Mississippi*. Mississippi Geological Survey: 351.

Heath, R.C. (1984) *Ground-water regions of the United States*. U.S. Geological Survey Water-Supply Paper 2242. 78 p.

Heath, R.C. (1987) *Basic ground-water hydrology*. U.S. Geological Survey Water-Supply Paper 2220. 84 p.

Heath, R.C. (1988) Hydrogeologic settings of the regions, chapter 3, pp. 15–23. In: Back, W., Rosenshein, J.S. & Seaber, P.R. (eds.) *Hydrogeology, The Geology of North America*. Volume O-2. 524 p. Boulder, Colorado, the Geological Society of America.

Hem, J.D. (1959) *Study and interpretation of the chemical characteristics of natural water*. U.S. Geological Survey Water-Supply Paper 1473. 269 p.

Hickey, J.J. (1982) *Hydrogeology and results at waste-injection test sites in Pinellas County, Florida*. U.S. Geological Survey Water-Supply Paper 2183. 42 p.

Hickey, J.J. & Vecchioli, J. (1986) *Subsurface injection of liquid waste with emphasis on injection practices in Florida*. U.S. Geological Survey Water-Supply Paper 2281. 24 p.

Hubbert, M.K. (1940) The theory of ground water motion. *Journal of Geology*, 48, 785–944.

Jacob, C.E. (1940) On the flow of water in an elastic artesian aquifer. *American Geophysical Union Transactions*, 21, 564–586.

Jacob, C.E. (1943) Correlation of ground-water levels and precipitation on Long Island, New York. *American Geophysical Union Transactions*, 24 (pt. 2), 564–573.

Jacob, C.E. (1946) Radial flow in a leaky artesian aquifer. *American Geophysical Union Transactions*, 27, 198–208.

Jacob, C.E. (1950) Flow of ground water, *Hunter rouse engineering hydraulics*. New York, Wiley and Sons [chapter 5].

Karplus, W.J. (1958) *Analog simulation*. New York, McGraw-Hill. 434 p.

King, F.H. (1899) *Principles and conditions of the movements of ground water*. U.S. Geological Survey 19th Annual Report, pt. 2: 59–294.

Kohout, F.A. (1960) Cyclic flow of salt water in the Biscayne aquifer of southeastern Florida, pp. 102–110. In: Back, W. & Freeze, R.A. (eds.) (1983) *Chemical Hydrogeology, Benchmark Papers in Geology*. Volume 73. Stroudsburg, Pennsylvania, Hutchinson Ross Publishing Company.

Kohout, F.A. (1961) Case history of salt water encroachment caused by a drainage canal in the Miami area, Florida. *American Water Works Association Journal*, 53 (11), 1406–1416.

Konikow, L.F. & Bredehoeft, J.D. (1973) *Simulation of hydrologic and chemical-quality variations in an irrigated stream–aquifer system; a preliminary report*. Colorado Conservation Board Water Resources Circular 17: 43.

Lee, C.H. (1912) *An intensive study of the water resources of a part of Owens Valley, California*. U.S. Geological Survey Water-Supply Paper 294. 135 p.

LeGrand, H.E. (1965) Patterns of contaminated zones of water in the ground. *Water Resources Research*, 1, 83–95.

Lusczynski, N.J. & Swarzenski, W.V. (1966) *Salt water encroachment in southern Nassau and southeastern Queens Counties, Long Island, New York*. U.S. Geological Survey Water-Supply Paper 1613-F.

Maxey, B. (1986) The Meinzer Era of hydrogeology in the United States, 1910–40, pp. 45–50. In: Rosenshein, J.S., Moore, J.E., Lohman, S.W. & Chase, E.B. (eds.) U.S. Geological Survey Open-File Report 86-480. p. 110.

McCallie, S.W. (1898) A preliminary report on the artesian well system of Georgia. *Georgia Geological Survey Bulletin*, 7, 214.

McDonald, M.G. & Harbaugh, A.W. (1988) *A modular three dimensional finite-difference ground-water flow model*. U.S. Geological Survey Techniques of Water-Resources Investigations Book 6, Chapter A1.

McGuinness, C.L. (1963) *The role of ground water in the national water situation*. U.S. Geological Survey Water-Supply Paper 1800. 1121 p.

Meinzer, O.E. (1923) *The occurrence of ground water in the United States, with a discussion of principles*. U.S. Geological Survey Water-Supply Paper 489. 321 p.

Meinzer, O.E. (1928) Compression and elasticity of artesian aquifers. *Journal of Economic Geology*, 23 (3), 263–291.

Meinzer, O.E. (1932) *Outline of methods for estimating ground-water supplies*. U.S. Geological Survey Water-Supply Paper 638-C. pp. 99–144.

Meinzer, O.E. (1939) *Ground water in the United States, a summary of ground-water conditions and resources, utilization of water from wells and springs, methods of scientific investigations, and literature relating to the subject*. U.S. Geological Survey Water-Supply Paper 836-D.

Mendenhall, W.C., Dole, R.B. & Stabler, H. (1916) Ground water in the San Joaquin Valley, California. U.S. Geological Survey Water-Supply Paper 398. 310 p.

Meyer, G., Davis, G. & LaMoreaux, P.E. (1988) Historical perspective, chapter 1, pp. 1–8. In: Back, W., Rosenshein, J.S. & Seaber, P.R. (eds.) Hydrogeology – the geology of North America. Volume O-2. Boulder, Colorado, Geological Society of America.

Meyers, F.W. (1989) *Hydrogeology, ground-water movement, and subsurface storage in the Floridan aquifer system in Southern Florida.* U.S. Geological Survey Professional Paper 1403-G.

Moore, J.E. & Hanshaw, B.B. (1987) Hydrogeological concepts in the United States: a historical perspective. *EPISODES*, 1 (4), 314–322.

Narasimhan, T.N. (2005) Hydrogeology in North America: past and future. *Hydrogeology Journal*, 13, 7–24.

Neuman, S.P. & Witherspoon, P.A. (1969) Applicability of current theories of flow in leaky aquifers. *Water Resources Research*, 5, 817–829.

Neuman, S.P. (1975) Analysis of pumping test data from anisotropic unconfined aquifers considering delayed gravity response. *Water Resources Research*, 11, 329–342.

Norton, W.H. (1897) Artesian wells of Iowa. Iowa Geological Survey Annual Report, volume 6: 123.

Palmer, C. (1911) The chemical interpretation of water analysis, pp. 12–17. In: Back, W. & Freeze, R.A. (eds.) (1983) *Chemical hydrogeology, benchmark papers in geology* Volume 73. Stroudsburg, Pennsylvania, Hutchinson Ross Publishing Company.

Pinder, G.F. & Bredehoeft, J.D. (1968) Application of the digital computer for aquifer evaluation. *Water Resources Research*, 4 (5), 1069–1093.

Pinder, G.F. & Cooper, H.H. (1970) A numerical method technique for calculating the transient position of the salt-water front. *Water Resources Research*, 6 (3), 875–882.

Pinder, G.F., Bredehoeft, J.D. & Cooper, H.H. (1969) Determination of aquifer diffusivity from aquifer response to fluctuations in river stage. *Water Resources Research*, 5 (4), 850–855.

Piper, A.M. (1944) A graphic procedure in the geochemical interpretation of water analysis, pp. 50–59. In: Back, W. & Freeze, R.A. (eds.) (1983) Chemical Hydrogeology, Benchmark Papers in Geology. Volume 73. Stroudsburg, Pennsylvania, Hutchinson Ross Publishing Company.

Poland, J.F. & Davis, G.H. (1969) Land subsidence due to withdrawal of fluids. *Geological Society of America Reviews Engineering Geology*, 2, 187–269.

Prickett, T.A. & Lonnquist, C.Q. (1971) Selected digital computer techniques for groundwater resource evaluation. *Illinios State Water Survey Bulletin*, 55, 62.

Reilly, T.E. (2004) A brief history of the contributions to ground-water hydrology by the U.S. Geological Survey. *Ground Water*, 42 (4), 625–631.

Rorabaugh, M.I. (1956) *Ground water in northeastern Louisville, Kentucky, with reference to induced infiltration.* U.S. Geological Survey Water-Supply Paper 1360-B. 169 p.

Rorabaugh, M.I. (1964) Estimating changes in bank storage and ground-water contribution to stream flow. *International Association Scientific Hydrology Publication No.* 63, 432–441.

Rosenshein, J.S. & Bennett G.D. (eds.) (1984) Groundwater hydraulics. *American Geophysical Union Water Resources Monograph* 9, 407 p.

Rosenshein, J.S., Moore, J.E., Lohman, S.W. & Chase, E.B. (eds.) (1988) *Two hundred years of hydrogeology in the United States (1776–1976).* U.S. Geological Survey Open-File Report 86-480: 110 p. (Republished by the National Water Well Association).

Sayre, A.N. & Bennett, R.R. (1942) *Drainage–basin evolution and aquifer development in a karstic limestone terrane South-Central Texas.* U.S. Geological Survey Open-File Report. 35 p.

Skibitzke, H.E. (1960) Electronic computers as an aid to the analysis of hydrologic problems. Publication 52, International Association Scientific Hydrology, Commission Subterranean Waters, Gentbrugge, Belgium. pp. 347–358.

Stallman, R.W. (1956a) Numerical analysis of regional water levels to define aquifer hydrology. *American Geophysical Union Transactions*, 37 (4), 451–460.

Stallman, R.W. (1956b) The use of numerical methods for analysing data on ground-water levels. *International Association of Scientific Hydrology Publication* 41, Tome II, 227–231.

Stallman, R.W. (1962) Numerical analysis. In: Ferris, J.G., Knowles, D.B., Brown, R.H. & Stallman, R.W. (eds.) *Theory of aquifer tests*. U.S. Geological Survey Water-Supply Paper 1536E. pp. 135–139.

Stallman, R.W. (1963) Notes on the use of temperature data for computing ground-water velocity. In: Bentall, R. (ed.) *Methods of collecting and interpreting ground-water data*. U.S. Geological Survey Water-Supply Paper 1544-H.

Stallman, R.W. (1964) *Multiphase fluids in porous media—a review of theories pertinent to hydrologic studies*. U.S. Geological Survey Professional Paper 411. pp. E1–E51.

Stiff, H.A. (1951) The interpretation of chemical water analysis by means of patterns, pp. 60–62. In: Back, W. & Freeze, R.A. (eds.) (1983) *Chemical hydrogeology, benchmark papers in geology*. Volume 73. Stroudsburg, Pennsylvania, Hutchinson Ross Publishing Company.

Stringfield, V.T. (1936) *Artesian water in the Florida Peninsula*. U.S. Geological Survey Water-Supply Paper 773-C.

Sun, R.J. & Johnston, R.H. (1994) *Regional Aquifer-System Analysis Program of the U.S. Geological Survey, 1978–1992*. U.S. Geological Survey Circular 1099. 126 p.

Taylor, G.J. (1976) *Historical review of the international water-resources program of the U.S. Geological Survey 1940–70*. U.S. Geological Survey Professional Paper 911. 146 p.

Theis, C.V. (1935) The relation between the lowering of the piezometric surface and the rate and duration of discharge of a well using ground-water storage. *American Geophysical Union Transactions*, 16, 519–524.

Theis, C.V. (1940) The source of water derived from wells. *Civil Engineering*, 10 (5), 277–280.

Theis, C.V. (1941) The effect of a well on the flow of a nearby stream. *American Geophysical Union Transactions*, 22, 734–738.

Thomas, H.E. (1951) *The conservation of ground water*. McGraw-Hill, New York, 327 p.

Thomas, H.E. (1952) *Ground water regions of the United States – their storage facilities*. U.S. 82nd Congress, House Interior and Insular Affairs Commission, The Physical and Economic Foundation of Natural Resources. Volume 3. 78 p.

Todd, D.K. (1959) *Ground water hydrology*. Wiley and Sons. 336 p.

Toth, J. (1963) A theoretical analysis of groundwater flow in small drainage basins. *Journal of Geophysical Research*, 68 (16), 4795–4812.

Trescott, P.C. (1973) *Iterative digital model for aquifer evaluation*. U.S. Geological Survey Open-File Report: 63 p.

Veatch, A.C. (1906) *Fluctuations of the water level in wells, with special reference to Long Island, New York*. U.S. Geological Survey Water-Supply Paper 155. 83 p.

Walton, W.C. (1962) Selected analytical methods for well and aquifer evaluation. *Illinois State Water Survey Bulletin*, 49, 81.

Walton, W.C. & Prickett, T.A. (1963) Hydrologic electric analog computer. *American Society Civil Engineers Proceedings, Hydraulics Division Journal*, 89 (HY6), 67–91.

Wenzel, L.K. (1942) *Methods for determining permeability of water bearing materials*. U.S. Geological Survey Water-Supply Paper 887.

Wisler, C.O. & Brater, E.F. (1948) *Hydrology*. New York, Wiley and Sons.

Wood, L.A. & Gabrysch, R.K. (1965) Analog model study of ground water in the Houston district, Texas, with a section on design, construction, and use of analog models by E. P. Patten, Jr. *Texas Water Commission Bulletin*, 6508, 103.

Zody, H.A.R., Eaton, G.P. & Maby, D. (1974) *Application of surface geophysics to ground-water investigations*. U.S. Geological Survey Techniques of Water-Resources Investigations Book 2, Chapter D1.

SERIES IAH International Contributions to Hydrogeology (ICH)

Volume 1–17 Out of Print

18. Shallow Groundwater Systems
    Edited by: Peter. Dillon & Ian Simmers
    1998, HB: 90-5410-442-2

19. Recharge of Phreatic Aquifers in (Semi-) Arid Areas
    Edited by: Ian Simmers, J.M.H. Hendrickx, G.P. Kruseman & K.R. Rushton
    1998, HB: 90-5410-694-8
    1998, PB: 90-5410-695-6 (out of print)

20. Karst Hydrogeology and Human Activities – Impacts, consequences and implications
    Edited by: David Drew and Heinz Hötzl
    1999, HB: 90-5410-463-5 (out of print)
    1999, PB: 978-90-5410-464-3

21. Groundwater in the Urban Environment – Selected City Profiles (1999)
    Edited by: John Chilton
    1999, HB: 978-90-5410-924-2
    1999, PB: 978-90-5809-120-8 (out of print)

22. Managing Water Well Deterioration
    Author: Robert G. McLaughlan
    2002, HB: 978-90-5809-247-2

23. Understanding Water in a Dry Environment
    Edited by Ian Simmers
    2003, HB: 978-90-5809-618-0

24. Urban Groundwater Pollution
    Edited by: David Lerner
    2003, HB: 978-90-5809-629-6

25. Introduction to Isotope Hydrology
    Author: Willem G. Mook
    2005, HB: 978-0-415-39805-3

26. Methods in Karst Hydrogeology
    Edited by: Nico Goldscheider & David Drew
    2007, HB: 978-0-415-42873-6

27. Climate Change Effects on Groundwater Resources: A Global Synthesis of Findings and Recommendations
    Edited by: Holger Treidel, Jose Luis Martin-Bordes & Jason J. Gurdak
    2011, HB 978-0-415-68936-6